Mott Insulators

Physics and applications

Sindhunil Barman Roy

UGC-DAE Consortium for Scientific Research, Indore, India

IOP Publishing, Bristol, UK

ISBN   978-0-7503-1596-8 (ebook)
ISBN   978-0-7503-1594-4 (print)
ISBN   978-0-7503-1595-1 (mobi)

DOI   10.1088/2053-2563/ab16c9

Version: 20190701

IOP Expanding Physics
ISSN 2053-2563 (online)
ISSN 2054-7315 (print)

British Library Cataloguing-in-Publication Data: A catalogue record for this book is available from the British Library.

Published by IOP Publishing, wholly owned by The Institute of Physics, London

IOP Publishing, Temple Circus, Temple Way, Bristol, BS1 6HG, UK

US Office: IOP Publishing, Inc., 190 North Independence Mall West, Suite 601, Philadelphia, PA 19106, USA

To my late parents Subodh and Binapani Barman Roy, and my beloved wife Gopa Barman Roy (nee Dasgupta).

Contents

Preface

The valence electrons in solids, which exist in the presence of the periodically placed ion-cores, have two competing tendencies: a desire to remain localized at the atomic sites to minimize the Coulomb repulsion between them, and a counter desire to become delocalized and reduce the cost of kinetic energy arising out of spatial confinement. In most of the bulk solids the tendency of an electron to become delocalized and behave more like a wave rather than a particle dominates, because this freedom of movement can compensate the Coulomb penalty. The valence electrons in crystalline solids, however, cannot exist in the same energy state, nor can they have arbitrary energies. The specific arrangement of the ion-cores within the solid determines the ranges of allowed energies or energy-bands for the electrons. The electrons systematically occupy the available energy-bands starting from the energetically lowest one, and the Pauli exclusion principle ensures that each energy state accommodates two electrons of opposite spin. The solid is an insulator if all the allowed energy-bands are occupied by the electrons. There is an energy gap between the uppermost filled band (known as the valence band) and lowest empty band (known as the conduction band). In contrast, a metal is a material with partially filled energy bands. In this nomenclature the semiconductors are insulators with relatively narrow energy band gap and with a small but finite number of electrons transferred from the uppermost filled energy band to the lowest unoccupied energy band at a finite temperature.

This development of electron band theory is based on the crucial assumption of independent electron (oblivious of Coulomb repulsion) approximation, which allows to map a many electron problem to the problem of one electron in the potential created by the rest of the electrons and ions in a solid. This reduces the need of computational resources to a reasonable limit, and today all practical quantitative methods depend on the inputs obtained from independent electron calculations. Almost in parallel to this purely scientific development, the electron band theory of solids also played a very important role in the evolution of semiconductor devices from transistor to IC to LSI/VLSI leading to the current complementary metal-oxide-semiconductor (CMOS) technology. The coverage of this scientific and technological development over the last seven decades is now an essential part of any modern solid state/condensed matter physics textbooks, several monographs/reference books, and numerous specialized books on semiconductor physics/technology.

However, even in the 1930s there were some reports that many transition-metal oxides with a partially filled d-electron band were found to be poor conductors and indeed in many a case were insulators. A typical example of such a material is NiO. Nevill Mott and P W Anderson, through their seminal works in the 1940–50s, took the pioneering steps towards understanding how electron–electron correlations could explain the insulating state in certain classes of materials, including technologically important materials like ferrites and garnets, which as per the electron band theory were expected to be metals. This insulating state of solids arising out of strong electron–electron correlation is now termed the Mott insulator. These elegant works of Mott–Anderson, however, did not generate very significant interest in the solid state/condensed matter physics community, that is, until the discovery of high T_C superconductivity (HTSC) in several oxide materials in late 1980s. In all these HTSC materials the parent compounds are identified to be a Mott insulating antiferromagnet. With this and the subsequent discovery of colossal magnetoresistance in another class of oxide compounds, there was a spurt of activity, both theoretical and experimental, in the field of Mott insulators. All these events provided stimulation towards a deeper understanding of such highly correlated electron materials, and the efforts in these directions are still being continued. These efforts in turn lead to the recognition of new emergent phenomena in many known insulating oxides and identification of newer Mott insulating systems.

Unlike the 1950s Bell Laboratories in the USA or the Atomic Energy Research Establishment in the UK, the corporate and government research laboratories nowadays expect the condensed matter physicists to remain engaged solely in applied or 'targeted basic research' leading to some tangible social benefits. Trendy basic physics research is pursuing Higgs boson or other exotic particles foreseen in 'the standard model' of particle physics or detecting gravitational waves generated by collapsing neutron stars, some about 130 million years ago, and that kind of research is to be left to the elite academic institutions and a few mega science facilities. This 'reductionist approach' for the quest for the fundamental particles and force fields still rules supreme in the field of physics in general, and very few pays attention to the fact that the concept of spontaneously broken gauge symmetry and Higgs mechanism (rather Anderson–Higgs mechanism) for generating particle mass arose in the first place in the condensed matter physics, to explain among other things the Meissner effect in superconductors, which also include the dense fermionic fluids like a neutron star, and that the concept of the asymptotic degree of freedom is already there in the so called heavy fermion materials at very much accessible temperatures. In fact, during the last 50 years or so there has been a silent continuous revolution in the field of condensed matter physics with the recognition of the idea of 'emergent phenomena'. This idea is nicely exemplified in a rather cult paper 'More is different' by P W Anderson published in *Science* in 1973 and in a book *A different Universe* by Robert Laughlin in early 2000. It says that everything we observe at one level, which while obeying the physical laws at a lower length scale or more microscopic level cannot necessarily be deduced from that level. When certain systems reach a sufficient level of higher length scale or complexity, new phenomena 'emerge' that do not exist for the elementary components of that system. A formal definition of emergent phenomena now exists: 'An emergent behaviour of a physical system is a qualitative property that can only occur in the limit that the number of microscopic constituents tends to infinity' (S Kivelson and S Kivelson 2016 *NPJ Quantum Materials* **1** 16024).

In this backdrop I started my tenure in an Indian national laboratory, focusing on the flux-lines or vortices in superconductors (irrespective of their microscopic origin and transition temperature), how to pin those down and make the superconductor suitable for some social benefits in the form of say medical applications (cheap and affordable MRI magnets) or energy applications (levitated trains or even thermo-nuclear fusion reactors) or even accelerating high energy particles with super-conducting RF cavities. Fortunately for some of us, the growing wave of the 'emergent phenomena' silently arrived in this field of applied superconductivity in the form of vortex matter physics and the rather ubiquitous first order phase transition. Soon a correlation emerged between various classes of functional (not merely academic) superconducting, magnetic and ferroelectric materials even in the phenomenological or macroscopic level. Over the period we realized that Mott insulators with their first order metal–insulator transition can provide a nice platform to study the interplay between such first order phase transitions and the associated functional properties. Simultaneously, during last two decades, a new direction has opened in the field of Mott insulators generating considerable interest, that is the possible technological applications of Mott insulators. It is now well recognized that current CMOS technology is fast approaching the fundamental limits, and there is a driving interest in the newer kind materials with newer physics concepts leading to the devices for information processing and storage. In this direction resistive switching based random access memory (RRAM) devices are emerging as a possible alternative to the current Flash metal-oxide-semiconductor field-effect transistor (MOSFET) technology devices. A purely electronically-driven resistive switching transition in the Mott insulators gives rise to the possibility for fast switching devices with long endurance and low dissipation. These possibilities have stimulated a significant amount of research activities on the device physics involving various classes of Mott insulators.

With all these recent developments in the area of physics and materials science of Mott insulators, especially their recognition as emergent materials for important and futuristic device applications, the time has now come to make the subject of Mott insulator (both physics and applications) accessible to a wider community. The only books currently available on this subject are the seminal book of Sir Nevil Mott and the subsequent addition to the field by Florian Gebhard, which are mostly used by advanced physics graduate students and professionals engaged in the theoretical studies of strongly correlated electron systems. In any case both these books do not contain much information on the experimental aspects of Mott-insulators and devices. The present book will be aimed more towards the advanced undergraduate and graduate students of physics, chemistry, materials science and electrical/electronics engineering. This will be a hybrid of a textbook and a reference book. The main aim of the book is to being this very interesting area of science and emerging technology to the attention of the young and fresh minds. The book (with a part review feature and lots of references) will also be useful for the professional researchers in academic institutions and industries engaged in programmes of correlated electron materials and devices.

I express my gratitude to Alak Majumdar, Abhijit Mookerjee, Greg Stewart and especially Bryan Coles under whose tutelage I developed my interest in correlated electron systems and condensed matter physics in general. It was while working with Bryan Coles as a post-doctoral research associate that I had the opportunity to meet Nevill Mott a couple of times. It was indeed a great experience to meet such an iconic figure. I am thankful to the anonymous referee for careful reading of all the chapters and very useful comments and constructive criticisms, which definitely helped to improve the presentation in the book. I am also thankful to Alak Majumdar and Amlan Majumdar for careful reading of some of the chapters and useful comments. I am thankful to P Chaddah and various other collaborators in my scientific works related to the first order phase transition in functional magnetic and superconducting materials; all those works eventually led me to the interesting subject of Mott insulators. I am thankful to L Sharath Chandra and M K Chattopadhyay for helping me to get some experimental results, which I have presented in chapter 1 of the book. I thank Sudip Pal and Rohit Sharma for help in preparing some of the figures, and library staff members of RRCAT, Indore for help in the literature survey. I thank the Director(s) of UGC-DAE Consortium for Scientific Research, Indore for kind hospitality and necessary support for completion of the book. I also thank various faculty members, especially Alok Banerjee, and students of UGC-DAE CSR, Indore for many stimulating discussions on correlated electron systems. I thank the editorial team of IOP especially Daniel Heatley for enthusiastically providing the necessary support at various stages of the book.

Author biography

Sindhunil Barman Roy

Sindhunil Barman Roy received his BSc (Hons) degree in Physics (in 1977) from North Eastern Hill University, Shillong, and MSc degree in Physics (in 1980) from Jadavpur University, Kolkata. He obtained his PhD degree from the Indian Institute of Technology, Kanpur in 1985. Subsequently, between 1986 and 1991, he was a postdoctoral research associate in Imperial College, London and the University of Florida, Gainesville. He was a staff member of the Raja Ramanna Centre for Advanced Technology, Indore from 1992 to 2018, and also a professor of the Homi Bhabha National Institute, Mumbai between 2007 and 2018. Presently he is an emeritus professor at the UGC-DAE Consortium for Scientific Research, Indore. His research interests span both basic and applied aspects of magnetism and superconductivity and he has published more than 200 research papers in international peer-reviewed journals. He served on the editorial board of Superconductor Science and Technology between 2008 and 2014. He is a recipient of the Homi Bhabha Science and Technology Award of the Department of Atomic Energy, India and a fellow of the Institute of Physics, UK.

Symbols

γ Coefficient of electronic heat capacity (mJ mol^{-1} K^{-2})
ϵ Permittivity
κ Dielectric constant
λ Wavelength (m)
ρ Electrical resistivity (ohm metres)
B Magnetic field (T)
C Molar heat capacity (J kg^{-1} K^{-1})
f Frequency
k Thermal conductivity (W m^{-1} K^{-1})
R Ideal gas constant (8.31 J mol^{-1} K^{-1})

Introduction

Matter in the form of a solid is mostly viewed as a collection of a large number of strongly interacting atoms or molecules. The outermost or valence electrons of the constituent atoms are at the root of such interactions between atoms, and technically this phenomenon gives rise to chemical bonding. The interactions of the electrons amongst themselves and with the ions in a solid dictate the physical and chemical properties of the concerned solid material. One of the main successes of quantum mechanics has been to be able to explain the physical properties of solid, and it also provides the basis for accurate calculation of such properties. The first application of quantum mechanics to understand the electronic properties of solid matter was in the form of the free electron model of Sommerfeld [1]. While this model could provide plausible explanations of many properties of metallic solid, it could not, however, say why some solids were insulating in nature. The concept of the electron energy band then came into the picture with the idea of electrons moving as a modulated plane wave i.e. Bloch wave in the background of the periodically distributed ionic potential in crystalline solid [1]. Subsequently, a clear cut distinction between the crystalline solids showing metallic and insulating behaviour was made within the model introduced by Wilson [2], with insulators as materials having energy bands either completely occupied or completely empty [1]. There is an energy gap between the uppermost filled band (known as the valence band) and lowest empty band (known as the conduction band). In contrast, a metal is a material with partially filled energy bands. In this nomenclature the semiconductors are insulators with relatively narrow energy band gap, and having a small but finite number of electrons transferred from the highest filled energy band to the lowest unoccupied energy band at any finite temperature. The development of electron band theory is based on the crucial assumption of independent electron approximation, which allows mapping of a many electron problem to the problem of one electron in the potential created by the rest of the electrons and ions in a solid. This reduces the need of computational resources to a reasonable limit, and today all practical quantitative methods of electronic properties calculations depend on the inputs obtained from independent electron approximations. In parallel to this purely scientific development, such studies of the electron theory of solids (while neglecting the role of strong electron–electron correlation) have played very important roles in various areas of evolving technologies since the 1950s, one prominent example being the evolution of semiconductor devices from transistor to integrated circuit (IC) to large-scale integration/very-large-scale integration (LSI/VLSI) leading to the current complementary metal-oxide-semiconductor (CMOS) technology. This field of electron band theory grew continuously over the last eight decades culminating into more than 10 000 research papers per year in the last three decades.

In the backdrop of this all-round success of electron band theory, however, there were some reports from even the 1930s that many transition-metal oxides with a partially filled d-electron band were found to be poor conductors and indeed, in many a cases, insulators [3]. A typical example of such a material is NiO. This possibly prompted Peierls [4] to point out the importance of the electron–electron correlation in a solid, and he suggested that strong Coulomb repulsion between electrons could be the source of such insulating behavior. Nevill Mott through his seminal works during the 1940–50s took the pioneering steps towards understanding how such electron–electron correlations could explain the insulating state in certain classes of materials, which as per the prediction of electron band theory should have been metals. Materials with such insulating state arising out of strong electron–electron interaction are now termed as Mott insulators. Mott showed that in the array of atomic potentials in a crystalline solid with one electron per atom and a Coulomb interaction between the electrons, the ion cores would be screened for sufficiently high electron density or small lattice spacing, and the crystalline solid would be metallic in nature [5, 6]. However, Mott also argued that the screening would break down for lattice spacing larger than a critical value, and the crystalline solid would become an insulator. It is known that almost any magnetic material has to have an open shell of atomic wave functions at each site, and these should normally overlap in a solid to form a metallic band. Mott physics thus provides a natural explanation for the occurrence of paramagnetic insulators with unfilled electron shells.

In early 1950s pioneering neutron scattering work of Shull and collaborators revealed the antiferromagnetic ordering in the insulating material MnO [7]. There was an attempt to understand the insulating nature of the antiferromagnetic MnO within the electron band theory by none other than John Slater, a pioneer in the field of electron band theory. Slater argued that the insulating nature of antiferromagnetic MnO was caused by the splitting of electron bands by the doubled periodicity of Hartree–Fock fields caused by antiferromagnetism [8]. However, it has been observed that MnO is a paramagnetic insulator at room temperature, hence without any periodic exchange field; this is in contradiction with the Slater band-splitting picture. Subsequently P W Anderson introduced the idea of super-exchange, which allowed the coupling of spins of relatively distant manganese ions through overlap of the wave functions with those of intervening oxygen atoms, and to explain the observed antiferromagnetic ordering in MnO [9]. The emergent antiferromagnetic state in MnO was naturally linked with the Mott magnetic insulating state. Anderson argued that the insulating antiferromagnetic state was the residue of the frustrated band forming tendency of the electrons. Thus Mott physics explained the occurrence of the magnetic insulator and Anderson super-exchange showed how antiferromagnetism followed from Mott physics. Nevill Mott and P W Anderson shared the Nobel Prize in 1977, which was partly attributed to these pioneering works. However, for some rather strange sociological reasons the actual impact of these elegant works of Mott–Anderson remained confined mostly within a relatively small section of condensed matter physics community interested in strongly correlated electron systems [10]. To quote P W Anderson here: 'It is the tragedy of Mott that although he almost certainly won his Nobel Prize for the Mott insulator, Slater …won the publicity battle' [11]. The most plausible reason being that the electron band theory dealing with the weakly interacting or independent electron picture was popular, and being followed extensively in the scientific community because of its practical ability in explaining many experimental results, which was of interest during the period 1950–80, and that too in the context of many technological developments in that period including the ever growing semiconductor/electronics industry. The only book covering the development of Mott–Anderson physics in that period was the seminal work of Nevill Mott [6], which was first published in the early 1970s and then followed by the second edition published in the early 1990s, and that too with most of the emphasis on the aspects of metal–insulator transition. In fact the interest of Mott himself in later years was more on this metal–insulator phase transition and the extremely interesting correlated electron metals in the vicinity of this phase transition [10]. And there was that seminal review article of P W Anderson published in *Solid State Physics* Vol 13 [9].

With the discovery of high T_C superconductivity (HTSC) in several oxide materials in the late 1980s, there was a revival in the interest in Mott insulator. In all these HTSC materials the parent compounds were identified to be a Mott insulating antiferromagnet. With this, and also the subsequent discovery of colossal magnetoresistance in another class of oxide compound, there was a spurt of activity, both theoretical and experimental, in the field of strongly correlated electron systems including Mott insulators. All these events provided some stimulation towards a deeper understanding of such strongly correlated electron materials, and the efforts in these directions still continue. These efforts in turn led to the identification of new emergent phenomena in many known insulating oxides and also identification of newer Mott insulating systems. However, the renewed interests in Mott physics still remained mainly confined with a section of condensed matter researchers mostly in academic institutions, and were covered mostly in a page or two in the standard condensed matter/solid state physics textbooks with perhaps a few exceptions [12–15]. The only dedicated books presently available on the subject are (1) the renewed edition of the seminal book of Nevill Mott (which covers the initial stages of HTSC research) [6] and (ii) the book by Florian Gebhard [16], which mostly emphasized on the theories to understand physics of the Mott metal–insulator transition. The later book is more suited for researchers (at least in the graduate student level) interested in the theoretical aspects of Mott physics, with rather little coverage on the experimental aspects concerning Mott insulators. There is of course a very readable review article [17] covering both the theoretical and experimental aspects of Mott physics known until the late 1990s.

A new direction, however, has opened up in the field of Mott insulators generating considerable interest during the last decade. This concerns the possible technological applications of Mott insulators. It is now well recognized that current CMOS technology is fast approaching the fundamental limits, and there is a driving interest in the newer kinds of materials with newer physics concepts leading to devices for information processing and storage. To this end resistive switching based random access memory (RRAM) devices are emerging as a possible alternative to the current metal-oxide-semiconductor field-effect transistor (MOSFET) technology devices [18, 19]. Resistive switching is a physical phenomenon, which can originate from a sudden and non-volatile change of the electrical resistance in a suitable material due to the application of electric field. A purely electronically-driven resistive switching transition in the Mott insulators gives rise to the possibility for newer kind of fast devices with long endurance and low dissipation. Mott memory is based on the electronic phase transition between metallic and insulating states of correlated electron materials. These possibilities have stimulated a significant amount of research activity on the device physics involving various classes of Mott insulators [19, 20]. For example the Mott field-effect transistor (Mott-FET) with a similar structure to conventional semiconductor FET, with the semiconductor channel materials being replaced by correlated electron materials i.e. Mott insulator is now a distinct possibility. A Mott insulator can undergo an insulator-to-metal phase transitions under an applied electric field due to electrostatically doped carriers. The Mott-FET will utilize this phase induced by a gate voltage as the fundamental switching paradigm. All these possibilities have now been recognized and recorded in the International Technology Roadmap for Semiconductors [19].

In addition to the electric field excitation, the Mott insulator–metal transition can also be triggered by photo- and thermal-excitations. This opens up possibilities of even further applications like optical and thermal switches, thermo-chromic devices, gas sensors and even solar cell applications. A considerable amount of information has now been generated in the literature based on the ongoing efforts in this rapidly emerging field. And there is a lesser known interesting case of nuclear fuel material UO_2. This material has recently been recognized as a classic case of Mott–Hubbard insulator involving 5f electrons [21]. It is also not quite well recognized that this material has enormous potential for technological applications in various other areas including thermo-electric devices, piezomagnetism and magnetoelastic memory.

With all the recent developments in the area of physics and materials science of Mott insulators, especially their recognition as emergent materials for important and innovative device applications, the time has now come to make this subject of Mott insulator (physics and applications) accessible to a wider community. The present book is aimed more towards the advanced undergraduate and fresh graduate students of physics, chemistry, materials science, and electrical and electronics engineering. This will be an admixture of an advanced undergraduate level textbook and a monograph.

Starting with this general introduction the book is presented in two parts. Part I of the book will deal with the science of Mott insulators and part II of the book will give an exposition of how Mott physics can play a very important role in various emerging and innovative technologies. Since some of the targeted readers of the book are advanced undergraduates in physics, chemistry, materials science and electrical engineering, a pedagogical exposition of the currently established electron band theory of solids is provided in the first two chapters. In chapter 1 starting from the Drude–Sommerfeld free electron theory, the concept of electronic band structure will be introduced through the framework of electrons in the periodic potential of ions in a solid. The crucial roles of Pauli exclusion principle and independent electron approximation will be highlighted and metals, insulators and semiconductors will be defined, and that will be followed by a discussion on the Peierls transition and charge density wave. A brief discussion on the effect of disorder in crystalline solid leading to Anderson localization will also be provided. The independent electron picture is very successful in explaining many fundamental properties of solid materials, but it is also realized that in order to explain many important materials phenomena one needs to take into account the many body effects, namely the interactions between electrons. This necessity of going beyond the independent electron approximation along with the evolution of the important theoretical ideas and formalism will be systematically discussed in chapter 2. The concepts of 'exchange' and 'correlation', two important building blocks of solid materials, will be introduced and explained. Some flavor will be given on the highly successful density functional theory and the computational techniques. The materials presented in chapters 1 and 2 can be found in many standard solid state/condensed matter physics book, but will be necessary here to appreciate the next steps to understand the importance of the many body physics of strong electron–electron interaction in the context of the applications in real life materials. Chapter 3 provides a basic introduction to Mott physics in strongly correlated electron solids. It will be pointed out that in contrast to the dominating behaviour of wave nature of electrons in metals, band insulators and semiconductors studied in earlier chapters, the particle nature of electrons along with their mutual Coulomb interaction will now come to the forefront. Chapter 4 will deal with magnetic insulators within the framework of Mott physics with a few examples of real materials. Chapters 5 and 6 will be devoted to theoretical and experimental studies, respectively, of the Mott metal–insulator transition. The real solid materials chosen for the discussion are not really exhaustive but will be an admixture of some well studied Mott-insulators as well as some newer materials. The main aim is to give a flavour of the experimental results on the Mott metal–insulator transition, which will be helpful in the discussions on the Mott-insulator based device application presented in part II.

In part II of the book, in chapter 7 a short introduction will be provided on the current CMOS technologies where the electron band theory of solids (summarized in chapters 1 and 2 of part I of the book) plays a big role. It will be discussed how the existing CMOS based technology is reaching its saturation, hence the need for conceptually new and innovative technologies, and also the ongoing efforts in this latter direction. Chapter 8 will provide an introduction to the exciting field of Mott memory and logic devices. Many interesting device concepts supported by actual experiments and device fabrications have emerged in this field during the last two decades. This chapter will also narrate the possibilities of various other applications involving the Mott insulator, namely temperature, infrared and gas sensors, cathode materials for modern Li-ion batteries, solar cells and magnetic refrigeration.

Efforts have been made to present the theoretical works in a form accessible to advanced undergraduate students, fresh graduate students and experimental scientists keeping the number of equations to bare minimum, but at the same without compromising on the rigor of the actual physics. However, for interested students two appendices on second quantization and Green's function technique have been provided, which are actually essential for a complete understanding and appreciation of Mott physics. An appendix on some experimental techniques relevant to the context of the present book has also been provided. This might enthuse some of the students to become a bit more adventurous, and they may find that such experimental facilities to explore the interesting world of Mott physics are many a times readily available in their institutions. There will also be a set of self-assessment questions and exercises at the end of each chapter. These are designed to cater to the taste of a wide range of readers comprising of advanced undergraduate students of physics, chemistry, materials science and electrical/ electronics engineering and fresh graduate students of physics. The answers to most of the questions are there in the corresponding chapters itself. However, there are few exercises which will need some more involvement including calculations and maybe some literature study.

References

[1] Ashcroft N W and Mermin N D 1976 *Solid State Physics* (Philadelphia, PA: Saunders College)
[2] Wilson A H 1931 *Proc. R. Soc. Lond.* A **133** 458
    Wilson A H 1931 *Proc. R. Soc. Lond.* A **134** 277
[3] de Boer J H and Verway E J W 1937 *Proc. R. Soc. Lond.* A **49** 59
[4] Peierls R 1937 *Proc. Phys. Soc. Lond.* A **49** 72
[5] Mott N F 1968 *Mod. Phys. Rev.* **40** 67
[6] Mott N F 1990 *Metal Insulator Transition* 2nd edn (London: Taylor and Francis)
[7] Shull C G 1951 *Phys. Rev.* **83** 333
[8] Slater J C 1951 *Phys. Rev.* **82** 538
[9] Anderson P W 1962 *Solid State Physics* vol 13 ed G T Rado and H Shull (New York: Academic) p 99
[10] Anderson P W 2011 *More and Different: Notes from a Thoughtful Curmudgeon* (Singapore: World Scientific)
[11] Anderson P W and Baskaran G 1997 *A Critique of 'A Critique of Two Metals'* arXiv: cond-mat/9711197
[12] Madelung O 1978 *Introduction to Solid State Theory* (Berlin: Springer)
[13] Martin R M, Reining L and Ceperley D M 2016 *Interacting Electrons: Theory and Computational Approaches* (Cambridge: Cambridge University Press)
[14] Fulde P 1995 *Electron Correlations in Molecules and Solids* (Berlin: Springer)
[15] Simon S H 2013 *Solid State Basics* (Oxford: Oxford University Press)
[16] Gebhard F 1997 *Metal Insulator Transition: Models and Methods* (Berlin: Springer)
[17] Imada M *et al* 1998 *Rev. Mod. Phys.* **70** 1039
[18] Sawa A 2008 *Mater. Today* **11** 28
[19] International Technology Roadmap for Semiconductors 2.0: 2015 Edition: Emerging Research Devices (Semiconductor Industries Association) https://www.semiconductors.org/wp-content/uploads/2018/06/0_2015-ITRS-2.0-Executive-Report-1.pdf
[20] Janod E *et al* 2015 *Adv. Funct. Mater.* **25** 6287
[21] Yu S W *et al* 2011 *Phys. Rev.* B **83** 165102

Part I

Physics of Mott insulators

IOP Publishing

Mott Insulators
Physics and applications
Sindhunil Barman Roy

Chapter 1

Electrons in crystalline solids

The atomic nature of matter has been firmly established right from the beginning of the 20th century, with the advent of quantum mechanics and newer experimental probes like x-ray diffraction and optical spectroscopy. At very high temperatures matter exists in a gaseous state, which ideally is an assembly of atoms in a uniform, disordered, uncorrelated and isotropic state. In the language of science the gaseous state of matter has complete translational and rotational symmetry. With a decrease in temperature, fundamental force laws between atoms start countering the disruptive effect of thermal or kinetic energy, which then leads to the formation of molecules in the case of elements like oxygen. Some other elements like inert gas argon, neon etc and various metals, however, continue to retain their atomic identity. With further decrease in average kinetic energy (with reduction in temperature) the role of inter atomic (molecular) interaction i.e. potential energy becomes more and more important. Eventually formation of clusters takes place in this assembly of atoms and molecules with an overall lowering in energy as well as entropy.

This competition between kinetic and potential energy has been exemplified with the beautiful example of water-H_2O in a well known textbook by Chaikin and Lubensky [1]. All of us are quite familiar with the three states of water, namely water vapour, liquid water and ice. Attraction in the form of hydrogen bonding between the water molecules would encourage the molecules to spend more time in the regions where there are other molecules; this in turn will enhance density fluctuations in the water vapour state. The density of water vapour, however, will still be uniform when averaged over a large region of space and over a long time interval. Eventually the intermolecular interaction will lead to the liquid phase of water below a particular temperature, known as the vapour to liquid transition temperature. Both water vapour and liquid water phases look irregular with random arrangement of water molecules (see figure 1.1(a)). However, these phases have a complete rotational and translational symmetry; figure 1.1(a) will look the same if the container of water vapour or liquid water is tipped or shaken. The liquid phase of

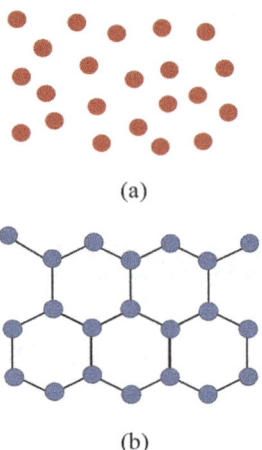

(a)

(b)

Figure 1.1. Schematic representation of (a) the random arrangement of water molecules in liquid water, (b) periodic arrangement of water molecules in solid water (i.e. ice crystal phase) with broken symmetry.

water will continue to have the translational and rotational symmetry of the higher temperature vapour phase, but its density is distinctly larger than the vapour phase. Thus the physical quantities distinguishing the liquid and gas phases in water are their density and compressibility. The positions of the atoms (molecules) in the liquid state, however, are relatively more correlated, and the distance between them is set by the competition between the attractive and repulsive parts of inter-atomic (inter-molecular) interactions [1].

Liquid water turns into ice crystal on further cooling. In fact water can exist in 13 different crystalline forms, of which at atmospheric pressure ordinary hexagonal ice is stable between 72 and 273 K [2]. The ice crystal does not flow like water; it is rigid and it resists shear. Microscopically, the water molecules in ice are arranged in a uniform repetitive way (see figure 1.1(b)), and the probability of finding a water molecule at a particular position in ice crystal will depend on the positions of distant molecules in the crystal. In other words the short range correlations amongst the molecules in liquid water turn into a long range order in the ice crystal with a periodic density. In general, the atoms (molecules) in a solid are arranged in a uniform repetitive way in a periodic lattice. A solid, however, is not homogeneous and isotropic like a liquid, and it is not invariant with respect to arbitrary translational and rotation. The transition from the liquid to the solid state breaks the symmetry of the liquid state, and the solid will remain invariant only under those operations, which leaves the periodic lattice of a solid unchanged. For example, the two-dimensional picture of ice presented in figure 1.1(b) clearly breaks the rotational invariance: it can be rotated only by 120° or 240°. This is also invariant only under translation by a lattice spacing that leaves the periodic arrangement unchanged. This broken symmetry phase of solid matter with periodic lattice structure is at the root of various interesting and useful physical properties of solids [1]; some such examples of interesting physical properties will be discussed in the later part of this book. In fact a proper understanding of the physical properties of solid matters would have been practically impossible if the most stable structure for most solids was not a regular periodic crystal lattice [3]. It would not be totally out of context to mention here that the concept of broken symmetry across a phase transition, which originated from the study of condensed matter phases, has now made profound impact in the fields well beyond the realm of condensed matter, especially in the modern day particle physics. In fact the whole concept of Higgs mechanism and Higgs boson actually emerged from the effort of understanding various interesting aspects of the superconducting state in certain classes of solid materials, in particular the concept of spontaneously broken gauge symmetry across the phase transition from normal conducting state to superconducting state [4].

Another event that gave a big impetus to the understanding of the physical properties of matter was J J Thomson's discovery of the electron in 1897. The existence of these minute particles with electrical charge provided immediate explanation of large electrical and thermal conductivity in a class of solid materials: the metals. Subsequently, quantum mechanics helped to recognize another very important attribute of the electron, namely its intrinsic spin; apart from carrying electric charge the electron also acts like a tiny magnetic dipole. Furthermore, within the framework of quantum physics, two electrons are not allowed to occupy the same energy state. This is known as the Pauli exclusion principle, which introduces some correlation in the motion of electrons with parallel spin configuration. These three important concepts: (1) periodic lattice structure of crystalline solid, (2) existence of electrons with tiny electrical charge $(1.6022 \times 10^{-19}$ C) and magnetic moment $(9.274 \times 10^{-24}$ A m^2 or J T$^{-1})$, and (3) Pauli exclusion principle, together form the corner stone of modern solid state physics and materials science, and provide the basis of understanding of all the important physical properties of solid crystalline matters.

Historically, the early scientists first used these three important concepts to understand the properties of metals. This choice was natural because the existence of the electron provided an obvious mechanism for large electrical conductivity in metals. It may, however, be noted that the majority of the solid matter we encounter in our day to day life, and also the core materials of interest in the present book namely the *Mott insulators*, are actually non-metals. We will see in the sections below that those initial efforts to study the metallic properties of the solid were not really out of the present context. In fact, that laid the foundation of a comprehensive quantum theory of solids encompassing all kinds of materials, which is now expanding rapidly and enabling one to understand the material properties better, and then tune their functionalities.

The present author has been significantly influenced in his study by the classic textbooks by Chaikin and Lubensky [1], Ziman [3], Ashcroft and Mermin [5], Ibach and Lüth [6] and the relatively recent books by Singleton [7], Hoffman [8], Simon [9], and that will be reflected in the narratives in the sections below and the rest of the book.

1.1 Periodic structure of solids and Bragg scattering

The microscopic arrangement of atoms/molecules in a solid matter cannot be observed with human eyes. One can, however, test the existence of this periodic arrangement of atoms (molecules) in the microscopic state of solid matter, in the same way one tests (theoretically) the periodicity in anything, that is by taking a Fourier transform and looking for discrete peaks in its spectrum [1]. Scattering of x-rays from matter is the experimental way of taking the Fourier transform, and the existence of a discrete spatial Fourier spectrum distinguishes a solid crystal from a liquid.

The well known example of scattering, which provides structural information in a solid, is the Bragg scattering of an incident wave from a set of equi-spaced and partially reflecting parallel planes. The set of parallel planes will diffract an incident wave, and the intensity of the wave will be modulated by constructive/destructive interference. The reflection of the wave for an infinite set of such parallel planes will survive, if there is a constructive interference between waves reflected by each set of neighbouring planes. In this case the path difference between waves reflected from the neighbouring planes separated by a distance d needs to be an integral multiple of the wavelength λ of the incident wave (see figure 1.2). This leads to *Bragg's law*:

$$2d \sin \theta = n\lambda \tag{1.1}$$

where θ is the angle as shown in figure 1.2 and n is an integer. The angle between the incident and the scattered wave is 2θ, and the scattered intensity at angle 2θ shows a fluctuation or inhomogeneity of the system with periodicity $(\lambda/2 \sin \theta)$ [1].

Scattering is most useful as a tool to probe the structure of a solid when the scattering particles interact only weakly with the solid under study. Bragg's law in equation (1.1) also indicates that the wavelength of the particle to be scattered needs to be less than twice the nearest neighbour distance i.e. $\lambda < 2d$. The usual solid materials have inter-particle spacing in the scale of angstrom (Å), and hence the energies of the scattered particles must correspond to the wavelengths in this

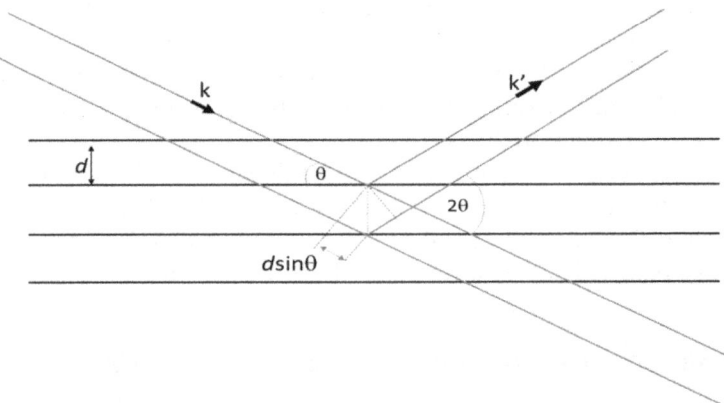

Figure 1.2. Schematic diagram of the Bragg scattering originating from equi-spaced parallel planes.

length scale. The dispersion relation corresponding to the energy and wave number $k = 2\pi/\lambda$ is given by:

$$\epsilon = \hbar\omega = \hbar ck = hc/\lambda \tag{1.2}$$

where $\hbar$ is Planck's constant, h divided by 2π, and c is velocity of light. Photons of visible light have energy in the range $\epsilon \sim 1$ eV and $\lambda = 0.4\text{--}0.7 \times 10^4$ Å, and thus they are suitable for studying structures in the scale of micro-meter. In this scale the scattering occurs due to the variations in the dielectric constant or refractive index in the matter. Investigation of the periodic lattice structure of a solid in the angstrom scale will require photons with energy in the range of 10^4 eV, and scattering events are due to variations in the electronic charge density. In the electromagnetic spectrum the x-rays have energies in this range, and can penetrate up to a millimeter of solids. Thus, x-rays are most widely used probe for studies of the microscopic bulk structure in solid materials.

We shall now study the scattering of a particle from a periodic lattice of a solid in a more rigorous way, by assuming that the particle is represented quantum mechanically by a plane wave of certain energy and momentum. Assuming $|\mathbf{k}\rangle$ and $|\mathbf{k}'\rangle$ to be the incident and final plane wave state of the scattered particle with momenta $\hbar\mathbf{k}$ and $\hbar\mathbf{k}'$, respectively, we shall now calculate the transition probability between these plane wave states of the scattered particle. We assume that the scattered particle interacts weakly with the scattering medium via a potential U so that only the lowest order scattering event needs to be considered for the entire sample. Then, by Fermi's golden rule, the transition rate between $|\mathbf{k}\rangle$ and $|\mathbf{k}'\rangle$ is proportional to the square of the matrix element $M_{\mathbf{k},\mathbf{k}'}$, where

$$M_{\mathbf{k},\mathbf{k}'} = \langle \mathbf{k}|U|\mathbf{k}'\rangle = \int d^d x\, e^{-i\mathbf{k}\cdot\mathbf{x}} U(\mathbf{x}) e^{i\mathbf{k}'\cdot\mathbf{x}}. \tag{1.3}$$

Here $U(\mathbf{x})$ is the scattering potential in the coordinate representation of the scattered particle, and $\mathbf{x}$ is a vector in a d-dimensional space. In most of the physical systems d is three, two or one. The differential cross-section $d^2\sigma/d\Omega$ per unit solid angle of the final wave vector $\mathbf{k}'$ is expressed as [1]:

$$\frac{d^2\sigma}{d\Omega} \sim \frac{2\pi}{\hbar} |M_{\mathbf{k},\mathbf{k}'}|^2. \tag{1.4}$$

Equation (1.4) represents a static cross-section, which can be obtained experimentally through x-ray diffraction by integrating over all possible energy transfers to the medium.

The scattering potential in multi particle systems, is the sum of terms arising from each of the individual atoms in the system [1]:

$$U(\mathbf{x}) = \sum_\alpha U_\alpha(\mathbf{x} - \mathbf{x}_\alpha) \tag{1.5}$$

where $\mathbf{x}_\alpha$ is the position of the atom arbitrarily labeled α. The matrix element in the differential cross-section then takes the form:

$$M_{\mathbf{k},\mathbf{k'}} = \sum_\alpha \int e^{-i\mathbf{k}\cdot\mathbf{x}} U_\alpha(\mathbf{x} - \mathbf{x}_\alpha) e^{i\mathbf{k'}\cdot\mathbf{x}} d^d x. \tag{1.6}$$

With the substitution $\mathbf{R}_\alpha = \mathbf{x} - \mathbf{x}_\alpha$ equation (1.6) can be rewritten as:

$$\begin{aligned} M_{\mathbf{k},\mathbf{k'}} &= \sum_\alpha \int e^{-i\mathbf{k}\cdot(\mathbf{x}_\alpha+\mathbf{R}_\alpha)} U_\alpha(\mathbf{R}_\alpha) e^{i\mathbf{k'}\cdot(\mathbf{x}_\alpha+\mathbf{R}_\alpha)} d^d R_\alpha \\ &= \sum_\alpha \left[\int e^{-i\mathbf{q}\cdot\mathbf{R}_\alpha} U_\alpha(\mathbf{R}_\alpha) d^d R_\alpha \right] e^{-i\mathbf{q}\cdot\mathbf{x}_\alpha} \\ &= \sum_\alpha U_\alpha(\mathbf{q}) e^{-i\mathbf{q}\cdot\mathbf{x}_\alpha}. \end{aligned} \tag{1.7}$$

Here $\mathbf{q} = \mathbf{k} - \mathbf{k'}$ is called the scattering wave vector, and $U_\alpha(\mathbf{q}) = \int e^{-i\mathbf{q}\cdot\mathbf{R}_\alpha} U_\alpha(\mathbf{R}_\alpha) d^d R_\alpha$ the atomic from factor or Fourier transform of the atomic potential. The differential scattering cross-section for this multiparticle system can be written as:

$$\frac{d^2\sigma}{d\Omega} \sim \frac{2\pi}{\hbar} \sum_{\alpha,\alpha'} U_\alpha(\mathbf{q}) U_{\alpha'}^*(\mathbf{q}) e^{-i\mathbf{q}\cdot\mathbf{x}_\alpha} e^{i\mathbf{q}\cdot\mathbf{x}_{\alpha'}}. \tag{1.8}$$

Equations (1.4) and (1.8) correctly give the scattering cross-section if the positions of the atoms in the system are fixed. In real systems there will be atomic movement due to thermal and quantum fluctuations, and hence some ensemble average of the ideal scattering cross-section is required [1]. If the detector in the x-ray diffractometer records all particles scattered by a certain wave vector independent of their energy change, then each scattering event obtains a snapshot of the sample. In the standard experiments, data are collected over a period of time that is long enough compared to thermal equilibration time, hence the different snapshots correspond to a time average over many sample configurations.

For a system with identical atoms equation (1.8) for scattering cross-section is rewritten as:

$$\frac{d^2\sigma}{d\Omega} \sim \frac{2\pi}{\hbar} |U_\alpha(\mathbf{q})|^2 I(\mathbf{q}) \tag{1.9}$$

where the function $I(\mathbf{q})$ is:

$$I(\mathbf{q}) = \left\langle \sum_{\alpha,\alpha'} e^{-i\mathbf{q}\cdot(\mathbf{x}_\alpha-\mathbf{x}_{\alpha'})} \right\rangle. \tag{1.10}$$

$I(\mathbf{q})$ is called the *structure function*, and it depends only on the positions of the atoms in the material and not on the nature of the interaction between atoms and the scattering probe particle. In a system with N atoms, $I(\mathbf{q})$ is a sum of N^2 complex numbers with phases determined by the positions of all N atoms [1]. We can see now that the differential scattering cross-section of plane-wave states provides direct information about the spatial structure of many-particle systems.

1.2 The direct lattice and the reciprocal lattice

All the elements except helium and many solid materials formed by combining those elements exist in a crystalline solid phase at ambient pressure at some temperatures. An ideal crystal consists of an array of identical copies of a single structural unit, which are repeated periodically. The structural unit contains a single atom in the most simple case. Figure 1.3 shows an example of a two-dimensional square lattice with a single atom at each lattice site. The repeated structural unit of the crystalline solid is called a *unit cell*, and the one with the smallest possible volume is known as a *primitive unit cell*. There are solid materials, where the lattice sites have many different atoms or a continuous variation of mass density around some mean value. When the unit cell has multiple atoms, the positions of the atoms relative to the center of the cell are termed the basis. In a d-dimensional perfect crystal, equivalent points in unit cells reside on a periodic lattice known as the Bravais lattice, which consists of a mathematical array of points. An integral linear combination of independent non-coplanar vectors, $\mathbf{a}_1, \mathbf{a}_2, \ldots, \mathbf{a}_d$, can specify any lattice in a d-dimensional space:

$$\mathbf{R}_n = n_1\mathbf{a}_1 + n_2\mathbf{a}_2 + \cdots + n_d\mathbf{a}_d \tag{1.11}$$

where $n_1, n_2, \ldots, n_d$ are components of a d-dimensional vector $\mathbf{n}$, which indexes a particular unit cell.

The Bravais lattice is completely defined by this set of noncoplanar vectors $\mathbf{a}_1, \mathbf{a}_2, \ldots, \mathbf{a}_d$, which are known as primitive lattice translation vectors. Equivalent points in the lattice are connected by a translation vector or lattice vector:

$$\mathbf{R} = \mathbf{R}_n - \mathbf{R}_{n'} \tag{1.12}$$

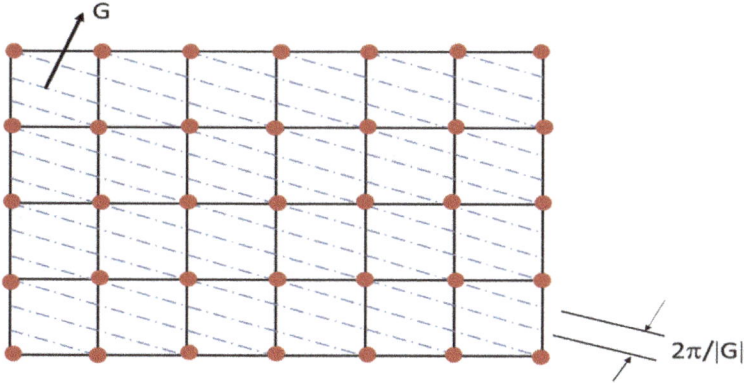

Figure 1.3. Schematic diagram of a two dimensional square lattice in real space represented by red coloured dots connected by bold black lines. Blue coloured dashed lines present a set of equi-spaced parallel planes containing all lattice sites of the square lattice. This set of planes is perpendicular to the vector **G**.

for any n and n'. The lattice consisting of points in coordinate space is termed the *direct lattice*. A physical crystalline solid is a material whose average mass density is a periodic function of space.

For a crystalline solid, a unit cell is defined as a region of space, the repetition of which can fill all the space. There can be an infinite set of choices of unit cell, and the minimum volume unit cells are known as primitive unit cells, which in three dimensions will have a volume:

$$V_{r3} = \mathbf{a}_1 \cdot (\mathbf{a}_2 \times \mathbf{a}_3). \tag{1.13}$$

This unit is a parallelepiped with edges as the primitive translation vectors. It is quite convenient, however, to choose a primitive unit cell, which is centered around a lattice point, and therefore explicitly reflects the symmetry of the underlying lattice. Such a primitive unit cell is called a *Wigner–Seitz cell*, which can be obtained by constructing perpendicular bisectors to all lattice vectors starting from a given lattice point. The smallest volume enclosed by bisector planes defines the Wigner–Seitz cell. The left-hand side of figure 1.4 shows an example of Wigner–Seitz cell for a representative two-dimensional periodic lattice. Apart from the translational invariance, a lattice may also be invariant under symmetry operations such as rotation through a certain angle, inversion and reflection. There are only 14 distinct types of Bravais lattice in three dimensions satisfying all these symmetry properties [5].

In any periodic lattice, there is a set of equi-spaced parallel planes containing all lattice points. This is shown in figure 1.3. A normal vector $\mathbf{G}$ defines each set of these planes (see figure 1.3). In a given plane perpendicular to $\mathbf{G}$, translation vectors $\mathbf{R}$ satisfy the condition $\mathbf{G} \cdot \mathbf{R} = $ constant. For this set of parallel planes all translation vectors $\mathbf{R}$ lie in some plane which satisfies the condition:

$$\mathbf{G} \cdot \mathbf{R} = 2\pi n \tag{1.14}$$

where n is some integer. The choice of the coefficient 2π leads to the relation: $e^{i\mathbf{G}\cdot\mathbf{R}} = 1$. In the nth plane associated with $\mathbf{G}$ any point $\mathbf{x}_n$ satisfies the relation $\mathbf{G} \cdot \mathbf{x}_n = 2n\pi$.

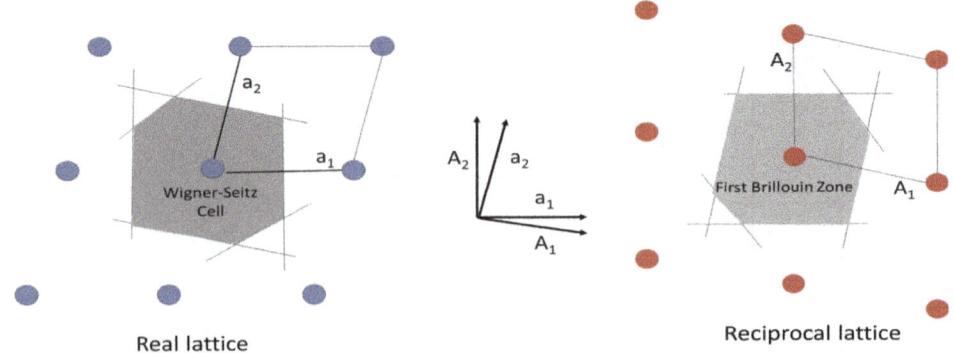

Real lattice   Reciprocal lattice

Figure 1.4. Wigner–Seitz cell for a representative two-dimensional periodic lattice and the first Brillouin zone of the corresponding two-dimensional reciprocal lattice. The vectors $\mathbf{A}_1$ and $\mathbf{A}_2$ of this two-dimensional representative reciprocal lattice are perpendicular to the primitive translation lattice vectors $\mathbf{a}_2$ and $\mathbf{a}_1$, respectively of the direct lattice.

The difference $(\mathbf{x}_n - \mathbf{x}_{n-1})$ between points in the adjacent planes satisfies the relation $\mathbf{G} \cdot (\mathbf{x}_n - \mathbf{x}_{n-1}) = 2\pi$. The component of $(\mathbf{x}_n - \mathbf{x}_{n-1})$ parallel to $\mathbf{G}$ is the distance l between the adjacent planes, and hence $l = 2\pi/|\mathbf{G}|$.

It is possible to have a set of reciprocal vectors $(\mathbf{A}_1, \mathbf{A}_2, \ldots, \mathbf{A}_d)$ corresponding to any set of primitive translation vectors $(\mathbf{a}_1, \mathbf{a}_2, \ldots, \mathbf{a}_d)$, satisfying the relation:

$$\mathbf{a}_i \cdot \mathbf{A}_j = 2\pi\delta_{ij}; \quad i, j = 1, 2, \ldots, d. \tag{1.15}$$

In three dimensions: $\mathbf{A}_1 = 2\pi(\mathbf{a}_2 \times \mathbf{a}_3)/[\mathbf{a}_1 \cdot (\mathbf{a}_2 \times \mathbf{a}_3)]$; $\mathbf{A}_2 = 2\pi(\mathbf{a}_3 \times \mathbf{a}_1)/[\mathbf{a}_1 \cdot (\mathbf{a}_2 \times \mathbf{a}_3)]$; $\mathbf{A}_3 = 2\pi(\mathbf{a}_1 \times \mathbf{a}_2)/[\mathbf{a}_1 \cdot (\mathbf{a}_2 \times \mathbf{a}_3)]$. Any vector that satisfies equation (1.14), can now be written as:

$$\mathbf{G} = m_1\mathbf{A}_1 + m_2\mathbf{A}_2 + \cdots + m_d\mathbf{A}_d \tag{1.16}$$

where $m_1, m_2, \ldots, m_d$ are positive or negative integers or zero.

The above exercise shows that the existence of a lattice in the real space or r-space also implies the existence of a lattice in the reciprocal space or k-space. The vectors $\mathbf{G}$ define the periodic reciprocal lattice with $\mathbf{A}_j$ as its primitive translation vectors.

At this point it is worth having a formal definition of a reciprocal lattice [5]. Let us consider a set of points $\mathbf{R}$, which constitutes a Bravais lattice and a plane wave $e^{i\mathbf{k} \cdot \mathbf{r}}$. Such a plane wave will not have the periodicity of the Bravais lattice for general $\mathbf{k}$. This will only happen for some special choices of wave vector. The set of all wave vectors $\mathbf{G}$ that gives rise to plane waves with the periodicity of a given Bravais lattice is known as its reciprocal lattice [5]. The vectors $\mathbf{G}$ will belong to the reciprocal lattice of a Bravais lattice of points $\mathbf{R}$, provided the following relation is satisfied:

$$e^{i\mathbf{G} \cdot (\mathbf{r}+\mathbf{R})} = e^{i\mathbf{G} \cdot \mathbf{r}} \tag{1.17}$$

for any $\mathbf{r}$, and all $\mathbf{R}$ in the Bravais lattice. If the term $e^{i\mathbf{G} \cdot \mathbf{r}}$ is canceled out from each side of the above equation, then the reciprocal lattice is characterized as the set of vectors $\mathbf{G}$ satisfying the relation:

$$e^{i\mathbf{G} \cdot \mathbf{R}} = 1 \tag{1.18}$$

for all $\mathbf{R}$ in the Bravais lattice. It may be noted that a reciprocal lattice is defined in reference to a particular Bravais lattice.

The k-space periodicity also implies that all information will be contained in the primitive unit cell of the reciprocal lattice. As in the case of the Wigner–Seitz cell in the direct lattice, a primitive unit cell can be constructed in the reciprocal lattice with the help of perpendicular bisectors to all lattice vectors emerging from a given reciprocal lattice point. The smallest volume thus obtained is known as the *first Brillouin zone*, and it has a k-space volume:

$$V_{k3} = \mathbf{A}_1 \cdot (\mathbf{A}_2 \times \mathbf{A}_3). \tag{1.19}$$

The right-hand side of figure 1.4 shows an example of the first Brillouin zone for a representative two-dimensional representative reciprocal lattice. The vectors $\mathbf{A}_1$ and $\mathbf{A}_2$ of this two-dimensional representative reciprocal lattice are perpendicular to the primitive translation lattice vectors $\mathbf{a}_2$ and $\mathbf{a}_1$, respectively of the corresponding direct lattice.

Any periodic function of position in $f(\mathbf{x}) = f(\mathbf{R} + \mathbf{x})$ in a real lattice can be expressed in terms of Fourier components with wave vectors in the reciprocal lattice:

$$f(\mathbf{x}) = \sum_G f_G e^{i\mathbf{G}\cdot\mathbf{x}}. \tag{1.20}$$

This can be shown by taking the general Fourier transform of $f(x)$ [1]:

$$f(\mathbf{q}) = \int e^{-i\mathbf{q}\cdot\mathbf{x}} f(\mathbf{x}) d^d x. \tag{1.21}$$

Now taking the integral over a unit cell, one can write:

$$f(\mathbf{q}) = \sum_T \int_0 e^{-i\mathbf{q}\cdot(\mathbf{x}+\mathbf{R})} f(\mathbf{x} + \mathbf{R}) d^d x$$
$$= \left(\sum_R e^{-i\mathbf{q}\cdot\mathbf{R}} \right) \int_0 e^{-i\mathbf{q}\cdot\mathbf{x}} f(\mathbf{x}) d^d x. \tag{1.22}$$

In the above equation zero lower limit in the integral sign indicates the integration over a unit cell, and the summation over $\mathbf{R}$ is equal to the number N of cells in the lattice if $\mathbf{q}$ is a reciprocal lattice vector or zero otherwise. Accordingly with V_0 being the volume of a unit cell equation (1.22) can be rewritten as:

$$f(\mathbf{q}) = NV_0 \sum_G \delta_{\mathbf{q},\mathbf{G}} f_G = \sum_G (2/\pi)^d \delta(\mathbf{q} - \mathbf{G}) f_G \tag{1.23}$$

where,

$$f_G = \frac{1}{V_0} \int f(\mathbf{x}) e^{-i\mathbf{G}\cdot\mathbf{x}} d^d x. \tag{1.24}$$

The scattering potential involved in equations (1.5) and (1.6) is a periodic function of position. Thus, for a periodic lattice with rigidly fixed scatterers at the lattice sites, the scattering matrix element in equation (1.6) can be rewritten as:

$$M_{\mathbf{k},\mathbf{k}'} = V \sum_G U_G \delta_{\mathbf{q},\mathbf{G}} \tag{1.25}$$

where $\mathbf{q} = \mathbf{k} - \mathbf{k}'$ is the difference between incident and scattered wave vectors. We can now write the scattering cross-section as:

$$\frac{d^2\sigma}{d\Omega} = V^2 \sum_G |U_G|^2 \delta_{\mathbf{q},\mathbf{G}}. \tag{1.26}$$

The above equation (1.26) indicates that the scattering pattern will have peaks at every reciprocal lattice vector. The intensity of the peak will be proportional to the square of the volume of the sample and to the square of the Fourier component of the scattering potential at wave vector $\mathbf{G}$. These are the Bragg scattering peaks

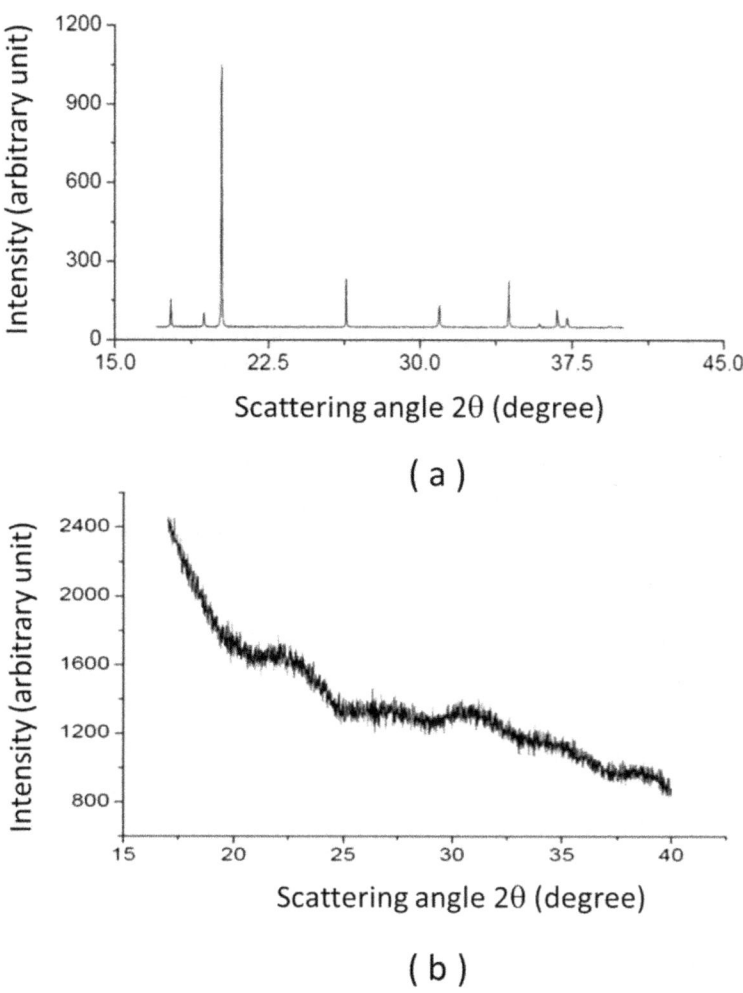

Figure 1.5. X-ray diffraction patterns of (a) elemental titanium (Ti), and (b) float-glass, obtained using a synchrotron radiation x-ray source. Characteristic Bragg peaks are observed in the Ti sample with periodic lattice structure, but not in aperiodic float-glass.[1]

of the solid material. Experimentally, x-ray diffraction study from crystalline material is the way of taking the Fourier transform. Figure 1.5(a) presents such an x-ray diffraction pattern obtained for elemental titanium (Ti) solid showing the existence of characteristic Bragg peaks. For comparison, we also show in figure 1.5(b) the x-ray diffraction pattern obtained from a non-crystalline solid material—float glass, where the structural units are arranged in a random manner. As expected, the x-ray diffraction pattern of float glass does not show the characteristic Bragg-peaks.

---

[1] These x-ray diffraction results were obtained using Indus-2 synchrotron radiation source at Raja Ramanna Centre for Advanced Technology, Indore, India.

The scattering into Bragg peaks discussed above is elastic in nature. Hence, the magnitude of the incident and the scattered wave vectors is the same i.e. $|\mathbf{k}| = |\mathbf{k}'|$. If we now set $\mathbf{q} = \mathbf{G}$, then $\mathbf{k}' = \mathbf{k} - \mathbf{G}$. From this one can write:

$$|\mathbf{k}'|^2 = |\mathbf{k}|^2 + |\mathbf{G}|^2 - 2\mathbf{k} \cdot \mathbf{G} \qquad (1.27)$$

or,

$$\mathbf{k} \cdot (\mathbf{G}/2) = |\mathbf{G}/2|^2. \qquad (1.28)$$

This is the well known *Laue condition*, which states that an incident wave, which lies on a Brillouin zone face in the reciprocal lattice, will be Bragg scattered across the zone to the opposite face. Substituting $|\mathbf{G}| = 2\pi d$ and $|\mathbf{k}| = 2\pi/\lambda$ into equation (1.27) (or (1.28)) and recalling that the angle θ in equation (1.1) representing Bragg's scattering law is the angle between $\mathbf{k}$ and the scattering planes, one can see that equations (1.1) and (1.27) (or (1.28)) are equivalent. It may be interesting to note that Max von Laue and co-workers recorded the very first diffraction x-ray diffraction image from a single crystal sample of $CuSO_4 \cdot 5H_2O$ [10], which is an interesting material in the context of the present book [11] (see chapter 4).

1.3 Metallic solid: a container of free electron gas

1.3.1 Drude model

A metal atom consists of a nucleus of charge eZ surrounded by the total number of Z electrons, where Z is the atomic number and e is the magnitude of electronic charge. Out of these Z electrons, a relatively small number Z_v of outer electrons or valence electrons remain more loosely bound. When the metal atoms are brought in together to form a crystalline metallic solid, the outermost electrons or valence electrons become dissociated from the individual atoms and wander freely through-out the solid, while the tightly bound inert-gas (He, Ar, Ne, Kr and Xe)-like ionic cores consisting of a nucleus of charge eZ and $(Z - Z_v)$ electrons remain intact and arrange themselves in a periodic lattice. Within three years of Thomson's discovery of electron, in 1900 Paul Drude introduced a theory to explain electrical and thermal conductivity of metals. The centre point of the Drude model is the idea of a gas of electrons moving freely in the periodic lattice of immobile ionic cores. Figure 1.6 presents a schematic representation of the Drude model of electrons in metals with a periodic lattice. This gas of electrons is responsible for the good electrical conductivity of metals. Drude applied the well established and highly successful kinetic theory of gases to this gas of conduction electrons, which has typically of the order of 10^{22} electrons per cubic centimeter of metal. It may be noted here that in contrast to the molecules of an ordinary gas, the electron gas in metals exist in an environment of periodic lattice of ionic cores. The main role of ions in the Drude model is to maintain the charge neutrality of the metal, and also to keep the electrons confined within the solid. The basic assumptions of the Drude model are the following:

(1) A collision within the electron gas represents the scattering of an electron only by an ionic core. Drude assumed that the electrons just bounce off the

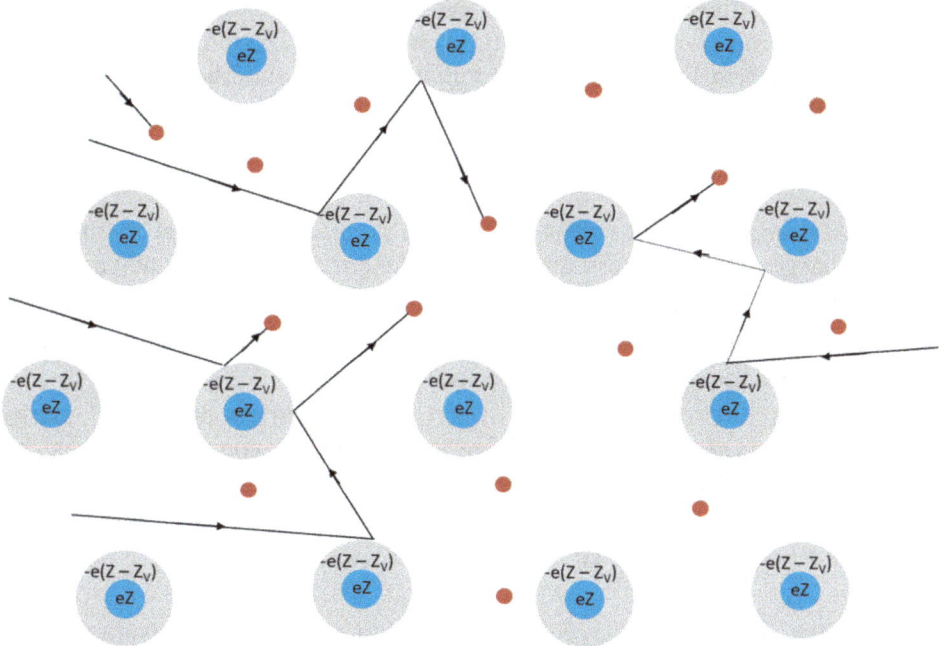

Figure 1.6. Schematic diagram of the Drude Model representing an electron gas in the background of periodically arranged ion-cores in metals. The bold lines represent typical trajectories of electrons in this electron gas.

impenetrable ion cores. This is an instantaneous event, which alters the velocity of an electron abruptly. A few such representative electron-trajectories are shown with bold lines in figure 1.6.

(2) In between collisions the electrons do not interact electromagnetically with the ionic cores. This is known as the *free electron approximation*.

(3) In between collisions the electrons do not collide or interact electro-magnetically with each other. This is known as the *independent electron approximation*.

(4) An electron experiences a collision with a probability per unit time $\frac{1}{\tau}$. This is known as the *relaxation time approximation*.

(5) The electrons in the conduction electron gas achieve thermal equilibrium with their surroundings only through collisions with the ionic cores.

A very well known electrical property of metals is that they follow Ohm's law, which states that the electrical current I flowing in a metallic wire is proportional to the potential drop U along the wire i.e. $U = IR$, where R stands for the electrical resistance of the metallic wire. The Drude model could explain this behavior and provided some idea on the magnitude of the electrical resistance. The electrical resistance R of the metallic wire depends on the dimension of the wire, but is independent of the magnitude of U and I. The dependence of R on the size and shape of the metallic wire can be eliminated by introducing a material dependent parameter called resistivity ρ, which is a proportionality constant between the electric field $\mathbf{E}$ at a point in the metal and the electrical current density $\mathbf{j}$ it induces:

$$\mathbf{E} = \rho \mathbf{j}. \tag{1.29}$$

The magnitude of current density $\mathbf{j}$, which is a vector parallel to the flow of electrical charge, can be expressed in terms of the uniform current I flowing through a metallic wire of length L and cross-sectional area A as $j = \frac{I}{A}$. The potential drop along the length of the wire is expressed as $U = EL$. This relation along with equation (1.29) leads to $U = \frac{I\rho L}{A}$, and in turn $R = \frac{\rho L}{A}$.

Electrons are moving at any point of metal with a variety of thermal energy in many directions. The net current density $\mathbf{j}$ thus can be expressed in terms of the average velocity $\mathbf{v}$ of electrons as,

$$\mathbf{j} = -ne\mathbf{v} = -\frac{ne}{m_e}\mathbf{p} \tag{1.30}$$

where n is the electron density, m_e is the electron mass, $\mathbf{p}$ is the average electron momentum and e is electronic charge.

In the absence of any applied electric field $\mathbf{E}$, the electrons are equally likely to move in any direction, hence $\mathbf{v}$ averages out to zero, and as a consequence there will be no net current density $\mathbf{j}$. In the presence of an applied electric field $\mathbf{E}$, the negatively charged electrons will experience a force $-e\mathbf{E}$ and acquire a mean velocity in the direction opposite to $\mathbf{E}$. This velocity can be estimated, starting with a representative electron, which has just undergone a collision at time $t = 0$. Starting with an instantaneous velocity $\mathbf{v}_0$ at that point of time the electron will gather an additional velocity $\frac{-e\mathbf{E}t}{m_e}$. The Drude model assumes that an electron emerges from a collision in a random direction and there is no contribution to the average electronic velocity $\mathbf{v}_{av}$ from $\mathbf{v}_0$. Hence, $\mathbf{v}_{av}$ will arise entirely from the average value of $\frac{-e\mathbf{E}t}{m_e}$. The average of t here is the relaxation time τ. Therefore,

$$\mathbf{v}_{av} = -\frac{e\mathbf{E}\tau}{m_e}; \quad \mathbf{j} = \left(\frac{ne^2\tau}{m_e}\right)\mathbf{E}. \tag{1.31}$$

This result is usually represented in terms of electrical conductivity $\sigma = \frac{1}{\rho}$:

$$\mathbf{j} = \sigma\mathbf{E}; \quad \sigma = \left(\frac{ne^2\tau}{m_e}\right). \tag{1.32}$$

Equation (1.32) establishes the linear dependence of $\mathbf{j}$ on $\mathbf{E}$, and also an estimate of electrical conductivity σ in terms of known parameters like electrical charge, mass and density of electrons in metals, except for the relaxation time τ. One may therefore use equation (1.32) and the experimentally determined value of electrical resistivities to have an estimate of the relaxation time:

$$\tau = \frac{m_e}{\rho n e^2}.$$

(1.33)

Room temperature resistivity of various metals at room temperature are usually of the order of microhm-centimeters [5]. This leads to typical values of relaxation time $\tau \approx 10^{-14}$–10^{-15} s at room temperature. To check whether this is a reasonable number or not, it may be instructive to estimate the electron mean free path $l = v_{av}\tau$, where v_{av} is the average electronic velocity. Electronic mean free path l is a measure of the average distance an electron travels between collisions. In the 'pre-quantum mechanics' period it was natural to estimate v_{av} from classical equipartition of energy: $\frac{1}{2}m_e v_{av}^2 = \frac{3}{2}k_B T$. Using the value of electron mass, the estimated value of v_{av} was of the order of 10^7 cm s^{-1} at room temperature. This in turn will indicate an electronic mean free path $l \approx 1$–10 Å for metals at room temperature. This distance is comparable to the spacing between the ions in metals, and is in consonance with the assumptions of the Drude model that collisions are due to electrons knocking into the large heavy ions.

We will, however, see shortly that the classical estimate of average electron velocity v_{av} is an order of magnitude too small at room temperature. In addition, electrical resistivity of metals is distinctly temperature dependent. Figure 1.7 shows the behaviour of resistivity in typical metals as a function of temperature. The resistivity in metals can easily decrease by a factor of 10 between room temperature and low temperature (say liquid helium temperature of 4.2 K). Thus, at low

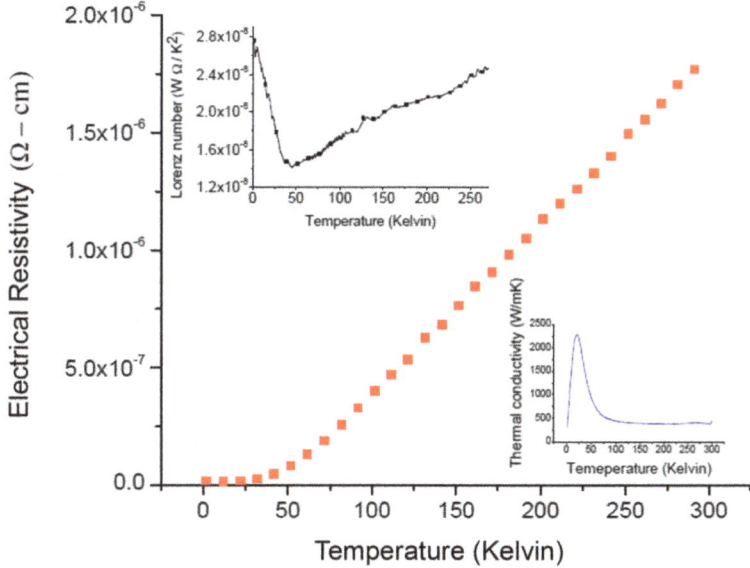

Figure 1.7. Electrical and thermal transport properties of National Institute of Standard and Technology (NIST) Copper standard sample. The main frame shows the temperature dependent electrical resistivity. The upper (lower) inset presents temperature dependence of Lorentz number (thermal conductivity).

temperatures τ can be an order of magnitude larger than at room temperature. All these factors can raise the low temperature electronic mean free path l to 10^3 Å or more, which is a thousand times the spacing between ions in a solid. In fact with careful sample preparations electronic mean free paths of the order of centimeters can be achieved by performing experiments at reasonably low temperatures. This clearly tells that Drude's assumption that electrons only collide with ions cannot be quite correct. In any case the observed temperature dependence of electrical resistivity (see figure 1.7) cannot be explained within the framework of the Drude model.

The most significant success of the Drude model was the explanation of the empirical law of Wiedemann and Franz, which states that the ratio of thermal conductivity to electrical conductivity κ/σ is directly proportional to the temperature T for a large number of metals. The proportionality constant, known as Lorenz number, is, to a fair accuracy, the same for all metals [5]. The assumption of the Drude model here is that the bulk of the thermal current in a metal is carried by the conduction electrons. The thermal conductivity κ is defined by the equations

$$\mathbf{J}^q = -\kappa \nabla T \tag{1.34}$$

where $\mathbf{J}^q$ is the heat-flux.

Thermal conductivity κ is expressed as

$$\kappa = \frac{1}{3} l v C_e = \frac{1}{3} v^2 \tau C_e \tag{1.35}$$

where l is the electronic mean free path, τ is the relaxation time, v^2 is mean square velocity of electron and C_e is electronic specific heat.

Combining equations (1.32) and (1.35) leads to the expression:

$$\frac{\kappa}{\sigma} = \frac{\frac{1}{3} C_e m v^2}{n e^2}. \tag{1.36}$$

Drude applied the classical ideal gas laws to evaluate both the electronic specific heat and mean square velocity. He assumed that the electron velocity distribution in the electron gas in equilibrium at a temperature T is represented by the Maxwell–Boltzmann distribution function, like that of an ordinary classical gas of density $n = \frac{N}{V}$. Thus the number of electrons per unit volume with velocities in the range dv about v is given by $f_B(v)dv$, where

$$f_B(v) = n \left(\frac{m}{2\pi k_B T} \right)^{\frac{3}{2}} e^{-\frac{m v^2}{2 k_B T}}. \tag{1.37}$$

With $C_e = \frac{3}{2} n k_B$ and $\frac{1}{2} m v^2 = \frac{3}{2} k_B T$, where k_B is Boltzmann constant, equation (1.36) can be written as,

$$\frac{\kappa}{\sigma} = \frac{3}{2} \left(\frac{k_B}{e} \right)^2 T. \tag{1.38}$$

The right side of the above equation (1.38) is proportional to T and depends on the universal constants k_B and e, which is in complete agreement with the empirical law of Wiedemann and Franz. Equation (1.38) gives a Lorenz number,

$$\frac{\kappa}{\sigma T} = \frac{3}{2}\left(\frac{k_B}{e}\right)^2 = 1.11 \times 10^{-8} \text{ W } \Omega \text{ K}^{-2}. \tag{1.39}$$

This is about half the typical values of Lorenz number obtained experimentally for various metals. However, it was quite puzzling that no electronic contribution comparable to $\frac{3}{2}nk_B$ had ever been observed experimentally. In fact the extraordinary success of the Drude model to explain the Wiedemann–Franz law came under a fortuitous circumstance, where two errors of about 100 arising due to the wrong application of the classical ideal gas laws to electron gas canceled each other. We will see later that the actual electronic contribution to the specific heat is about 100 times smaller than the classical ideal gas prediction, whereas the mean square electron velocity is about 100 times larger. The transport properties of solids along with heat capacity are relatively easy to measure (see appendix A) and we will see in this book that they can provide very useful information on the properties of materials under study.

Using the same classical statistical mechanics, an expression for the thermo-electric power can also be obtained [5]:

$$Q = -\frac{C_e}{3ne} = -\frac{k_B}{2e}. \tag{1.40}$$

This predicts a value of metallic thermoelectric power $Q = -0.42 \times 10^{-4} \text{ V K}^{-1}$. However, experimentally determined values of thermopower for metals at room temperature are of the order of micro-volts per degree, which is a factor of 100 smaller. In this case the same error of 100, which appeared twice in Drude's derivation of the Wiedemann–Franz law, remained uncompensated. All these observations were pointers towards the inadequacy of classical statistical mechanics in describing the behavior of electrons in a metal.

The Drude model has also been used to investigate various other electromagnetic properties of metals including the Hall effect, which occurs when a magnetic field $\mathbf{H}$ is applied perpendicular to the current density $\mathbf{j}$ flowing in a sample. A voltage drop namely Hall voltage appears in the direction perpendicular to both $\mathbf{j}$ and $\mathbf{H}$. Assuming that current flows parallel to the x direction and the magnetic field is applied to the z direction, the Hall voltage will then develop in the y direction. The Hall effect in materials is often represented by the Hall coefficient R_H:

$$R_H = \frac{E_y}{j_x B_H}. \tag{1.41}$$

The Drude model, however, made a rather surprising prediction that the Hall coefficient $R_H = -\frac{1}{nec}$. This is independent of both the magnitude of the applied magnetic field H and the relaxation time τ. While the experimentally determined

Hall coefficient of metals shows this order of magnitude values, they often depend on the temperature and the strength of applied magnetic field. In a typical example of aluminum, R_H never gets within a factor of 3 of the Drude model value, depends strongly on the magnetic field, and does not even show the expected sign in high magnetic fields. In addition the Drude model predicts that in a metallic wire with current flowing in a perpendicular direction to a uniform magnetic field, the electrical resistance does not depend on the strength of the magnetic field. This is in contrast to what has been observed experimentally. In fact in the cases of noble metals like copper, silver and gold, the electrical resistance actually increases without a limit with the increase in the strength of magnetic field. Also in most of the metals the behavior of electrical resistance depends on the orientation of the sample with respect to the magnetic field direction.

Another important property of metals are their optical response or ac-conductivity. The response of a metal in the presence of a time varying electric field depends on how electron momentum in the free electron gas changes on average as a function of time. Assuming one-dimensional current flow in the x-direction, the relation between the current density and ac-electric field can be written as:

$$\mathbf{j} = \sigma(w)\mathbf{E}(x). \tag{1.42}$$

Here $\sigma(w)$ is complex electrical conductivity and $E(x) = E_e^{ik_E x - wt}$, where k_E and w are wavevector and frequency, respectively, of the applied ac-electric field. A local response to the electric field is assumed here, which ignores any wave vector dependence of electrical conductivity. Such an approximation is justified when the electronic mean free path l is much smaller than the wavelength of the electric field i.e. $l \ll \lambda/2\pi = 1/k_E$. If the average momentum of electrons in the homogeneous electron gas at time t is $p(t)$ then using equation (1.30) we can write:

$$\mathbf{j}(t) = -\frac{en\mathbf{p}(t)}{m_e}. \tag{1.43}$$

We now have to calculate the momentum to first order at time $t + dt$ in the presence of the ac-electric field $\mathbf{E}$. The probability of an electron undergoing a collision is dt/τ and the probability of no collision is $(1 - dt/\tau)$, where the average time between electron collisions is τ. In the event of no collision, the electron gains a momentum $-e\mathbf{E}dt$ from the electric field. The momentum at time $t + dt$ to a first order is expressed as:

$$\mathbf{p}(t + dt) = \left(1 - \frac{dt}{\tau}\right)(\mathbf{p}(t) - e\mathbf{E}dt + O(dt^2)). \tag{1.44}$$

The contribution from electrons that had undergone a collision is of second-order in nature, and is not considered here.

Equation (1.44) can be rewritten as:

$$\frac{\mathbf{p}(t + dt) - \mathbf{p}(t)}{dt} = -\frac{\mathbf{p}(t)}{\tau} - e\mathbf{E}(t). \tag{1.45}$$

In the limit $t \to 0$, $\mathbf{p}(t + dt) - \mathbf{p}(t) \to dp(t)$ and the average change in momentum can be written as:

$$\frac{d\mathbf{p}(t)}{dt} = -\frac{\mathbf{p}(t)}{\tau} - e\mathbf{E}(t). \tag{1.46}$$

If the electric field which drives the ac-current is of the form $\mathbf{E}(t) = \mathbf{E}_0 e^{-iwt}$, then for a linear response, $\mathbf{p}(t) = Re(\mathbf{p}(w)e^{-iwt})$, and equation (1.46) can be rewritten as:

$$-iw\mathbf{p}(w) = -\frac{\mathbf{p}(w)}{\tau} - e\mathbf{E}(w) \tag{1.47}$$

or,

$$\mathbf{p}(w) = \frac{-e\tau\mathbf{E}(w)}{1 - iw\tau}. \tag{1.48}$$

Now from equation (1.30) we know that $\mathbf{j} = -ne\mathbf{p}/m_e$, hence the current density can be written as:

$$\mathbf{j}(t) = Re(\mathbf{j}(w)e^{-iwt}) \tag{1.49}$$

with,

$$\mathbf{j}(w) = -\frac{ne\mathbf{p}(w)}{m_e} = \frac{ne^2\tau}{m_e}\frac{\mathbf{E}(w)}{1 - iw\tau}. \tag{1.50}$$

Customarily ac-current density $\mathbf{j}(w)$ is expressed as [5]:

$$\mathbf{j}(w) = \sigma(w)\mathbf{E}(w). \tag{1.51}$$

Here $\sigma(w)$ is the frequency dependent conductivity or as conductivity, which is expressed as:

$$\sigma(w) = \frac{\sigma_0}{1 - iw\tau} \tag{1.52}$$

where $\sigma_0 = ne^2\tau/m_e$ is the zero-frequency or dc-conductivity, as we have derived earlier within the framework of the Drude model (see equation (1.32)). The ac-conductivity $\sigma(w)$ is a useful quantity since it is related to the dielectric constant $\epsilon_r(w)$ through the relation [12]:

$$\epsilon_r(w) = 1 + \frac{i\sigma(w)}{\epsilon_0 w} \tag{1.53}$$

where ϵ_0 is the permittivity of free space. Thus, ac-conductivity can be accessed in a wide frequency range through the optical measurement of the dielectric constant. In the measured optical response of the metals, there is, however, more subtle frequency dependence than can be explained within the framework of the simple free electron dielectric constant.

All these drawbacks of the Drude model are narrated here only to highlight the need to go beyond the simple *free electron model* in order to explain the physical properties of metals and in turn towards the development of a complete understanding of the solid crystalline materials in general. At this point one, however, needs to appreciate the great pioneering efforts of Drude, who made the first attempt to understand metals on a microscopic level.

1.3.2 Pauli exclusion principle, Fermi electron gas and quantum statistical mechanics of electrons

It took more than twenty five years after Drude's time for the next significant step to take place towards overcoming the inadequacies of the Drude model. With the progress in the field of the quantum theory, the statistics of electron motion was described in terms of waves. It was recognized that the electrons belonged to a class of particles called 'fermion', where the interference between identical wave-packets was destructive. This property restricts two electrons from residing in the same quantum state. This is the famous Pauli exclusion principle. Accordingly, the classical electron gas of the Drude model needs to be replaced with an electron gas obeying quantum statistical mechanics, namely Fermi gas. The Pauli exclusion requires that in the lowest energy state i.e. for $T \rightarrow 0$ K of the Fermi electron gas, the electrons occupy the available energy levels progressively starting with the lowest energy and ending at some highest limit. This highest energy limit, which separates the occupied states of the electron from the unoccupied ones at $T \rightarrow 0$ K is called the Fermi energy. One interesting consequence of the Pauli exclusion principle is that in contrast with the classical gas of electrons, Fermi electron gas has a finite internal energy at $T = 0$ K, and this value lies many orders of magnitude above the internal energy of a classical gas at $T = 300$ K [6]. The probability that a quantum mechanical state is occupied by an electron is given by the Fermi–Dirac distribution function (see appendix B). In the context of the present study, the Maxwell–Boltzmann distribution function (equation (1.37)) for electron gas in metals needs to be replaced by the following distribution function [5]:

$$f(v) = \left(\frac{\left(\frac{m}{\hbar} \right)^3}{4\pi^3} \right) \frac{1}{\exp\left[\left(\frac{1}{2}mv^2 - k_B T_0 \right)/k_B T \right] + 1}. \tag{1.54}$$

Here $\hbar$ is Planck's constant divided by 2π and T_0 is a temperature that is determined by the normalization condition:

$$n = \int d\mathbf{v} f(\mathbf{v}). \tag{1.55}$$

T_0 is typically of the order of tens of thousands of degrees. Figure 1.8 presents a comparison between the Maxwell–Boltzmann and Fermi–Dirac distribution functions for typical metallic densities at temperature $T \sim 0.01 T_0$. This shows that there is a drastic difference between the Maxwell–Boltzmann distribution and the

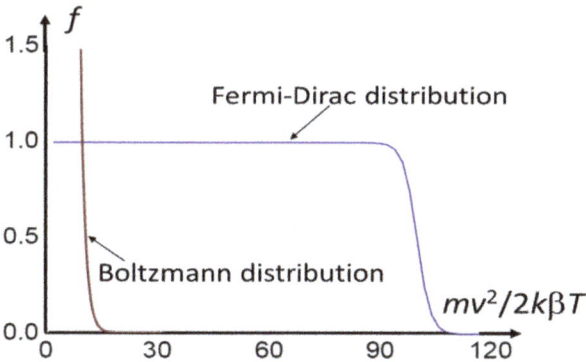

Figure 1.8. The comparison of the Maxwell–Boltzmann distribution function and the Fermi–Dirac distribution function for typical metallic densities at temperature $T \sim 0.01 T_0$.

Fermi–Dirac distribution for electrons at temperatures of present interest, namely T less than 1000 K (see figure 1.8).

1.3.3 Sommerfeld model

Sommerfeld applied this quantum statistical mechanics to the free electron gas of metals and resolved most of the anomalies of thermal properties of the early Drude model. However, the arrangement of positively charged metal ions in a periodic lattice was still ignored, except that it provided a mean potential $-U_0$ to keep the electron gas confined to the solid.

It is quite instructive to examine the ground state (i.e. $T = 0$ K state) of the electron gas in metals. It turns out that for the electron gas at metallic densities, the room temperature is actually a very low temperature, and many of the electronic properties of a metal are hardly different from their values at $T = 0$ K. In the Sommerfeld model, a solid metal crystal is treated as a three-dimensional square potential well with an infinite barrier at the surfaces (see figure 1.9). The electrons are unable to escape from the solid metal crystal, which is of course an over simplification especially when the work function value for typical metals is about 5 eV. The time independent Schrödinger equation in the one-electron approximation in an infinite square well of size L is:

$$-\frac{\hbar^2}{2m}\nabla^2\psi(\mathbf{r}) + U(\mathbf{r})\psi(\mathbf{r}) = E'\psi(\mathbf{r}) \tag{1.56}$$

where the potential $U(\mathbf{r})$ is given by:

$$U(x, y, z) = U_0 = \text{constant} \tag{1.57}$$

for $0 \leqslant x,\, y,\, z \leqslant L$; otherwise

$$U(x, y, z) = \infty. \tag{1.58}$$

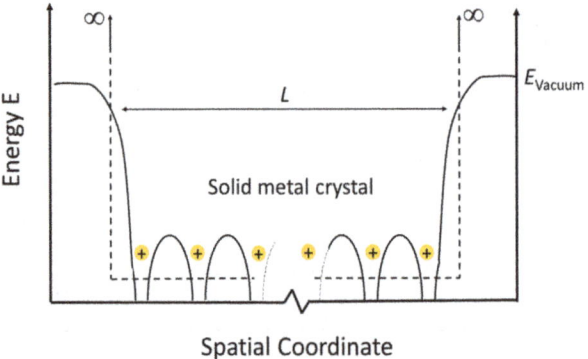

Figure 1.9. Schematic representation of the potential for an electron in a periodic lattice of positive ion cores in a crystalline solid.

With $E = E' - U_0$, this yields:

$$-\frac{\hbar^2}{2m}\nabla^2\psi(\mathbf{r}) = E\psi(\mathbf{r}). \tag{1.59}$$

To solve this Schrödinger equation (1.59) we now need to choose an appropriate boundary condition. We assume that the size of the metal is sufficiently large so that its bulk properties are unaffected by the detailed configuration of its surface. This leads to the popular choice of a cube of metal with a side $L = V^{1/3}$. Setting a boundary condition $\psi = 0$ in the boundaries, as in the standard quantum mechanical particle-in-a-box problem, is not quite appropriate for a metal, since it will imply that the surfaces of a large (but not infinite) metallic sample are perfectly reflecting for electrons. This condition will give rise to standing wave solutions to equation (1.59) and it would not be possible to probe the metallic state by passing a current as that would need traveling electron waves. A more useful choice would be the one that avoids the surface altogether. This is realized by imagining that each face of the cube could be joined to the face opposite to it. In this situation an electron coming to the metallic surface will not be reflected back. The electron will leave the metallic cube but simultaneously re-enter at a corresponding point on the opposite surface of the cube. Thus, in this one-dimensional metal the line 0 to L in which the electron is confined will be replaced by a circle of circumference L. This one-dimensional circular model of metal requires the boundary condition $\psi(x + L) = \psi(x)$, which in a three-dimensional metal cube takes the form:

$$\begin{aligned}
\psi(x, y, z + L) &= \psi(x, y, z) \\
\psi(x, y + L, z+) &= \psi(x, y, z) \\
\psi(x + L, y, z) &= \psi(x, y, z).
\end{aligned} \tag{1.60}$$

Equation (1.60) is known as the *Born–von Karman* or *periodic* boundary condition.

Without invoking any boundary condition, the Schrödinger equation (equation (1.59)) gives the solution:

$$\psi_k(\mathbf{r}) = \frac{1}{\sqrt{V}} e^{i\mathbf{k} \cdot \mathbf{r}} \qquad (1.61)$$

with energy,

$$E(k) = \frac{\hbar^2 k^2}{2m} \qquad (1.62)$$

where $\mathbf{k}$ is any position independent vector. The normalization constant in equation (1.61) is chosen such that the probability of finding the electron in the whole volume V is unity:

$$1 = \int |\psi(\mathbf{r})|^2 \, dr. \qquad (1.63)$$

The Born–von Karman boundary condition (equation (1.60)) will permit only certain discrete values of $\mathbf{k}$, since equation (1.60) will be satisfied by the general wave function (equation (1.61)) only if,

$$e^{ik_x L} = e^{ik_y L} = e^{ik_z L} = 1. \qquad (1.64)$$

Now, $e^x = 1$ only if $x = 2\pi n$, where n is an integer. Hence, the components of k must be of the form:

$$k_x = \frac{2\pi}{L} n_x, \quad k_y = \frac{2\pi}{L} n_y, \quad k_z = \frac{2\pi}{L} n_z, \qquad (1.65)$$

where n_x, n_y, n_z are integers.

The possible states of an electron in a three-dimensional infinite square potential well can be counted according to their quantum numbers n or k. The allowed values in the three dimensional wave-vector space or k-space results in constant energy surfaces, $E = \frac{\hbar^2 k^2}{2m} = $ constant, which are spherical in nature. The allowed values of k in a region (assuming that the region is enormous on the scale of $2\pi/L$) of k-space is the volume of k-space contained within the region divided by the volume of k-space per point in the network of the allowed values of k. The points in this network representing a cubic sample of metal are separated by a distance of $(2\pi/L)$, hence the volume of k-space per point is $(2\pi/L)^3$. It is therefore concluded that a region of k-space of volume Ω will contain:

$$\frac{\Omega}{(2\pi/L)^3} \qquad (1.66)$$

allowed values of k. In another way, it can be said that the number of allowed k-values per unit volume of k-space or the density of states is:

$$\frac{V}{8\pi^3}. \qquad (1.67)$$

In practice the k-space region one deals with is quite large (with $\sim 10^{22}$ points) and regular, and for all practical purposes equations (1.66) and (1.67) are regarded as exact.

In an N-electron system the lowest possible electron-energy state i.e. the ground state is with $\mathbf{k} = 0$. Within the framework of quantum mechanics, two electrons with opposite spin, can be placed in the ground state as permitted by the Pauli exclusion principle. Addition of more electrons will lead to the successive filling of the one-electron levels of lowest energy that are not already occupied. When the number of electrons N is very large, the region occupied by the electrons will be a filled-up sphere of radius $\mathbf{k}_F$ with a volume $\Omega = 4\pi k_F{}^3/3$ (see figure 1.10). The radius $\mathbf{k}_F$ is known as the Fermi wave vector. Using equation (1.66) we can write for the total number of allowed-$\mathbf{k}$ values as:

$$\left(\frac{4\pi k_F^3}{3}\right)\left(\frac{V}{8\pi^3}\right) = \frac{k_F^3}{6\pi^2} V. \tag{1.68}$$

Taking into account of the $\pm\frac{1}{2}\hbar$ spin degeneracy of the electrons, in order to accommodate N electrons in the allowed k-levels one must have:

$$N = 2\frac{k_F^3}{6\pi^2} V = \frac{k_F^3}{3\pi^2} V. \tag{1.69}$$

Thus, if there are N electrons in a volume V with an electronic density $n = N/V$, then the ground state of this N-electron system is created by occupying all single particle levels with $\mathbf{k} < \mathbf{k}_F$ and leaving all the states with $\mathbf{k} > \mathbf{k}_F$ unoccupied.

Equation (1.69) can be rearranged to give an expression for the *Fermi wavevector* in terms of electronic density $n = N/V$:

$$k_F = (3\pi^2 n)^{\frac{1}{3}}. \tag{1.70}$$

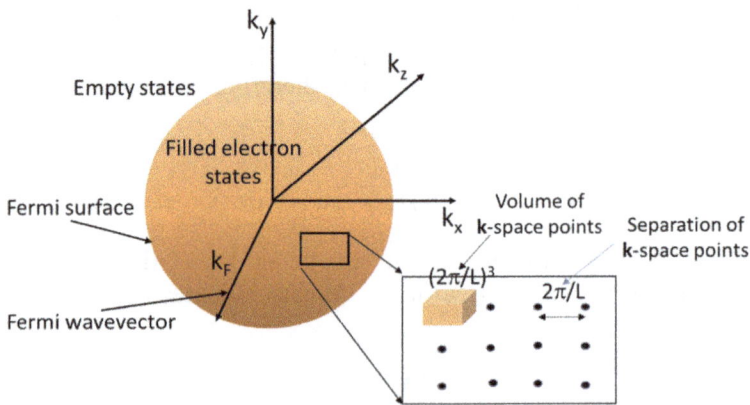

Figure 1.10. Schematic representation of the Fermi sphere distinguishing occupied free electron states from empty states at $T = 0$ K.

The corresponding energy is called the *Fermi energy*, and is expressed as:

$$E_F = \frac{\hbar^2 k_F^2}{2m_e} = \frac{\hbar^2}{2m_e}(3\pi^2 n)^{\frac{2}{3}}. \tag{1.71}$$

Fermi energy E_F provides the conceptual idea of a constant energy surface, namely *Fermi surface*, which at $T = 0$ K separates the filled electron energy states of the electron gas from the empty states. The presence of the Fermi surface has many important consequences for the properties of metals.

For typical electron densities $\approx 10^{22}$–10^{23} cm^{-3} in metals, equation (1.71) gives values of E_F in the range ≈ 1.5–15 eV. This in turn provides a value of the characteristic Fermi temperature $T_F = \frac{E_F}{k_B}$ in the range 20–100 $\times$ 10^3 K, which is much greater than room temperature. The striking difference between the classical electron gas of the Drude model and the quantum Fermi gas of electrons in the Sommerfeld model is visible here. To have the same energy the electrons in the Drude model need to be at 20–100 $\times$ 10^3 K. Typical values of the Fermi wave vector k_F are of the order of the reciprocal atomic spacing, and the velocities of electrons at the Fermi surface are $v_F = \frac{\hbar k_F}{m_e} \approx 0.01c$ (c is the velocity of light). This indicates that the electrons are very energetic even in the ground state (i.e. $T = 0$ K) of the electron gas.

In order to proceed further to investigate the metallic properties, the Fermi–Dirac distribution function (see appendix B) can be rewritten in the form to give the probability of occupation of a state of energy E:

$$f = \frac{1}{e^{\frac{(E-\mu)}{k_B T}} + 1} \tag{1.72}$$

where μ is the *chemical potential*.

Figure 1.11 shows the Fermi–Dirac distribution for electron gas at $T = 0$ K and at finite temperatures $T \ll E_F/k_B$, which include room temperature. At $T = 0$ K, $E_F \equiv \mu$ ($T = 0$ K). From figure 1.11 it is evident that f varies appreciably only in the temperature range approximately within $k_B T$ of the chemical potential μ. This implies that in metals, only the electrons within the energy range $\approx k_B T$ of μ will be able to contribute to thermal processes, transport properties etc. Electrons further below in energy will not be able to acquire sufficient thermal energy to be excited

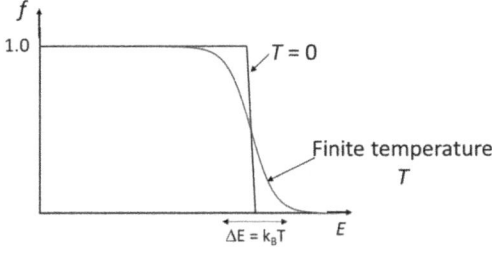

Figure 1.11. The Fermi–Dirac distribution function at $T = 0$ and at a finite temperature $T \ll T_F$.

into empty states above the Fermi surface. Furthermore, for a typical metal $\mu \sim E_F \equiv \mu \ (T = 0 \ \text{K})$ at all accessible temperatures until the metal melts.

At this stage it would be useful to introduce the concept of the electronic density of state $g(E)$, which will be useful for studying thermal properties of metal like electronic heat capacity. The number of energy states in the interval $E(k)$ and $E(k) + dE$ can be estimated by taking the volume of a thin shell of the octant bounded by the energy surfaces $E(k)$ and $E(k) + dE$ and dividing this by the volume V_k associated with a single k-point:

$$dZ = \left(\frac{1}{8}4\pi k^2 dk\right) \bigg/ (\pi/L)^3 \tag{1.73}$$

$$dE = (\hbar^2 k/m_e)dk. \tag{1.74}$$

Hence, the number of states per unit volume of the solid crystal:

$$dZ = \frac{(2m_e)^{3/2}}{4\pi^2 \hbar^3} E^{1/2} dE. \tag{1.75}$$

Taking into account $\pm\frac{1}{2}$ spin degeneracy of electron, every k-space point corresponds to two possible electron states. Thus, the electron density of states $g(E) = \frac{dZ}{dE}$ can be expressed as:

$$g(E) = \frac{1}{2\pi^2}\left(\frac{2m_e}{\hbar^3}\right)^{\frac{3}{2}} E^{1/2}. \tag{1.76}$$

A rigorous mathematical derivation of electronic heat capacity involving electron density of states and Fermi–Dirac distribution can be found in textbooks like Ashcroft and Mermin [5] and Singleton [7]. However, since one of the aims of this book is to make it accessible to a wider audience of scientists (including chemists and electrical engineers), we will rather follow a path of reasoning (with minimal usage of equations) to get a fairly accurate estimate of the electronic specific heat C_{el}.

As discussed above, at all temperatures of practical interest $k_B T$ is much smaller than E_F, so it can be assumed that only the electrons in an energy range $\approx k_B T$ on either side of $\mu \equiv E_F$ will take part in thermal processes. The number density of such electrons will be $\approx k_B T g(E_F)$. In the free-electron approximation, the Fermi energy is given by $E_F = \frac{\hbar^2}{2m_e}(3\pi^2 n)^{\frac{2}{3}}$ (see equation (1.71)). The natural logarithm of E_F leads to:

$$ln(E_F) = \frac{2}{3}\ln(n) + \text{constant}. \tag{1.77}$$

Differentiation of equation (1.77) gives:

$$\frac{dE_F}{E_F} = \frac{2}{3}\frac{dn}{n}. \tag{1.78}$$

This can be rearranged to obtain:

$$\frac{dn}{dE_F} \equiv g(E_F) = \frac{3}{2}\frac{n}{E_F}. \tag{1.79}$$

Each electron will be excited to an energy state $\approx k_B T$ above its ground-state ($T = 0$ K) energy. This can lead to a reasonable estimate of the thermal energy of the electron gas [7]:

$$U_{th}(T) - U_{th}(0) \approx (k_B T)^2 g(E_F) = \frac{3}{2}n k_B T\left(\frac{k_B T}{E_F}\right). \tag{1.80}$$

Differentiating equation (1.80) with respect to T leads to:

$$C_{el} = 3n k_B\left(\frac{k_B T}{E_F}\right). \tag{1.81}$$

The value of electronic specific heat C_{el} thus obtained is within a factor 2 of the more accurate method to estimate C_{el} [5, 7]. From equation (1.81) it is evident that the Sommerfeld model leads to a significant improvement over the Drude model. It successfully explains the temperature dependence of C_{el}. Furthermore, the magnitude of C_{el} is a factor $\approx\frac{k_B T}{E_F}$ smaller than the Drude value, as is observed in experiments on the alkali metals.

1.3.4 Inadequacies of the Drude and Sommerfeld models

The predictions of Sommerfeld theory, however, are less accurate for the specific heat in noble metals, and quite poor for the transition metals such as iron and manganese (much too small a prediction), and also for bismuth and antimony (much too large a prediction). In addition there is clearly a T^3 term in the experimentally observed temperature dependent specific heat, the presence of which cannot be explained by the Sommerfeld model.

The great success of the free electron theory of Drude and Sommerfeld is also rather limited to room temperature and very low temperatures of a few degrees of Kelvin. It fails to explain the temperature dependence of the electrical resistivity, thermal conductivity and Lorenz number (see figure 1.7). Furthermore, the difficulties concerning the free electron transport coefficients, namely the Hall coefficient, magneto-resistance, and also thermoelectric power, ac conductivity and the optical properties still remain. Then there are the fundamental questions:

1. It is assumed that all valence electrons in a metallic atom become conduction electrons in a solid metal, while the rest of the electrons remain bound to the ionic cores. The question remains, why this should be, and also how this is to be rationalized in the case of elements, like iron, which show more than one chemical valence?
2. Why are some of the elements metals, and the others are insulators and semiconductors?

All these observations call for a revisit to the basic assumptions of the Drude–Sommerfeld models, especially the *free electron approximation*. The various simplifications of the *free electron approximation* are the following:

1. The effect of the ionic cores on the dynamics of the electrons is ignored between collisions.
2. The exact role the ionic cores play as a source of collisions, remains unspecified.
3. The ions themselves can possibly contribute as independent dynamical entities to the physical properties like specific heat and thermal conductivity. This possibility is totally ignored.

To start with it may be difficult to distinguish between assumptions (i) and (ii) above, since it is far from clear that the effect of the ions on the electrons can be unambiguously resolved into *collisional* and *noncollisional* parts. In the next section we shall first go beyond the Drude–Sommerfeld free electron theory where the electrons are considered to be moving not in an empty space, but in the environment of a specified static potential originating from a fixed array of stationary ions. This is called the 'static ion approximation', in which collisions are entirely absent. We will see that this 'static ion approximation' provides a wealth of information and is a significant step towards the formation of a theory of solids encompassing both metals and insulators. Only after that one can examine the consequences of the dynamical deviations of the ionic positions from the static periodic lattice. The actual role of ions as a source of electron collision can only be understood at that stage.

1.4 Electron waves in the periodic ionic-potential of a solid: Bloch's theorem

The most important fact regarding the ionic cores in crystalline solid is that they are arranged in a regular periodic lattice. This important feature of the crystalline solids was first realized by x-ray diffraction experiments in the early 20th century, which was subsequently confirmed by other microscopic techniques including neutron diffraction and electron microscopy. This concept of a periodic lattice of ions in a solid lies at the heart of modern solid state physics.

In crystalline solid state of materials one is interested in the condensed state whose binding energy is of the order of 1–10 eV per atom. This is much smaller than the binding energy of electrons to the nucleus in an atom, hence very different from the energy scale to describe the formation of atoms or ions. During the process of the formation of solids, only the electron orbitals or the shells that are not completely filled will be changed significantly. Hence, to know the solid state properties it is sufficient to focus on the valence electrons with a binding energy of up to 10 eV and their interactions with the ions, with themselves, and of course the ion–ion interaction.

The major aim of crystalline solid state physics is to study the motion of electrons in the electrostatic field of the ions in a periodic lattice, in the presence of electron–electron interactions. In any system with more than one electron, the electron–electron interaction affects the energy and leads to correlation between the electrons. Electrons are correlated with one another through the Pauli exclusion principle and they interact via Coulomb repulsion. Heisenberg and Dirac identified this electron–electron interaction as the underlying cause of magnetism in terms of the exchange energy, which depends on the spin state of the electron and the change of sign of the electron wavefunction when two electrons are exchanged. In the field of chemistry the prominent roles of Pauli exclusion principle and Coulomb interaction between electrons are exemplified in the covalent bond and the formation of hydrogen molecule. However, one cubic centimeter of solid metal for example will contain $\sim 10^{23}$ numbers of conduction electrons. While in the Sommerfeld model some part of electron correlation has been taken into account through the Pauli exclusion principle, it obviously becomes a difficult (if not impossible) many body problem when one deals with the Coulomb interactions between such a huge number of electrons.

In order to make further progress, there are two distinct approaches. The first approach is to design effective single-electron theories, where the electron–electron interaction is approximated in such a way that the resulting problem becomes an effective single-electron problem. In the second approach the simplified model systems are studied for the electronic many-body problem. The first approach is popularly known as one-electron band theory or band structure calculations, in which there are three common schemes:

1. The independent electron approximation in which the electron–electron interaction is ignored. Electrons are independent except for the requirement of the Pauli exclusion principle.
2. The Hartree–Fock approximation in which only the interaction of an electron with the average electron density representing the many electron or N-electron system is considered.
3. The density-functional theory with local electron density approximation and its refinements, which provide the basis of modern electron band theory.

In the rest of this chapter, after studying the motion of electron in the periodic ion-potential of the solid, we will discuss the independent electron approximation leading to the formation of electron band structure. The Hartree–Fock approximation and density functional theory will be the subject of discussion in chapter 2. The effects of strong electron–electron interaction (which is the core subject of the present book) will then be covered in chapters 3–6.

We recall our discussions in section 1.2 that a crystalline solid is not invariant with respect to arbitrary translation and rotation. It will remain invariant only under those operations, which leaves the periodic lattice of a solid unchanged. The underlying periodicity of the crystalline solid is defined by the 'lattice translation vectors':

$$\mathbf{R} = n_1\mathbf{a}_1 + n_2\mathbf{a}_2 + n_3\mathbf{a}_3 \tag{1.82}$$

where n_1, n_2 and n_3 are integers and $\mathbf{a}_1$, $\mathbf{a}_2$ and $\mathbf{a}_3$ are three noncoplanar vectors called the primitive lattice translation vectors.

In order to proceed beyond the Drude–Sommerfeld free electron model, one must now consider the interactions between the ionic cores and the electrons. In other words one needs to introduce the effects of a periodic ion potential $U(\mathbf{r})$. In order to be $U(\mathbf{r})$ must satisfy the relation:

$$U(\mathbf{r} + \mathbf{R}) = U(\mathbf{r}). \tag{1.83}$$

We have learnt in section 1.2 that any periodic function of position in $f(x) = f(R + x)$ in the real lattice can be expressed in terms of Fourier components with wave vectors in the reciprocal lattice (see equation (1.20)). Hence, the periodic potential $U(\mathbf{r})$ may also be expressed as a Fourier series with wave vectors in the reciprocal lattice:

$$U(\mathbf{r}) = \sum_G U_G e^{i\mathbf{G}\cdot\mathbf{r}}. \tag{1.84}$$

In this equation U_G are Fourier coefficients and $\mathbf{G}$ are a set of reciprocal lattice vectors:

$$\mathbf{G} = m_1\mathbf{A}_1 + m_2\mathbf{A}_2 + m_3\mathbf{A}_3 \tag{1.85}$$

where the m_j are integers, and the $\mathbf{A}_j$ are primitive translation vectors of reciprocal lattice (see section 1.2). We shall see below that the periodicity of the reciprocal lattice also implies that the primitive unit cell of the reciprocal lattice, i.e. the first Brillouin zone, will contain all important information.

In order to study the motions of electrons through the periodic ion potential in a crystalline solid, one needs an appropriate set of functions, which should reflect the translational symmetry of the periodic lattice. To this end one can start with a plane wave:

$$\psi(\mathbf{r}) = e^{i(\mathbf{k}\cdot\mathbf{r} - wt)} \tag{1.86}$$

subject to the boundary conditions, which include the symmetry of the crystal:

$$\psi(\mathbf{r} + N_j\mathbf{a}_j) = \psi(\mathbf{r}) \tag{1.87}$$

where $j = 1, 2, 3$ and N_j is the number of unit cells in the jth direction. The periodic boundary conditions involved here is the *Born–von Karman boundary conditions*, which was introduced earlier in section 1.3.3 (see equation (1.60)). This boundary condition leads to:

$$e^{iN_j\mathbf{k}\cdot\mathbf{a}_j} = 1 \tag{1.88}$$

for $j = 1, 2, 3$.

Comparison of this equation (1.88) with equations (1.14) and (1.18) suggests that the allowed wave vectors are:

$$\mathbf{k} = \sum_{j=1}^{3} \frac{m_j}{N_j} A_j. \tag{1.89}$$

A new state is generated each time when all of the m_j change by one. Hence, the volume of k-space occupied by one state is:

$$\frac{\mathbf{A_1}}{N_1} \cdot \frac{\mathbf{A_2}}{N_2} \times \frac{\mathbf{A_3}}{N_3} = \frac{1}{N}\mathbf{A_1} \cdot \mathbf{A_2} \times \mathbf{A_3}. \tag{1.90}$$

Now, from equation (1.19) we know that the volume $\mathbf{A_1} \cdot \mathbf{A_2} \times \mathbf{A_3}$ stands for the primitive unit cell of the reciprocal lattice—the first Brillouin zone. Thus, in conjunction with equation (1.19) equation (1.90) indicates that the Brillouin zone always contains the same number of k-states as the number of primitive unit cells in the crystal [7].

With all this background information we shall now study the behavior of electrons in a static potential originating from an array of ions in a perfect periodic lattice. The justification of the static lattice potential is made with the adiabatic or Born–Oppenheimer approximation. This is based on the argument of very different mass of ions and electrons. The massive ions respond very slowly to any change of configuration of electron cloud. On the other hand, the electrons with much less mass move faster and respond adiabatically to any change of positions of the ions. So, for any practical purpose the static potential seen by the electrons originates from an array of stationary ions whose positions are fixed in a periodic lattice. It may be pointed out here that a perfect periodicity is actually an idealization. In real solids there are always impurity atoms and a temperature dependent probability of finding displaced or missing ions, which will of course destroy the perfect translational symmetry. In addition, at finite temperatures the ions are actually not stationary; they undergo thermal vibrations about their equilibrium positions. In fact, such imperfections are all of great importance for many physical properties of solid. However, the progress is best made by separating the problem in two parts: (i) the ideal perfect crystal, in which the potential is genuinely periodic, and (ii) the effects of all the deviations from perfect periodicity on the properties of a hypothetical perfect crystal, which are treated as small perturbations.

We shall now study the general properties of a single electron of mass m_e in the static periodic potential $U(\mathbf{r})$ with the help of the Schrödinger equation:

$$\left(-\frac{\hbar^2 \nabla^2}{2m_e} + U(\mathbf{r})\right)\psi = E\psi. \tag{1.91}$$

We rewrite the potential $U(\mathbf{r})$ as a Fourier series as in equation (1.84) in terms of the reciprocal lattice vectors $\mathbf{G}$. The wave function ψ can be expressed as a sum of plane waves obeying the Born–von Karman boundary conditions:

$$\psi(\mathbf{r}) = \sum_k C_k e^{i\mathbf{k}\cdot\mathbf{r}}. \tag{1.92}$$

Substitution of the wave function from equation (1.92) and the potential from equation (1.84) into the Schrödinger equation (equation (1.91)) leads to:

$$\sum_k \frac{\hbar^2 k^2}{2m_e} C_k e^{i\mathbf{k}\cdot\mathbf{r}} + \left[\sum_G U_G e^{i\mathbf{G}\cdot\mathbf{r}}\right]\left[\sum_k C_k e^{i\mathbf{k}\cdot\mathbf{r}}\right] = E\sum_k C_k e^{i\mathbf{k}\cdot\mathbf{r}}. \tag{1.93}$$

The potential energy term can be rewritten as:

$$U(\mathbf{r})\psi = \sum_{G,k} U_G C_k e^{i(\mathbf{G}+\mathbf{k})\cdot\mathbf{r}}. \tag{1.94}$$

The sum is over all possible values of $\mathbf{G}$ and $\mathbf{k}$, hence equation (1.94) can be written as:

$$U(\mathbf{r})\psi = \sum_{G,k} U_G C_{k-G} e^{i\mathbf{k}\cdot\mathbf{r}}. \tag{1.95}$$

The Schrödinger equation (equation (1.91)) now takes the form:

$$\sum_k e^{i\mathbf{k}\cdot\mathbf{r}} \left[\left(\frac{\hbar^2 k^2}{2m_e} - E \right) C_k + \sum_G U_G C_{k-G} \right] = 0. \tag{1.96}$$

Now the Born–von Karman plane waves are an orthogonal set of functions. Hence, the coefficient of each term in the sum above must vanish [7]:

$$\left(\frac{\hbar^2 k^2}{2m_e} - E \right) C_{\mathbf{k}} + \sum_G U_G C_{k-G} = 0. \tag{1.97}$$

Since all the useful information regarding the k-space is contained in the first Brillouin zone, it may be convenient to deal just with solutions in the first Brillouin zone. Accordingly, one can write $\mathbf{k} = (\mathbf{q} - \mathbf{G}')$, where $\mathbf{q}$ lies in the first Brillouin zone and $\mathbf{G}'$ is a reciprocal lattice vector. Equation (1.97) can now be rewritten as:

$$\left(\frac{\hbar^2(\mathbf{q} - \mathbf{G}')^2}{2m_e} - E \right) C_{q-G'} + \sum_G U_G C_{q-G'-G} = 0. \tag{1.98}$$

Changing the variables further so that $\mathbf{G}'' \to \mathbf{G} + \mathbf{G}'$, one gets the equation of coefficients in the form:

$$\left(\frac{\hbar^2(\mathbf{q} - \mathbf{G}')^2}{2m_e} - E \right) C_{q-G'} + \sum_{G''} U_{G''-G'} C_{q-G''} = 0. \tag{1.99}$$

It may be noted here that equations (1.98) and (1.99) are nothing but restatement of the original Schrödinger equation (equation (1.91)) in momentum space, simplified by the fact that because of periodicity of the potential, U_G will not vanish only when $\mathbf{q}$ is a vector of the reciprocal lattice.

Equation (1.99) of coefficient specifies the form the wave function ψ will take in equation (1.92), and it only involves coefficients $C_{\mathbf{k}}$ in which $\mathbf{k} = \mathbf{q} - \mathbf{G}$ with $\mathbf{G}$ being the general reciprocal lattice vectors. This implies that for each distinct value of $\mathbf{q}$, there is a wave function $\psi_{\mathbf{q}}(\mathbf{r})$, which takes the form:

$$\psi_q(\mathbf{r}) = \sum_G C_{q-G} e^{i(\mathbf{q}-\mathbf{G})\cdot\mathbf{r}}. \tag{1.100}$$

This equation is obtained by substituting $\mathbf{k} = \mathbf{q} - \mathbf{G}$ into equation (1.92), and can be rewritten further as [7]:

$$\psi_q(\mathbf{r}) = e^{i\mathbf{q} \cdot \mathbf{r}} \sum_G C_{q-G} e^{-i\mathbf{G} \cdot \mathbf{r}} = e^{i\mathbf{q} \cdot \mathbf{r}} u_{j,q}. \tag{1.101}$$

The above equation represents a wave function, which is essentially a plane wave with wave vector $\mathbf{q}$ within the first Brillouin zone modulated by a function $u_{j,q}$ with the periodicity of the lattice of the crystalline solid. This leads to *Bloch's theorem*, which states: 'the eigenstates ψ of a one-electron Hamiltonian $H = -\frac{\hbar^2 \nabla^2}{2m_e} + U(\mathbf{r})$, where $U(\mathbf{r} + \mathbf{R}) = U(\mathbf{r})$ for all Bravais lattice translation vectors $\mathbf{R}$ can be chosen to be a plane wave multiplied by a function with the periodicity of the Bravais lattice'. The index j appears in Bloch's theorem because for given $\mathbf{q}$ there are many solutions to the Schrödinger equation.

1.5 Independent electron approximation and electronic band structure

At this stage it may be pointed out that the framework of electrons in a periodic potential is not confined only to cases of metals. This framework actually applies to all crystalline solids, and will have an important role in the classification of insulators and semiconductors. The problem of electrons in a solid is in principle a many-electron problem. One is dealing here with not only the one-electron potentials describing the interactions of the electrons with the massive ionic-cores, but also pair potentials describing the electron–electron interactions. However, in the *independent electron approximation* these interactions are represented by an effective one-electron potential $U(\mathbf{r})$. If the crystal is perfectly periodic, it must satisfy the condition $U(\mathbf{r} + \mathbf{R}) = U(\mathbf{r})$ irrespective of the form the one-electron effective potential might be having. It may also be noted that Bloch's theorem makes no assumptions on the strength of the ionic potential.

Independent electrons, each of which obey a one-electron Schrödinger equation with a periodic potential, will be called *Bloch electrons*. Each set of $u_{j,q}$ results in a set of electron states with a particular energy dispersion relation $E(\mathbf{q})$. This leads to a description of the energy levels of an electron in a periodic potential in terms of a set of continuous functions $E_j(\mathbf{q})$, each with the periodicity of the reciprocal lattice. This forms the basis of the idea of *electronic band structure* of the solid. The number of possible wave functions in this electron band is given by the number of distinct $\mathbf{q}$, that is the number of Born–von Karman wavevectors in the first Brillouin zone. Therefore the number of the electron states in each band is 2 times the number of primitive cells in the solid crystal. The factor two here arises from $\pm\frac{1}{2}\hbar$ spin-degeneracy of electron.

The wave vector $\mathbf{k}$ used in the Bloch wavefunction plays the same fundamental role in the general problem of the motion of electron in a periodic potential of crystalline lattice, that the free electron wave vector plays in the Sommerfeld theory. The free electron wave vector is simply $\mathbf{p}/\hbar$, where $\mathbf{p}$ is the momentum of the electron. However, the Bloch electron wavevector **k** is not proportional to the electron momentum. It is known as quasi momentum or crystal momentum of the electron, and it is a quantum number describing the electron's state within an energy band. Thus **k** of a Bloch electron can be viewed as a quantum number characteristic of the translational symmetry of a periodic potential in a crystalline lattice, where **k** = **p**/$\hbar$ of the Drude–Sommerfeld electron is a quantum number characteristic of the complete translational symmetry of free space.

It can be shown [5] that an electron in a level specified by band index j and wave vector **k** has a finite mean velocity, given by:

$$\mathbf{V}_j(k) = \frac{1}{\hbar}\nabla_k E_j(k). \tag{1.102}$$

This asserts that there are stationary energy levels for an electron in a periodic potential, in which the electron moves forever without any degradation of its mean velocity in spite of the interaction between the electron with the fixed periodic lattice of ions. In the periodic array of ions in a solid, the electron wave can propagate without attenuation because of the coherent constructive interference of the scattered electron waves. This is clearly in contrast with the Drude model, where the electrons were supposed to be colliding with every ion.

1.6 Nearly free electron model versus tight binding model

It is particularly instructive to study the limiting case of a vanishingly small periodic potential in a crystalline solid. This is widely known as the *nearly free electron model*. This rather unrealistic model surprisingly gives results quite close to the experimental observations in many solids. The metals whose atomic structure consists of s and p electrons outside a close-shell noble gas configuration are potential candidates, where the conduction electrons can be considered as moving in an almost constant potential. These metallic elements are often termed as *nearly free electron* metals, and they were actually the starting point of the study of the Drude–Sommerfeld free electron gas model. There are some fundamental reasons behind the near free electron nature of such metals, in spite of the presence of strong electron–electron and electron–ion interactions:

1. The Coulomb interaction between electron and ion is strongest at small separation. However, the immediate neighbourhood of the ions is already occupied by the core electrons. Hence the conduction electrons are prevented from coming to this region by the Pauli exclusion principle.
2. In the region where the conduction electrons have access, the mobility of the electrons again somewhat screen the field of positively charged ions, and this effect further diminishes the net potential experienced by an individual electron.

We shall first consider the energy states in the limiting case of the free electrons in the Sommerfeld model i.e. in an infinite square well potential. One may recall here that the free electron can be described by the energy parabola $E = \frac{\hbar^2 k^2}{2m_e}$ (see figure 1.12(a)). Even

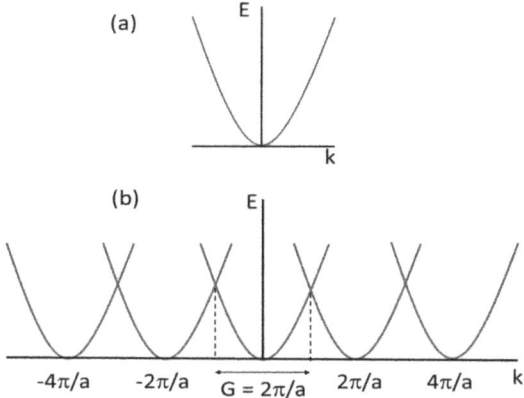

Figure 1.12. (a) Parabolic energy curves of a free electron. (b) Parabolic energy curves of a free electron in a one-dimensional lattice with vanishingly small potential, which is periodically continued in the reciprocal space.

in such a case where the Fourier coefficients U_G of the potential $U(\mathbf{r}) = \sum_G U_G e^{i\mathbf{G}\cdot\mathbf{r}}$ in a periodic lattice vanishes, one needs to take the symmetry requirement of the periodicity into consideration. This requirement will be of much importance even for the smallest non-vanishing potential. This periodicity of the crystal lattice will imply that the possible electron states are not restricted to a single parabola in k-space, but can be found with equal probability on parabolas shifted by any reciprocal lattice vector $\mathbf{G}$ i.e.

$$E(\mathbf{k}) = E(\mathbf{k} + \mathbf{G}) = \frac{\hbar^2}{2m}\,|\mathbf{k} + \mathbf{G}|^2. \tag{1.103}$$

The parabolic energy curves for a free electron where $\mathbf{G} \to G = n\frac{2\pi}{a}$ (a is lattice constant in real space and n is an integer) in one dimension is periodically continued in reciprocal space (see figure 1.12(b)). The behavior of $E(k)$ is periodic in k-space, hence it is adequate to represent this only in the first Brillouin zone. One can simply displace the part of the energy parabola of interest by the appropriate multiple of $G = \frac{2\pi}{a}$. The $E(k)$ bands are more complicated in three dimensions even in the case of a vanishing potential, since there will be $\mathbf{G}$ contributions in $E(k)$ bands in all three coordinate directions.

In the one dimensional problem there is a degeneracy of the energy values when two energy parabolas intersect at the edges of the first Brillouin zone i.e. at $\frac{G}{2} = \frac{\pi}{a}$ and $-\frac{G}{2} = -\frac{\pi}{a}$. The state of an electron with such k values will be a superposition of at least two corresponding plane waves. Such plane waves are $e^{\frac{iGx}{2}}$ and $e^{\frac{-iGx}{2}}$. We will see shortly that the degeneracy at the intersection points of the parabolas is lifted by the perturbing non-zero lattice potential $U(\mathbf{r})$.

The Schrödinger equation (equation (1.93)) in reciprocal space implies that one must also consider the G-values larger than $2\frac{\pi}{a}$. However, in the first approximation such contributions from the other reciprocal lattice vectors may be neglected. We will now try to see the effect of vanishingly small periodic potential with a study of the Schrödinger equation using perturbation theory, and write the equation as:

$$(H_0 + \alpha H_1)\psi = E\psi \tag{1.104}$$

where $H_0 \equiv -\frac{\hbar^2}{2m_e}\nabla^2$ and $\alpha H_1 \equiv U$. We consider the unperturbed state to be the limiting case of the free electrons in the Sommerfeld model. In reference to that we shall now calculate the energy and wave function in the presence of the vanishingly small perturbation potential U representing the nearly free electron condition in the periodic lattice. To make further progress we will now make a series expansion in the powers of α:

$$E_k = E_0 + \alpha E_1 + \alpha^2 E_2 + \cdots$$
$$\psi_k = \psi_0 + \alpha\psi_1 + \alpha^2\psi_2 + \cdots. \tag{1.105}$$

Since in the nearly free electron model the perturbation term is very small compared to H_0, then the series for E_k and ψ_k will converge rapidly. Hence, good approximations to the true eigenstates and eigenvalues of the new system can be made with calculation involving only the lowest order terms in the series expansion. In such perturbative calculation, E_1 can be expressed as:

$$E_1 = \frac{\int \psi_0^* U\psi_0 dx}{\int \psi_0^* \psi_0 dx}. \tag{1.106}$$

This equation (1.106) is valid only if the unperturbed system is non-degenerate. However, the present case is degenerate since the energy dispersion is quadratic in nature i.e. $E_0 = \hbar^2 k^2 / 2m_e$. Hence, $E_{0k} = E_{0(-k)}$, and the corresponding eigenstates ψ_{0k} and $\psi_{0(-k)}$ are proportional to e^{ikx} and e^{-ikx}, respectively. In such a situation of two-fold degeneracy, it is possible to use a modification of equation (1.106) with the replacement of ψ_{0k} by a new function that is orthogonal linear combinations of ψ_{0k} and $\psi_{0(-k)}$. This new function is designated as ϕ_{0k} and this must satisfy the condition:

$$\int \phi_{0k}^* U\phi_{0(-k)} dx = 0 \tag{1.107}$$

Noting that U is a periodic function, which can be expressed as a Fourier series (see equation (1.94)), only linear combinations of $e^{ikx} + e^{-ikx}$ that satisfy equation (1.107) are $\sin(kx)$ and $\cos(kx)$. Replacing ϕ_{0k} with $\sin(kx)$ and designating $U(x)$ as:

$$U(x) = -\sum_{n=1}^{\infty} U_n \cos\left(\frac{n2\pi x}{a}\right) \tag{1.108}$$

equation (1.106) can be rewritten as:

$$E_{1k} = \frac{-\sum_{n=1}^{\infty} U_n \int \sin^2(kx)\cos\left(\frac{n2\pi x}{a}\right)dx}{\int \sin^2(kx)dx}$$

$$= \frac{-\sum_{n=1}^{\infty} U_n \int \frac{1}{2}[1 - \cos(2kx)]\cos\left(\frac{n2\pi x}{a}\right)dx}{\int \frac{1}{2}[1 - \cos(2kx)]dx}. \tag{1.109}$$

From equation (1.109) $E_{1k} = \frac{U_n}{2}$ for $k = \pm\frac{n\pi}{a}$ and zero otherwise. If the above calculation is repeated with ϕ_{0k} as $\cos(kx)$, one gets $E_{1k} = -\frac{U_n}{2}$ for $k = \pm\frac{n\pi}{a}$ and zero otherwise. So, within the framework of first order perturbation theory, the solutions of the Schrödinger equation in a weak periodic potential at special points $k = \pm n\pi/a$ are not plane waves but standing waves. We have studied in section 1.2 that the traveling waves with such wave-vector satisfy the Bragg condition and get Bragg reflected in the Brillouin zone boundary. Now, a superposition of two traveling electron waves moving in opposite directions create a standing wave. These standing waves possess zero at fixed positions in space. The probability densities corresponding to these standing waves represented by ϕ_{0k} and $\phi_{0(-k)}$ are:

$$\rho_+ = \phi_{0k}^* \phi_{0k} \sim \cos^2 \pi\frac{x}{a} \tag{1.110}$$

$$\rho_- = \phi_{0(-k)}^* \phi_{0(-k)} \sim \sin^2 \pi\frac{x}{a}. \tag{1.111}$$

These probability densities along with a qualitative representation of the periodic potential in a one-dimensional lattice is presented in figure 1.13. The charge density for an electron in the state ϕ_{0k} ($\phi_{0(-k)}$) is maximum (minimum) at the position of the positive ion cores and minimum (maximum) in between the ion cores. The standing wave ϕ_{0k} ($\phi_{0(-k)}$) thus have a higher (lower) energy on the energy parabola at the Brillouin zone boundary than that of traveling plane wave e^{ikx} of free electron. This increase and decrease in energy at the Brillouin zone boundary represents a deviation from the free electron energy parabola, which is actually a good approximation to the solution in the region away from the zone boundary (see figure 1.14).

It is also clear that the electron waves will travel through the lattice unaffected unless they satisfy the required matching condition with the lattice points. This implies that the conduction electrons in a metal feel the presence of ions in the solid lattice under certain special circumstances, and in turn provides a plausible explanation of the large values of electronic mean free path in pure metals. This is a consequence of the wave-particle duality of electrons.

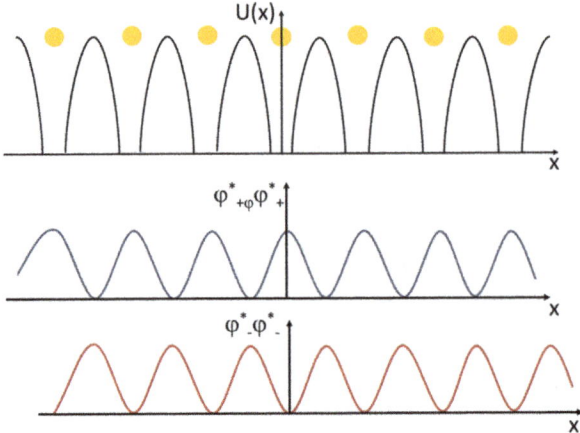

Figure 1.13. Schematic representation of the potential energy $U(x)$, probability density $\rho_+ = \phi_{0k}^*\phi_{0k}$ and $\rho_- = \phi_{0(-k)}^*\phi_{0(-k)}$ in a one-dimensional lattice.

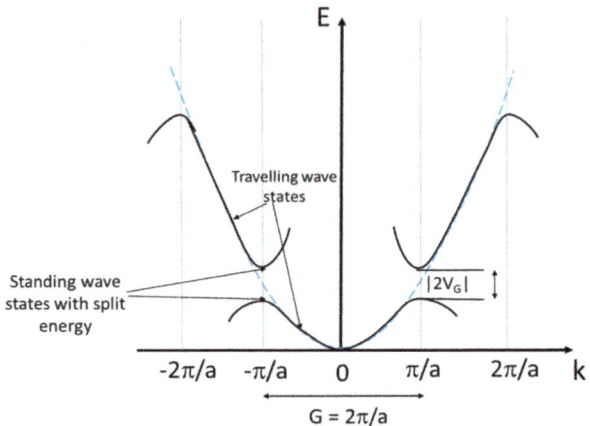

Figure 1.14. Schematic representation of the formation of the band gap by splitting of the energy parabola of the free electron at the boundary of the first Brillouin zone $\mathbf{k} = \pm\frac{\pi}{a}$ in a one-dimensional lattice.

With this information one can now proceed to get an idea of the magnitude of the band splitting or energy gap as shown in figure 1.14.

Starting from equation (1.97) one can write [6]:

$$C_{k-G} = \frac{\sum_{G'} U_{G'-G} C_{k-G'}}{E - \frac{\hbar^2}{2m_e}|\mathbf{k} - \mathbf{G}|^2}.\tag{1.112}$$

A first approximation to the calculations of $C_{k-G'}$ can be made for small perturbations by setting the true eigenvalue E to the free electron energy $\frac{\hbar^2 k^2}{2m_e}$. In this first approximation only the largest coefficient of $C_{k-G'}$ is of interest, and the largest deviation from the free electron behavior occurs when the denominator of equation (1.112) becomes zero. This will lead to:

$$\mathbf{k}^2 \approx |\mathbf{k} - \mathbf{G}|^2. \tag{1.113}$$

The strongest perturbation to the energy surface of the free electron i.e. the Fermi sphere in k-space occurs for the $\mathbf{k}$ vectors at the edge of the first Brillouin zone. It also follows from equation (1.98) that apart from $C_{k-G'}$, the coefficient C_k is equally important. For this approximation one needs to consider only two relations starting from equation (1.98):

$$\left(E - \frac{\hbar^2}{2m_e} k^2 \right) C_k - U_G C_{k-G} = 0 \tag{1.114}$$

and,

$$\left(E - \frac{\hbar^2}{2m_e} |\mathbf{k} - \mathbf{G}|^2 \right) C_{k-G} - U_{-G} C_k = 0. \tag{1.115}$$

These equations are simultaneous equations of the coefficients $C_{\mathbf{k}}$ and $C_{\mathbf{k}-\mathbf{G}}$, and the solutions are [6]:

$$E^{\pm} = \frac{1}{2} \left(E_{\mathbf{k}-\mathbf{G}}^0 + E_{\mathbf{k}}^0 \right) \pm \left[\frac{1}{4} \left(E_{\mathbf{k}-\mathbf{G}}^0 - E_{\mathbf{k}}^0 \right)^2 + |U_{\mathbf{G}}|^2 \right]^{\frac{1}{2}} \tag{1.116}$$

where $E_{\mathbf{k}}^0 = \frac{\hbar^2}{2m_e} |\mathbf{k}|^2$ and $E_{\mathbf{k}-\mathbf{G}}^0 = \frac{\hbar^2}{2m_e} |\mathbf{k} - \mathbf{G}|^2$ are the free electron dispersion curves.
The equation (1.116) can be written approximately as:

$$E^{\pm} = \frac{1}{2} \left(E_{k-G}^0 + E_k^0 \right) \pm |U_G|. \tag{1.117}$$

At the Brillouin zone boundary where the contributions from the two waves with C_k and C_{k-G} are equal and where $E_{k-G}^0 = E_k^0$, there is a symmetrical splitting of energy levels about the free electron value. This energy gap has a value:

$$\Delta E = E_+ - E_- = 2|U_G|. \tag{1.118}$$

Equation (1.117) describes the form of two electron energy curves that are separated by this energy gap near the Brillouin zone boundaries. Figure 1.14 illustrates this phenomenon for the one-dimensional lattice near the Brillouin zone boundary at $k = \frac{G}{2}$.

There is another model to represent the electrons in a crystalline solid, which is quite opposite to the *nearly free electron model*. The lattice potential here is so large that the electrons remain most of the time bound to the ionic cores except for some occasional visits to the other atoms. Let us consider the solid as a type of giant molecule and investigate the possible energy levels in such a molecule. As a simple case, we can imagine building a molecule from sodium (Na) atoms that have only one valence electron. For two Na atoms, the situation is similar to that of the hydrogen molecule, and as the atoms approach each other, bonding and antibonding molecular orbitals are formed. Each Na atom has one $3s$ electron and the two electrons can be accommodated in the bonding orbital, which have opposite spins in order to fulfill the requirements of the Pauli exclusion principle. The energy scale can be set to zero for a very large separation of the atoms. For the separation a, the bonding level reaches the lowest energy. Since only the bonding level is occupied by two electrons, the energy of the antibonding state is irrelevant and the energy gain is maximized for the separation a. What happens if more than two atoms are considered? The interaction of two atomic states leads to the formation of two levels that are delocalized over the entire molecule. The situation is similar for N atoms. The N atomic energy levels split up into N nondegenerate molecular levels. $N/2$ of these levels are then occupied by two electrons each. For a very large N, the same principles apply. There is a quasi-continuum of states between the lowest and the highest level, which is called an energy band. Now we can qualitatively see why Na should display metallic behavior. The energy band of the valence electrons is exactly half-filled. When an electric field is applied to a sample of Na, the electrons experience a force opposite to the field direction. In order to move in that direction, they have to increase their kinetic energy by some amount so that they can go into a state with a slightly higher energy. For the electrons in the highest occupied states, this is easily possible because there are plenty of unoccupied states available at slightly higher energies.

We now consider the framework that starts with atomic states and construct a Bloch wave function through a linear combination of the atomic orbitals. This method is known as the *tight binding approach*. The nearly free electron approach described earlier is a more natural starting point to describe metals, while the tight-binding approach is the obvious starting point for covalently bonded crystals, and also for more localized electrons in metals, such as the d electrons in transition metals. Eventually, both are mere approximations, and refining them will lead to the same result from both ends. But a discussion on the tight-binding approach here will give more insight into the meaning of the band structure of solids.

We describe the tight-binding approximation in its simplest form by starting with the Hamiltonian for the atoms, which constitute the crystalline solids. Considering only one kind of atom for simplicity, the Hamiltonian is given by:

$$H_{at} = -\frac{\hbar^2 p^2}{2m_e} + U_{at}(\mathbf{r}) \tag{1.119}$$

where U_{at} is the atomic one-electron potential. The atoms have different energy levels E_n and the associated wave functions. Each energy level turns into a band when the atoms come together to form a solid. Starting with the Na atom for example, the band is derived from the $3s$ state with the energy E_{3s} and the wave function $\phi_{3s}(\mathbf{r})$. With an atom on every point $\mathbf{R}$ of the Bravais lattice, the Hamiltonian for the solid is expressed as [8]:

$$H_{sol} = -\frac{\hbar^2}{2m_e}p^2 + \sum_R U_{at}(\mathbf{r} - \mathbf{R}). \tag{1.120}$$

The first term of equation (1.120) is the kinetic energy of the single electron, and the second term represents the sum of the atomic potentials of all the atoms in the solid; this potential has the periodicity of the lattice. This equation can be rewritten in terms of the potential of the atom at the origin plus the potential of the rest of the solid:

$$H_{sol} = -\frac{\hbar^2}{2m_e}p^2 + U_{at}(\mathbf{r}) + \sum_{R\neq 0} U_{at}(\mathbf{r} - \mathbf{R}) = H_{at} + U(\mathbf{r}), \tag{1.121}$$

where

$$U(\mathbf{r}) = \sum_{R\neq 0} U_{at}(\mathbf{r} - \mathbf{R}). \tag{1.122}$$

The equation (1.121) can be considered as the Hamiltonian for an atom placed at the origin with some correction potential arising from all the other atoms in the solid. Starting with the limiting case where the atoms are quite far from each other, one can try to use the atomic wave functions belonging to the atomic energy levels E_n to calculate the energy eigenvalues of the solid. In such a case the eigenvalue equation can be written down as [8]:

$$\int \phi_n^*(\mathbf{r}) H_{sol}\phi_n(\mathbf{r})dr = E_n + \int \phi_n^*(\mathbf{r}) U(\mathbf{r})\phi_n(\mathbf{r})dr = E_n - \beta. \tag{1.123}$$

Here, $-\beta$ represents a small shift of the atomic energy level due to the presence of all the other atoms in the solid. If the atoms are sufficiently separated from each other, the wave function $\phi_n(\mathbf{r})$ will go to zero before the potential $U(\mathbf{r})$ from the neighbouring atoms at $\mathbf{R} \neq 0$ becomes appreciably larger than zero, and in such cases $\beta = 0$.

The atomic wave function centered on any other lattice site $\mathbf{R}$ will also solve the Schrödinger equation, where one has to rewrite the Hamiltonian such that it is centered on the atom at $\mathbf{R}$ plus the potential from all the other atoms in the solid. For a solid of N atoms placed far apart so that they do not interact, one thus obtains N degenerate solutions for every energy eigenvalue of the atomic Hamiltonian. The resultant band structure will consist of a band at the energy E_n with no dispersion at all.

In the case where some interaction exists between the neighboring atoms in a solid, the wave function of the solid can be written as a linear combination of the atomic wave functions on every lattice site $\mathbf{R}$:

$$\psi_k(\mathbf{r}) = \frac{1}{\sqrt{N}} \sum_R C_{k,R}\phi_n(\mathbf{r} - \mathbf{R}) \tag{1.124}$$

where, $\frac{1}{\sqrt{N}}$ is a normalization factor. The coefficients of equation (1.124) $C_{k,R}$ will depend on the wave vector $\mathbf{k}$, and need to be determined. While it may not be entirely correct to use the atomic wave functions $\phi_n(\mathbf{r} - \mathbf{R})$ here because the proximity to the other atoms can influence these wave functions, this fact is ignored for simplicity. Now this wave function must have the character of Bloch wave if it is to satisfy the Schrödinger equation of the solid with periodic lattice. This condition is satisfied by choosing the coefficients $C_{k,R}$ such that the equation (1.124) turns into:

$$\psi_k(\mathbf{r}) = \frac{1}{\sqrt{N}} \sum_R e^{i\mathbf{k}\cdot\mathbf{R}} \phi_n(\mathbf{r} - \mathbf{R}) \tag{1.125}$$

where $\mathbf{k}$ takes the values permitted by the periodic boundary conditions i.e. $\mathbf{k} = (k_x, k_y, k_z) = \left(\frac{n_x 2\pi}{a}, \frac{n_y 2\pi}{a}, \frac{n_z 2\pi}{a} \right)$. This wave function fulfills the Bloch condition [8]:

$$
\begin{aligned}
\psi_k(\mathbf{r} + \mathbf{R}') &= \frac{1}{\sqrt{N}} \sum_R e^{i\mathbf{k}\cdot\mathbf{R}} \phi_n(\mathbf{r} - \mathbf{R} + \mathbf{R}') \\
&= \frac{1}{\sqrt{N}} e^{i\mathbf{k}\cdot\mathbf{R}'} \sum_R e^{i\mathbf{k}(\mathbf{R}-\mathbf{R}')} \phi_n(\mathbf{r} - (\mathbf{R} - \mathbf{R}')) \\
&= \frac{1}{\sqrt{N}} e^{i\mathbf{k}\cdot\mathbf{R}'} \sum_{R''} e^{i\mathbf{k}\cdot\mathbf{R}''} \phi_n(\mathbf{r} - \mathbf{R}'') = e^{i\mathbf{k}\cdot\mathbf{R}'} \psi_k(\mathbf{r})
\end{aligned}
\tag{1.126}
$$

where

$$\mathbf{R}'' = \mathbf{R} - \mathbf{R}'. \tag{1.127}$$

This wave function can be used to calculate the desired band structure $E(\mathbf{k})$. Assuming that the wave functions are already normalized so that:

$$
\begin{aligned}
E(\mathbf{k}) &= \int \psi_k^*(\mathbf{r}) H_{sol} \psi_k(\mathbf{r}) dr \\
&= \frac{1}{N} \sum_{R,R'} e^{i(\mathbf{k}\cdot(\mathbf{R}-\mathbf{R}'))} \int \phi_n^*(\mathbf{r} - \mathbf{R}') H_{sol} \phi_n(\mathbf{r} - \mathbf{R}) dr
\end{aligned}
\tag{1.128}
$$

where both summations run over all the lattice sites. We now consider a finite solid with Born–von Karman periodic boundary conditions. All the sums for a particular choice of $\mathbf{R}'$ are the same, and one can get rid of the double summation by recognizing that there are N such sums. Setting $\mathbf{R}' = 0$ arbitrarily in equation (1.128), leads to:

$$E(\mathbf{k}) = \sum_R e^{i(\mathbf{k}\cdot\mathbf{R})} \int \phi_n^*(\mathbf{r}) H_{sol} \phi_n(\mathbf{r} - \mathbf{R}) dr. \tag{1.129}$$

Using equation (1.123) the above equation (1.129) can be written as:

$$E(\mathbf{k}) = E_n - \beta + \sum_{R\neq0} e^{i(\mathbf{k}\cdot\mathbf{R})} \int \phi_n^*(\mathbf{r}) H_{sol} \phi_n(\mathbf{r} - \mathbf{R}) dr. \tag{1.130}$$

With equation (1.121) the integral in the above equation (1.130) can now be separated into:

$$
\begin{aligned}
\int \phi_n^*(\mathbf{r}) H_{sol} \phi_n(\mathbf{r} - \mathbf{R}) dr = E_n &\int \phi_n^*(\mathbf{r}) \phi_n(\mathbf{r} - \mathbf{R}) dr \\
&+ \int \phi_n^*(\mathbf{r}) U(\mathbf{r}) \phi_n(\mathbf{r} - \mathbf{R}) dr.
\end{aligned}
\tag{1.131}
$$

The first integral on the right-hand side of equation (1.131) can be neglected, because it contains two wave functions on different lattice sites, which have very little overlap. For the same reason the second integral on the right-hand side can also be small, but often not quite as small because the potential $U(\mathbf{r})$ falls less rapidly to zero when going away from $\mathbf{R}$, hence $U(\mathbf{r})\phi_n(\mathbf{r} - \mathbf{R})$ can increase in the region where it overlaps with $\phi_n^*(\mathbf{r})$. We define $\gamma(\mathbf{R})$ as:

$$\gamma(\mathbf{R}) = -\int \phi_n^*(\mathbf{r})U(\mathbf{r})\phi_n(\mathbf{r} - \mathbf{R})dr. \tag{1.132}$$

With the help of equation (1.132) the expression for band structure can now be written as:

$$E(\mathbf{k}) = E_n - \beta - \sum_{\mathbf{R}\neq 0} \gamma(\mathbf{R})e^{i(\mathbf{k}\cdot\mathbf{R})}. \tag{1.133}$$

This expression represents how the atomic energy level turns into a band when the atoms are placed in a crystalline solid with periodic lattice.

We shall now examine the case of a primitive cubic lattice of atoms with lattice spacing a such that $\mathbf{R} = (\pm a, 0, 0); (0, \pm a, 0); (0, 0, \pm a)$. For an atomic s-state equation (1.133) can be written as [8]:

$$\begin{aligned} E_s(\mathbf{k}) &\approx E_s - \beta_s - \gamma_s(e^{ika} + e^{-ika}) \\ &= E_s - \beta_s - 2\gamma_s(\cos k_x a + \cos k_y a + \cos k_z a) \end{aligned} \tag{1.134}$$

where β_s is the value of β calculated for this s-band. With the atoms being brought together to form a solid with primitive cubic lattice, the atomic energy levels E_i become electronic bands (see figure 1.15). The centre of gravity of such bands is lowered by an amount β_s with respect to E_i and the bandwidth is proportional to γ_s. In a solid containing N atoms, and thus N primitive cubic lattice, the atomic energy level E_i of the isolated atom will split into N states due to interaction with $N - 1$ atoms. These states form a quasi-continuous band, which can accommodate $2N$ electrons. The close-packed structure of metals with a high coordination number of the atoms thus

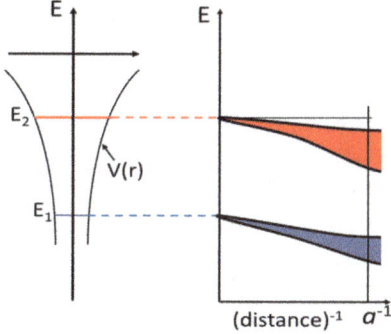

Figure 1.15. Schematic representation of how the nondegenerate electronic levels in atomic potential broaden into bands with the reduction in atomic spacing.

leads to a large band width. The value of γ_s is usually even more important for the width of the band because of the very rapid decay of the wave functions away from the nucleus. An atomic wave function, that is strongly localized near the nucleus for a given structure of the crystalline solid, will give rise to a significantly narrower band than a wave function that is less localized. For example, atomic $3d$ level gives rise to a much narrower band than an atomic $4s$ level, although these atomic levels are very similar in energy. The innermost ($1s$) level of a heavy atom is an extreme case of a localized wave function, which retains its atomic localized character.

It may also be noted that the tight binding wavefunction (equation (1.125)) is actually a wave $e^{ik\cdot R}$ superposed on sets of localized atomic orbitals, giving the phase and amplitude at each lattice site. Thus an electron in the tight binding level can be found with equal probability in any cell of the crystal since its wave function changes only by a phase factor as one moves from one cell to another at a distance $\mathbf{R}$ away. A further indication of the itinerant character of the tight-binding levels come from the fact that the mean velocity of an electron in a Bloch level with wave vector $\mathbf{k}$ and energy $E(\mathbf{k})$ is given by $v(\mathbf{k}) = \frac{1}{\hbar}\frac{\partial E}{\partial k}$. If E is independent of $\mathbf{k}$, $\frac{\partial E}{\partial k}$ is zero. This is consistent with the fact that in genuinely isolated atomic levels (i.e. zero bandwidth cases) the electrons are indeed localized at the individual atoms. When there is a non-zero overlap in the atomic wave functions, then $E(\mathbf{k})$ will not be constant throughout the crystal, and a small variation in $E(\mathbf{k})$ will imply a non-zero value of $\frac{\partial E}{\partial k}$ and hence a small but non-zero mean velocity. This means that as long as there is an overlap of atomic wave functions the electron will be able to move freely through the crystalline solid. The decrease in overlap only reduces the electron velocity, but it does not eliminate the electron motion.

1.7 Quantum chemistry and electron theory of solids

In the tight-binding model we have considered a crystalline solid with fixed atoms at the lattice sites with electrons moving between these atoms. The core concept of this model is actually very straightforward: the electrons do not like to remain confined. This aspect we can see easily from our earlier discussion in section 1.3.3 where the electrons in a metal were treated to be confined in a one-dimensional infinite square well with width L. The kinetic energy of the electron was expressed as $E_n = \frac{1}{2m}\left(\frac{n\pi}{L^2}\right)$ (see equations (1.62) and (1.65)). The smaller L is, the larger the kinetic energy. Hence, kinetic energy can be saved if the electrons are allowed to move in a larger volume. In the tight binding model, the electrons lower their kinetic energies by hopping from one lattice site to the other. The readers with a chemistry background might have already sensed the similarity of the tight-binding method with the famous valence bond theory for molecule formation. In early 20th century, G N Lewis had proposed the formation of chemical bond by the interaction of two shared bonding electrons. Later Heitler and London using the Schrödinger equation showed how two hydrogen atom wave functions in a hydrogen molecule join together to form a covalent bond. Subsequently Linus Pauling used the pair bonding ideas of Lewis together with the Heitler–London (HL) theory to introduce two other key concepts in valence bond theory, namely *resonance* and *hybrid-orbital*, and turned it into a hugely successful method for understanding an enormous range of chemical phenomena. The HL wavefunction does not take into account any ionic configuration in which two electrons can reside in the same atom, and it is assumed that there will always be one electron centred at atom 1 and the other at atom 2. The HL wave function can be expressed as:

$$\psi_{HL} = \frac{1}{2}[\phi_1(r_1)\phi_2(r_2) + \phi_2(r_1)\phi_1(r_2)](\alpha_1\beta_2 - \beta_1\alpha_2). \qquad (1.135)$$

Here $\phi_{1,2}(r)$ are centred at atoms 1 and 2, respectively. The terms α and β represent spin up and spin down states, respectively. In the HL approach the Coulomb repulsion between electrons is reduced by keeping the two electrons well separated, and a covalent bond is formed by taking advantage of the attraction of electrons to each of the nuclei. But this happens at the expense of the kinetic energy of electrons, which would have been lower in an ionic configuration. Thus, in the HL approach more importance is implicitly given to the Coulomb repulsion between electrons than the energy gain due to electron delocalization.

On the other hand, around the same time Hund and Mulliken introduced the method of *molecular orbitals*, where the electrons are placed in wave functions extending over the whole molecule. This is quite akin to the nearly free electron model of electron band theory of the solids. The Hund–Mulliken method was also a very successful method of understanding most of the observed facts of quantum chemistry. The drastic difference between the Heitler–London and Hund–Mulliken methods generated considerable controversies over the years in the field of quantum chemistry. In fact it was the early pioneers of the electron band theory of solids like Slater, Wigner and Van Vleck, who actually showed how to reconcile between those two extreme approaches, and in turn enabled a deeper understanding of the electron band theory of solids. Within the molecular orbital approach the bonding molecular orbital for hydrogen molecule is occupied by two electrons of opposite spin and the antisymmetric total wavefunction is expressed as:

$$\begin{aligned}\psi_M o = \frac{1}{2^{3/2}}[&\phi_1(r_1)\phi_1(r_2) + \phi_1(r_1)\phi_2(r_2) + \phi_2(r_1)\phi_1(r_2) \\ &+ \phi_2(r_1)\phi_2(r_2)](\alpha_1\beta_2 - \beta_1\alpha_2).\end{aligned} \qquad (1.136)$$

The independent electron approximation is very much prevalent here and the probability of finding both the electrons at the same site is 50%. The electron kinetic energy is optimized, while the Coulomb repulsion between the electrons remains considerably large.

In spite of these similarities one may realize that the electron theory of solid state, however, is a many-body problem. To describe the intrinsic properties of materials one must employ statistical concepts to the large system in thermodynamic limit. P W Anderson [13] in a short essay captured this cross fertilization of ideas between the fields of quantum chemistry and electron band theory of crystalline solid, and interestingly how those ideas in many ways influenced the development of the main subject matter of this book, namely strong electron correlation and Mott insulators.

In the one-electron band theory of solids, the wave property of electrons dominate. The electrons propagate in the solid, and in this process the optimization of kinetic energy compensates the effect of strong Coulomb repulsion between the electrons. However, we will see in the later chapters that the role of Coulomb energy cannot always be neglected, and this competition between kinetic energy gain and Coulomb repulsion plays a very important role in the electronic properties of many emerging classes of solid materials including of course the Mott insulators.

From the study of the molecular problem it is apparent that in general the antisymmetric wavefunctions are slightly higher in energy than the symmetric wavefunctions. This is because an electron in the antisymmetric state is excluded from a small region of space where the antisymmetric wave function changes its sign. An electron in the symmetric state, however, is allowed in this region. The exclusion from a region shortens the range of positions allowed for the electrons in the antisymmetric state, and through the uncertainty principle there is an increase in the kinetic energy for such electrons. This argument can be generalized in a qualitative manner for a solid where the Pauli exclusion principle ensures that for a collection of many electrons the wave function (involving both spatial spin parts) must be antisymmetric. This in turn will say that the total kinetic energy of a collection of electrons shared between the atoms in a solid will increase rapidly as a greater number of electrons is packed into a small area. Thus in the absence of Pauli's exclusion principle ensuring the antisymmteric state for electrons, a large collection of nuclei/ions and electrons in a solid would have their energy reduced by packing more tightly, and the solid material would have been quite unstable. Hence, we can see that the Pauli exclusion principle is instrumental for the stability of the solid matter. The quantitative argument behind this striking phenomena is, however, very complicated, and it took 40 years from the introduction of the Pauli exclusion principle to conclusively show that there is a lower limit in the energy of a collection of matter [14, 15].

1.8 Methods for calculating electron band structure

We have seen above that the electron band structure in solids can be studied through two opposite approaches: (i) treating the crystalline lattice structure as a weak perturbation to the free electron model, or (ii) starting from the essentially atomic single particle orbitals situated at each lattice site.

There are a number of methods that interpolate between these two extreme limits, which can be employed to study the real systems:

1. Cellular method: In this method solutions to the single-particle problem are obtained within a single Wigner–Seitz cell of the lattice in real space with appropriate boundary conditions on the cell, rather than the atomic boundary condition that the wave function vanishes at large distances. In this way a whole lattice may be built out of such cells. Instead of using exact ionic potential within the cell, a muffin-tin approach defines an atomic region in the neighbourhood of a particular ion and the rest of the cell. Within the atomic region a regular Coulomb atomic potential which diverges to ∞ at the atomic site is used, and the potential is set to a constant value in the interstitial region.

2. Plane wave method: In this method, the single-particle states are written as an expansion of plane waves, and the potential throughout the crystal is a superposition of free-atom potentials.

3. Augmented plane wave (APW) method: In the APW method (which is a computational improvement over the cellular and plane wave method), one uses the cellular idea of atomic and interstitial regions. An APW function is treated like a plane wave in the interstitial region. In the atomic region it must satisfy an atomic-like Schrödinger equation, while being continuous across the boundary between the atomic and interstitial region. The superpositions of APWs build the single-particle states, and a variational approach is followed for optimization of the states.

4. Orthogonalized plane wave (OPW) method: From the computational point of view the OPW method is a further improvement over the APW method. In this method while dealing with the single-particle states of the non-core electrons i.e. conduction electrons, instead of just plane waves one starts with superpositions of plane waves already chosen to be orthogonal to the core states of the ion potential.

5. Pseudopotential method: The effective potential seen by the conduction electrons is very different from the bare ion potential, because the core electrons act to screen the nuclear charge, and in addition the Pauli exclusion principle causes an effective repulsion between electrons. In the pseudopotential method, the potential to start with is one that is essentially zero in the core region, and a screened Coulomb potential $\frac{e^{-r/\lambda}}{r}$ outside the core, where λ is some screening length scale. In fact, the pseudopotential method also takes some initial step to tackle the major shortcomings of the single-electron models dealing with non-interacting electrons.

The readers are referred to the book by Ashcroft and Mermin [5] for more details on these methods of electron band structure calculations. Currently the most popular theory, however, is the density functional theory for the electronic ground state. This will be discussed in chapter 2.

1.9 Metals, insulators and semiconductors

Within the framework of the Bloch theory and the independent electron approximation, Wilson [16] in the early 1930s introduced a model that predicted at $T = 0$ a sharp distinction between crystalline solids showing metallic conduction and insulating behavior. According to the Wilson model, insulators are materials in which all energy bands are either completely occupied or completely empty. There is an energy gap E_g between the uppermost full band (known as the valence band) and lowest empty band (known as the conduction band). An insulator under the framework of conventional electron band theory without any explicit consideration of the electron–electron Coulomb interactions is usually called a band insulator or Bloch–Wilson insulator. This is to be contrasted with the main subject matter of this book, namely the Mott-insulator, where electron–electron interaction plays a very

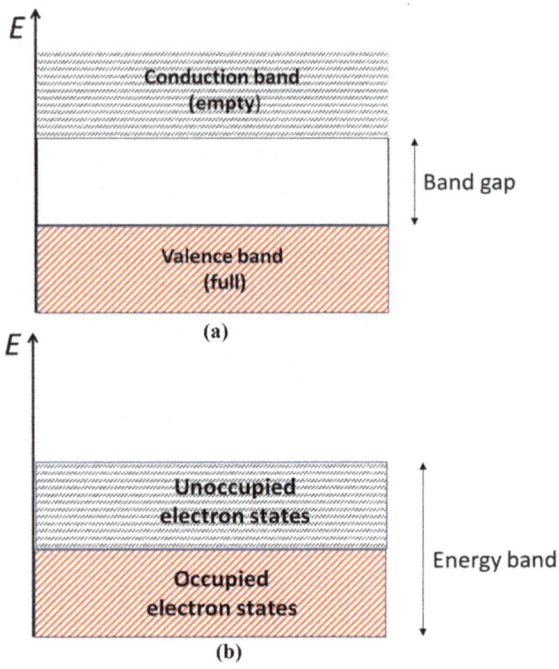

Figure 1.16. Schematic representation of the band picture of a solid: (a) insulator; (b) metal.

important role. A metal, on the other hand, is a material in which one or more energy bands are partly full. This band picture is schematically presented in figure 1.16.

Let us consider the case of alkali metals Na, K, etc residing in the first column or period of the periodic table (see figure 1.17), which have one electron per primitive cell. They are called monovalent metals, and all have just one half filled band. The closed shells in these elements are nearly inert, and there is hardly any interaction with the outer or conduction electrons. The true ground state of these elements will be symmetrical between $\psi_{\mathbf{k}}$ and $\psi_{-\mathbf{k}}$, and will carry no current. However, there are current carrying states in the band available in an infinitesimal distance in energy above the top of the occupied levels (see figure 1.16(b)). Hence, the conductivity will be high, which is a characteristic property of metals. These alkali atoms have a body-centred cubic (BCC) lattice, which is actually a cube with side a containing two lattice points, and hence two alkali metal atoms. Therefore the electron density is $n = 2/a^3$. We can determine the Fermi wavevector by substituting this value of electron density in equation (1.70), which gives $k_F = 1.24\frac{\pi}{a}$. For a BCC lattice the shortest distance to the Brillouin zone boundary is half the length of one of the primitive translation vectors $\mathbf{A}_j$ of the reciprocal lattice vector $\mathbf{G}$ and can be written in terms of the cube edge a as $1.41\frac{\pi}{a}$. This suggests that the free-electron Fermi surface reaches only to a distance $1.24/1.41 = 0.88$ of the path to the nearest Brillouin zone boundary. The filled electronic states will thus have wave vectors, which remain away from any of the Brillouin zone boundaries. Therefore in these monovalent alkali metals there is no distortion of the bands due to the bandgap, and their physical properties are quite close to the predictions of the

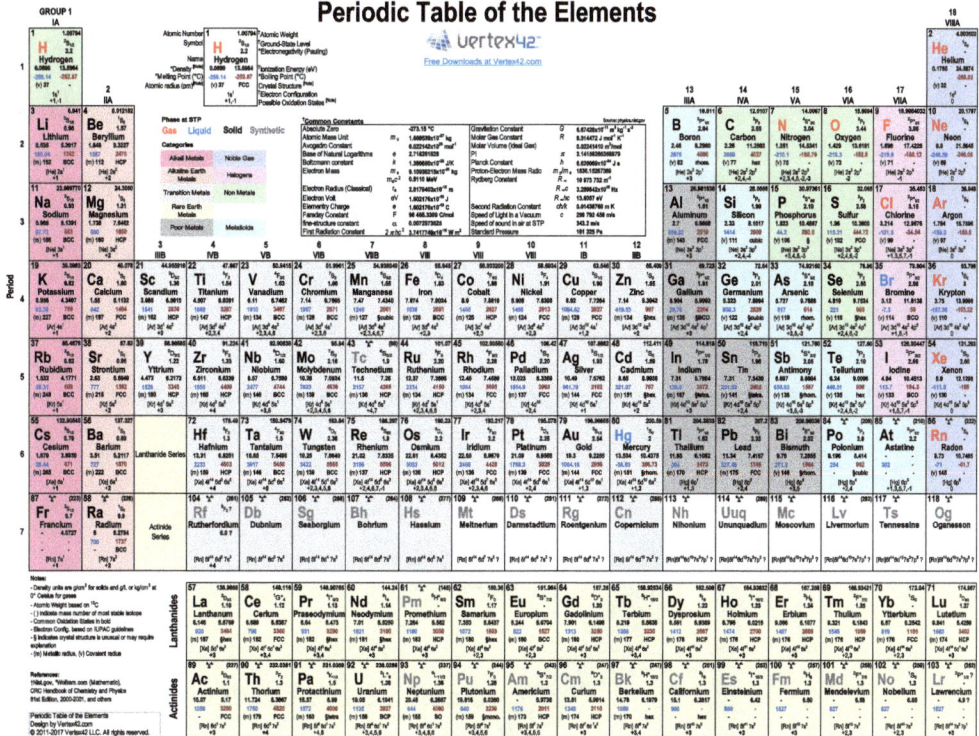

Figure 1.17. Periodic table of elements with their electronic configuration. (Reproduced with permission from https://www.vertex42.com.)

Sommerfeld free electron model. Similarly the other monovalent metals like noble metals Cu, Ag, Au occupying column 11 of the periodic table (see figure 1.17) will also have just one half filled band, or half filled Brillouin-zone. At this stage it is possible to make a generalized statement that a crystalline solid with an odd number of electrons per unit cell will be a metal. For example, for Al, Ga, In etc there are three electrons per atom, which can fill only one and a half band. However, it may be noted that As, Sb and Bi with five electrons per atom are actually semi-metals. These elements crystallize into a structure with two atoms per unit cell, so the odd electron rule does not apply here. The ten electrons almost exactly fill five bands, but the 5th band is not quite full. In fact there is a little overlap into the 6th band, and regardless of the temperature there are always some electrons available to carry a current.

The elements occupying the 2nd and 12th column of the periodic table with filled *s*-shell have all electrons in closed shells. In principle they are expected to be insulator. However, a solid with an even number electrons per unit cell is not necessarily an insulator, because the bands may overlap in energy. It may be energetically favourable to let some of the electrons go into the upper band, while leaving the lower band partially filled. This situation is pictured in figure 1.18 for a solid metal. There will be an energy gap where the Brillouin-zone boundary is crossed at any particular place, but the lowest part of the 2nd band may lie below the highest part of the 1st band. It actually happens that all divalent elements are metals.

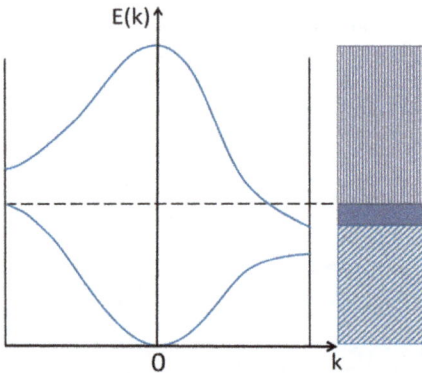

Figure 1.18. Schematic representation of the overlapping bands in a divalent metal.

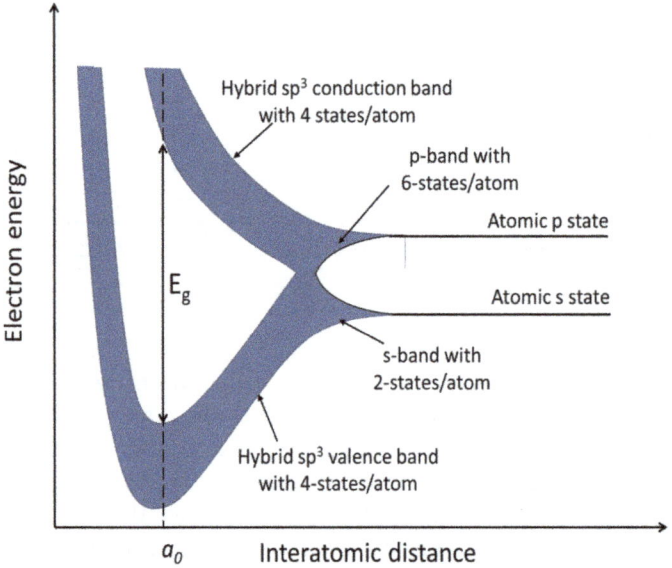

Figure 1.19. Schematic representation of the energy bands of tetravalent elements like C, Si and Ge as a function of atomic spacing. There is a forbidden energy gap of width E_g between the occupied (valence) and the unoccupied (conduction) bands, which results from sp^3 hybrid orbitals.

The energy gaps there are not large enough to keep the electrons within a single Brillouin-zone. However, some of these divalent metals like Sr and Ba are poor conductors, since the band overlap in such cases is probably small.

The tetravalent elements are either metals or insulators. Let us consider the formation of energy bands in carbon. The electronic configuration of isolated carbon atoms is $1s^2$, $2p^2$, $2s^2$. As the carbon atoms approach each other in the formation of solid, the atomic $2s$ and $2p$ states start overlapping leading to the formation of $2s$ and $2p$ bands (see figure 1.19). There will 2(6) states per atom available for filling the electrons in $2s$ ($2p$) bands. In solid carbon i.e. diamond crystals, however, there is an overlap of $2s$ and $2p$ wavefunctions with tetrahedral bonding. This in turn modifies the 2s and 2p bands and eventually leads to a sp^3 hybrid, which splits into two bands, each of which can contain four electrons. This feature is shown schematically in figure 1.19. The lower part of this split hybrid sp^3 band accommodates the four electrons of the atomic 2s and 2p states of carbon. The upper part of the band remains unoccupied, and there is an energy gap of width E_g between the two sp^3 sub-bands. The properties of Si and Ge can be explained in a similar way. The hybrid sp^3 band arises from 3s and 3p atomic states for Si, and 4s and 4p atomic states for Ge. When a voltage is applied, the electrons in the filled band cannot increase their kinetic energy by a small amount because there are no vacant states with energies slightly above the band. The band gap E_g of Si is about 1.1 eV. At finite temperatures a variety of processes enables the electrons to be excited into the conduction band leaving behind some empty states in the valence band. This event allows some electrical conduction in elements like Si and Ge, and they are known as semiconductors. The 14th column of the periodic table is occupied by these semiconducting elements. On the other hand, the energy gap in diamond is so large that at room temperature it as a perfect electrical insulator. But carbon again behaves like a semi-metal in the form of graphite. The tetravalent element Sn is also an interesting case; this is metallic in one crystalline phase and semiconducting in another.

The transition elements occupying 3rd to 10th column in the periodic table (see figure 1.17) are characterized by their incomplete inner d-shells, which get filled up while moving from left to right. For example in atomic Fe only 6 of the 10 states in the 3d-shell are filled, with 2 other electrons in the outer 4s-state. These outer valence electrons form a broad s-band in crystalline solid Fe, which is not very different from the nearly free electron conduction band in an ordinary metal. The inner d-electrons in some way still behave as if they are localized in the ion cores. However, there is finite overlap of d orbitals from the neighbouring atoms, and that leads to the formation of a narrow d-band capable of holding up to 10 electrons per atom. Hence, neither of the bands is full, but the conduction is mainly by the s-electrons.

1.10 Peierls transition and charge density wave

In the discussions above we have seen how electrons can interact with the periodic potential of the ions in a crystalline solid, giving rise to a gap in the electron energy spectrum. This is the origin of band insulators. Here we will see how the interaction of the electrons with static lattice deformations can also give rise to an energy gap in a class of materials called Peierls insulator. We will illustrate this phenomenon schematically taking the example of a one-dimensional metal at $T = 0$. Figure 1.20 presents a nearly-free electron band in a monovalent one-dimensional metal, where the band gap exists at the Brillouin-zone boundary at $\pm\frac{\pi}{a}$ and the Fermi surface (here actually a pair of points) is half the length of the Brillouin zone i.e. $k_F = \pm\frac{\pi}{2a}$. In the real (r)-space there is one electron per atom on average, and the electrons are free to move along the chain of atoms, placed at a distance a apart (see figure 1.20). However, a very simple distortion of the r-space lattice, where the neighbouring atoms moved together to form dimers, can lower the energy of the electron system.

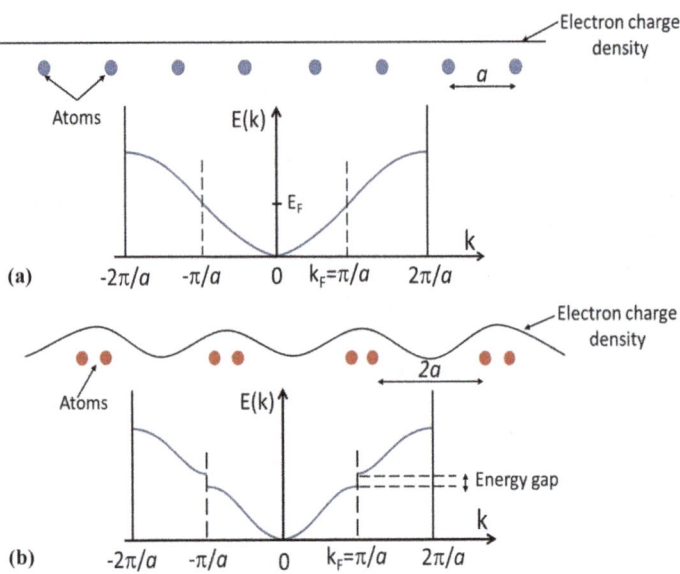

Figure 1.20. Schematic representation of Peierls distortion in a one-dimensional metal: (a) undistorted metal; (b) Peierls insulator.

This kind of distortion is shown in figure 1.20, where the unit cell length has doubled to $2a$ and the associated Brillouin zone is reduced in size by half. A band-gap now opens up at the new Brillouin-zone boundaries at $\pm k_F = \pm\frac{\pi}{2a}$ lowering the electronic energy, and turning the materials into an insulator. This kind of distortion results in the piling up of electronic charge on the dimers, and the resulting state is called a *charge density wave* (CDW).

Such Peierls transition are readily observed in materials, which have chain or layer structures. Organic conductor tetrathiafulvalenetetracyanoquinodimethane (TTF-TCNQ) is the prototype of one group of solids with CDW ground state, and other prominent groups of solids consists of inorganic layer compounds being the tri-chalcogenides ($NbSe_3$, NbS_3, $TaSe_3$, TaS_3). The dichalcogenide TaS_2 is an interesting compound that shows both commensurate and incommensurate CDW transition as a function of temperature. We will discuss more on TaS_2 in chapter 6 in the context of strong electron correlation and metal–insulator transition. Another prominent example of a Peierls insulator is the blue bronze $K_{0.3}MoO_3$, which undergoes a Peierls transition around 180 K, with concomitant resistivity and structural change. For more details on Peierls transition and CDW, the readers are referred to the review article by Gruner [17].

1.11 Effects of static disorder: Anderson localization

Within the framework of Bloch's theory a periodic potential formed by the ions in a lattice can be approximated by a regular array of potential wells (see figure 1.21(a)). The electron wave functions assume the form of plane waves modulated by a function with the periodicity of the lattice, and the wave functions spread

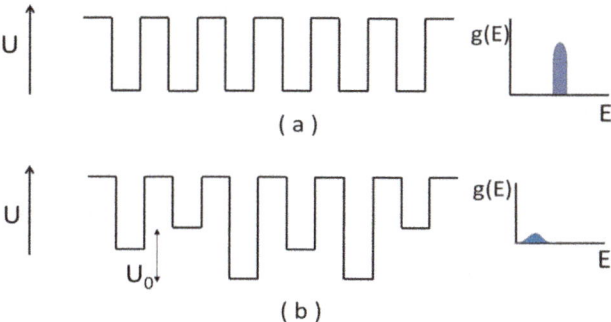

Figure 1.21. One-dimensional periodic potential approximated as: (a) a regular set of square wells representing a perfectly ordered system (upper frame), and (b) randomly varying potential wells representing a strongly disordered system (lower frame).

throughout the crystal giving rise to the normal electrical conductivity. We have also learnt from the Bloch theory that an electron in a perfectly periodic array of ions in a crystalline solid experiences no collision at all. Collisions can arise within the independent electron approximation only from deviations from perfect periodicity in a crystalline solid, which can be classified in two broad categories:

1. Deviations from periodicity in a crystalline solid, due to thermal vibrations of the lattice:

    At finite temperatures the ions are subjected to some thermal energy, and do not remain rigidly fixed at the periodic lattice points. The primary effect of this thermal energy is to cause the ions to undergo small vibrations about their lattice positions. This deviation of the ionic network from the perfect periodicity is the dominant scattering mechanism for electrons at room temperature, and the most important cause of the temperature dependence in the electrical resistivity.

2. Static impurities and crystal defects:

    Point defects like impurity atoms, missing ions at the lattice sites, and ions in the wrong places (say at the interstitials) provide a localized scattering centre. There may also be more extended defects like dislocations, in which the lattice periodicity is violated along a line, or even in an entire plane.

Continuing within the framework of Bloch's theory, the traditional view is that scattering by the random potential causes the Bloch waves to lose phase coherence on the length scale of the electron mean free path l. The wave function, however, remains extended throughout the sample. P W Anderson in his seminal paper in 1958 [18] pointed out that if the disorder is very strong, the wave function may become localized. We will study here briefly the effect of strong disorder of the second kind i.e. static disorder, on the electronic state in a crystalline solid. The narrative below will closely follow the treatment of the subject presented in the book by Mott and Davis [19]. To proceed further we need to consider a strongly varying potential in contrast with a perfect periodic potential. In contrast to the picture of the Bloch periodic potential, the size of the wells are far from uniform (see figure 1.21(b)). Starting from the Schrödinger equation, Anderson used the framework of tight-binding electrons, within which a crystalline array of ion potential wells gave rise to a narrow band of levels (see figure 1.21(a)) as in the case of the d band of a transition metal or to an impurity band produced by donors in a semiconductor. It is assumed that the potential wells are far apart that so the overlap between the atomic wavefunctions $\phi(r)$ on adjacent wells is small. Representing the nth potential well by the suffix n and $\mathbf{R}_n$ as its lattice site, the Bloch wavefunction for an electron in this crystalline is expressed as [19]:

$$\psi_k(x, y, z) = \sum_n e^{i\mathbf{k}\cdot\mathbf{R}_n}\phi(\mathbf{r} - \mathbf{R}_n). \tag{1.137}$$

The functions ϕ are considered to be spherically symmetrical. Taking E_0 as the energy level for an electron in a single well, the energy for an electron in a simple cubic lattice corresponding to the wavefunctions in equation (1.137) is expressed as [19]:

$$E = E_0 + E_k \tag{1.138}$$

where,

$$E_k = -2I(\cos(k_x a) + \cos(k_y a) + \cos(k_z a)). \tag{1.139}$$

In this equation (1.139) the term I is called the transfer integral and is expressed as:

$$I = \int \phi^*(\mathbf{r} - \mathbf{R}_n)H\phi(\mathbf{r} - \mathbf{R}_{n+1})dr \tag{1.140}$$

where H is the Hamiltonian. The transfer integral depends on the shape of the wells, and for the present purpose it is sufficient to write it as [19]

$$I = I_0 e^{-\alpha r}. \tag{1.141}$$

Here α is defined such that the wavefunction on a single well falls off with distance with the rate $e^{-\alpha r}$, and $\alpha = (2mE_0)^{1/2}/\hbar$. The effective mass m^* at the bottom of a band is given by:

$$m^* = \frac{\hbar^2}{2Ia^2} \tag{1.142}$$

and the bandwidth B is given by:

$$B = 2zI \tag{1.143}$$

where z is the coordination number [19].

Anderson considered a non-periodic potential U that took all values at random between $\pm U_0$ (see figure 1.21(b)); here U_0 is the spread of energies. A large electron mean free path l is introduced if U_0 is considered to be small and an application of the Born approximation gives [19]

$$\frac{1}{l} = \frac{2\pi}{\hbar} \left(\frac{1}{2\sqrt{2}} U_0 \right)^2 a^3 \frac{g(E)}{v} \tag{1.144}$$

where the electron energy E and velocity v are taken at the Fermi energy. Using equation (1.76) for $g(E)$ the equation (1.144) becomes:

$$\frac{a}{l} = \frac{(U_0/I)^2}{32\pi}. \tag{1.145}$$

Now, based on the idea of Ioffe and Regel [20] the lower limit for the electron mean free path l in a metal is the interatomic spacing, and that means a mean free path such that $kl < 1$ is impossible. With no disorder and with the potential energy illustrated in figure 1.21(a), $ka = \pi$ in the middle of the band. The Ioffe–Regel rule would then say that the shortest l arises when the Bloch wavefunction loses phase memory in going from atom to atom. So the Bloch wavefunction for an electron in the crystal now instead of equation (1.137) takes the form:

$$\psi_k(x, y, z) = \sum_n A_n \phi(\mathbf{r} - \mathbf{R}_n) \tag{1.146}$$

where the A_n possess random phases and amplitudes. Such wavefunction is shown schematically in figure 1.22(a). In this case one can write the magnitude of the mean free path to be $l \sim a$. One can see from equation (1.145) that this will happen when $\frac{U_0}{I} \simeq \sqrt{32\pi} \sim 10$. Alternatively, if the co-ordination number z is 6, $U_0/B \simeq 0.83$. With further increase in the value of U_0/B the random fluctuations of the amplitude and the phase of wavefunction become larger (see figure 1.22(b)). When U_0/B becomes very large the wavefunctions for each isolated well will be influenced by all the other wells, and will fall off exponentially with distance as in figures 1.22(c) and (d). Anderson showed that there is no diffusion of electron if U_0/B is greater than a constant that depends on the co-ordination number z. In this situation all the wavefunctions for an electron in such non-periodic solid are of the type shown in figure 1.22(c), decaying with distance r from the neighbourhood of some well. The wavefunctions of the localized electron states are of the form:

$$\psi_k = \sum_n A_n \phi(\mathbf{r} - \mathbf{R}_n) e^{-\alpha r} \tag{1.147}$$

with the coefficients A_n having random phases. This phenomenon is now formally known as Anderson localization, and this leads to a static disorder induced insulating state. The exponential decay of the envelope of the wave function from some point $\mathbf{R}_n$ in space is sometime expressed as $e^{-\frac{|\mathbf{r} - \mathbf{R}_n|}{\xi}}$ where ξ is the localization length.

  With some understanding of the two limits of perfect periodic crystalline solid and non-periodic solid with strong disorder, the interesting question now arises as to what happens to the state with intermediate disorder? It is found that even if the Anderson criterion for electron localization is not satisfied there can be a situation

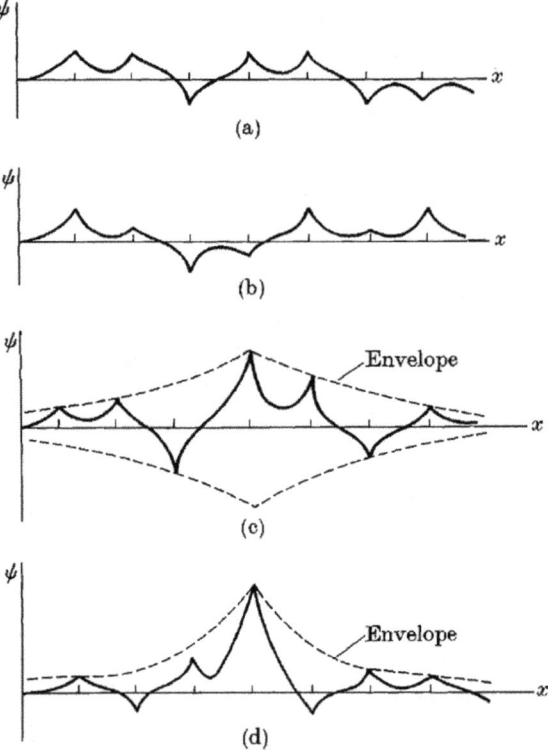

Figure 1.22. Anderson wavefunction: (a) when electronic mean free path $l \sim a$; (b) when states are just non-localized; (c) when states are just localized; (d) case of strong localization. (Reproduced from [19]. Copyright of OUP, Copyright 1992.)

where states are localized in one range of energies and not localized in another range. Electronic conductivity σ_E will then be finite in the middle of the band, but zero for energies near band extremities [19]. Any form of random potential, however small it may be, will introduce a range of localized states at the band tail. A critical energy E_c separates localized states from non-localized states, so that:

$$\langle \sigma \rangle = 0; \; E < E_c$$
$$\neq 0; \; E > E_C. \tag{1.148}$$

This is shown in figure 1.23(a) for a density of states resulting from the Anderson potential. $E_c(E_{C'})$ separates localized and non-localized states at the bottom (top) edge of the band. This critical energy at which this change form localized to delocalized state occurs is called the *mobility edge* [21]. The name mobility arises from the fact that, the conductivity at zero temperature would vanish if the Fermi energy lies in a region of localized states, whereas extended states will result into a finite conductivity at zero temperature. The transition between a metal and an insulator is thus marked by the mobility edge.

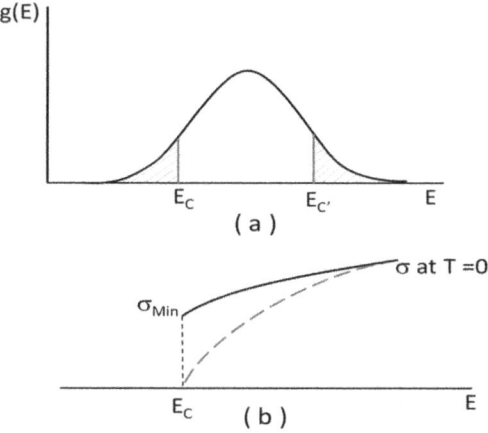

Figure 1.23. Schematic representation of: (a) the mobility edge, separating localized and extended electron states; (b) the two possibilities of a continuous or discontinuous transition with minimum metallic conductivity σ_{min}.

In normal metals electrical resistivity increases with the increase in temperature and that is attributed to the electron scattering due to the increasing deviation from the lattice periodicity due to thermal vibrations. The electronic mean free path l in metals decreases progressively with the increasing lattice vibration. Ioffe and Regel [20] in 1960 proposed that a metallic state does not survive this reduction in l indefinitely and it can never become shorter than the inter-atomic spacing a. At that instant a coherent electron motion vanishes and the concept of charge carrier (electron) velocity is lost. Hussey *et al* [22] have pointed out the universality of the Ioffe–Regel limit in a wide variety of metals. The precise numerical definition of the Ioffe-Regel limit is not generally agreed upon, and criteria ranging from $k_F l_{min} \approx 1$ through $l_{min} \approx a$ to $k_F l_{min} \approx 2\pi$ have all been employed [22].

Based on the idea of Ioffe–Regel limit, Mott addressed the important question whether the metal–insulator transition was continuous or discontinuous in nature (see figure 1.23(b)). He had argued for a discontinuous transition and introduced the concept of the minimum metallic conductivity [19], or the size of conductivity jump at the disorder-induced metal–insulator transition [23]:

$$\sigma_{min}^{3D} \simeq \left[\frac{1}{3\pi^2} \right] \left[\frac{e^2}{\hbar} \right] \left[\frac{1}{a} \right] \tag{1.149}$$

where a is some microscopic length scale in the problem, such as the inverse of the Fermi wave number, $a \approx k_F^{-1}$. Experimental support for the existence of Mott minimum metallic conductivity σ_{min} has been found for a large variety of systems, and the concept has also been extended to two dimensions, where

$$\sigma_{min}^{2D} \approx \frac{e^2}{\hbar}. \tag{1.150}$$

The scaling theory of localization [23], however, calls into question the existence of σ_{min} in both three and two dimensions. This theory predicts instead that the metal–insulator transition is a continuous transition in three dimensions, and that all states are localized in two dimensions. This debate is yet to be settled, but this subject matter is beyond the scope of this book.

Further reading

1. Chaikin P M and Lubensky T C 1995 *Principle of Condensed Matter Physics* (Cambridge: Cambridge University Press).
2. Ziman J M 1972 *Principles of the Theory of Solids* 2nd edn (Cambridge: Cambridge University Press).
3. Ashcroft N W and Mermin N D 1976 *Solid State Physics* (Philadelphia, PA: Saunders College).
4. Singleton J 2001 *Band Theory and Electronic Properties of Solids* (Oxford: Oxford University Press).
5. Ibach H and Lüth H 2003 *Solid State Physics* 3rd edn (Berlin: Springer).
6. Hofmann P 2015 *Solid State Physics* 2nd edn (New York: Wiley).
7. Simon S H 2013 *Solid State Basics* (Oxford: Oxford University Press).
8. Fox M 2001 *Optical Properties of Solids* (Oxford: Oxford University Press).
9. Mott N F and Davis E A 1977 *Electronic Processes in Non-crystalline Solids* (Oxford: Clarendon).

Self-assessment questions and exercises

1. The number density of conduction electrons in a typical metal is in the range 10^{22} cm^{-3}–10^{23} cm^{-3}, yet the independent electron approximation without considering Coulomb repulsion can explain various metallic properties quite well. Why is this effect of Coulomb interaction not felt in spite of the presence of such a huge number of electrons?
2. Resistivity ρ of a reasonably pure copper (Cu) sample at 300 K is approximately 1.8 μΩ cm. Assuming the conductivity of Cu metal is due to the free electrons, what is the mean free path for such conduction electrons? The lattice parameter for Cu is 3.72×10^{-8} cm.
3. What role does the Pauli exclusion principle play in the existence of crystalline solid materials?
4. A box of volume V contains a number of electrons occupying the lowest allowed energy levels. If the volume of the box is doubled while leaving the number of electrons unchanged, will there be a change in the maximum momentum of the electrons? If so, by what factor?
5. Within the realm of the Sommerfeld free electron model, what will be the separation of energy levels at the Fermi energy for (a) 1 cm^3, (b) 1 μm^3, (c) 10 nm^3 samples of a metal like Al?
6. Copper (Cu) forms in a face-centred cubic (FCC) crystal structure with the edge of the unit cell $a = 3.72 \times 10^{-8}$ cm. Assuming one conduction electron per atom in Cu metal, find the concentration of conduction electron in Cu.

Within the realm of the free electron gas model, will the Fermi energy depend on: (a) the mass of the crystal, (b) the concentration of conduction electron or (c) both? Derive an expression for Fermi energy (at $T = 0$) and check your answer.

7. The crystal potential of a two-dimensional square lattice with lattice spacing a is given by $U_{XY} = U_0\left(\cos\frac{2\pi x}{a} + \cos\frac{2\pi y}{a}\right)$, where U_0 is a constant. Find out approximately the energy gap at the midpoint of a side face of the first Brillouin zone.

8. Calculate the bandwidth of the lowest energy band of a one-dimensional solid of lattice spacing a represented by $U(x) = U_0 a \delta(x)$.

9. What is the ground state of a single-atom problem represented by a potential $U(x) = a U_0 \delta(x)$ in a one-dimensional solid with lattice spacing a? What is the bandwidth of the lowest energy band within the tight binding approximation?

10. If the material considered in the problem above is a divalent one, what will be the necessary condition for the material to be a metal?

11. What are the necessary conditions for the formation of Peierl's insulating state?

12. Anderson localization leads to the insulating behaviour in the concerned solid. Is the concept of Bloch electron as a propagating wave ignored altogether in such a situation?

References

[1] Chaikin P M and Lubensky T C 1995 Principle of Condensed Matter Physics (Cambridge: Cambridge University Press)

[2] Debenedetti P G and Stanley J E 2003 *Phys. Today* **56** 40

[3] Ziman J M 1972 *Principles of the Theory of Solids* 2nd edn (Cambridge: Cambridge University Press)

[4] Wilczek F 2005 *Nature* **433** 239

[5] Ashcroft N W and Mermin N D 1976 *Solid State Physics* (Philadelphia, PA: Saunders College)

[6] Ibach H and Lüth H 2003 *Solid State Physics* 3rd edn (Berlin: Springer)

[7] Singleton J 2001 *Band Theory and Electronic Properties of Solids* (Oxford: Oxford University Press)

[8] Hofmann P 2015 *Solid State Physics* 2nd edn (New York: Wiley)

[9] Simon S H 2013 *Solid State Basics* (Oxford: Oxford University Press)

[10] Friedrich W, Knipping P and Laue M 1912 *Proc. Bavarian Acad. Sci.* 303

[11] Mourigal M *et al* 2013 *Nat. Phys.* **9** 435

[12] Fox M 2001 *Optical Properties of Solids* (Oxford: Oxford University Press)

[13] Anderson P W 2011 *More and Different: Notes from a Thoughtful Curmudgeon* (Singapore: World Scientific)

[14] Dyson F and Lenard A 1967 *J. Math. Phys.* **8** 423

[15] Lenard A and Dyson F 1968 *J. Math. Phys.* **9** 698

[16] Wilson A H 1931 *Proc. R. Soc. Lond.* A **133** 458
    Wilson A H 1931 *Proc. R. Soc. Lond.* A **134** 277

[17] Gruner G 1988 *Rev. Mod. Phys.* **60** 1129

[18] Anderson P W 1958 *Phys. Rev.* **109** 1492

[19] Mott N F and Davis E A 1977 *Electronic Processes in Non-crystalline Solids* (Oxford: Clarendon)

[20] Ioffe A F and Regel A R 1960 *Prog. Semicond.* **4** 237

[21] Mott N F 1967 *Adv. Phys.* **16** 49

[22] Hussey N E, Takenaka K and Takagi H 2004 *Philos. Mag.* **84** 2847

[23] Lee P A and Ramakrishnan T V 1986 *Rev. Mod. Phys.* **57** 287

Chapter 2

Electron–electron interaction in crystalline solids

In the chapter 1 we have seen that within the *independent electron approximation*, each electron in a solid obeys a one-electron Schrödinger equation in a periodic potential $U(\mathbf{r})$:

$$-\frac{\hbar^2}{2m_e}\nabla^2\psi(\mathbf{r}) + U(\mathbf{r})\psi(\mathbf{r}) = E\psi(\mathbf{r}). \qquad (2.1)$$

In a crystalline solid, however, one is dealing with: (i) the one-electron potentials describing the interactions of the electrons with the massive ionic-cores, and (ii) pair potentials describing some aspects of the electron–electron interactions. All these interactions together are represented by an effective one-electron potential $U(\mathbf{r})$ in the independent electron approximation. In the case of a metallic solid it is really impossible to describe electrons correctly by the one-electron Schrödinger equation (2.1), however innovative ways one may define the one-electron potential $U(\mathbf{r})$. This is because of the various interesting but complex effects arising from the electron–electron interactions. Hence, there is a real need to go beyond the *independent electron approximation*. In this chapter we shall explore some mean-field methods to study interacting many electron systems, for example a metal. In this approach the main problem under study is replaced with a set of independent-particle problems with an effective potential representing electron–electron interactions. This subject has been covered earlier in various textbooks on solid state physics, and the present author has been particularly influenced by the books: Ashcroft and Mermin [1], Madelung [2], Martin [3] and Mahan [4].

2.1 Electron–electron interaction in metals; Hartree–Fock theory

An accurate picture of metals is ideally represented by a Schrödinger equation with an N-particle wave function $\Psi(\mathbf{r}_1\mathbf{s}_1, \mathbf{r}_2\mathbf{s}_2, \ldots, \mathbf{r}_N\mathbf{s}_N)$, which takes into account both position $\mathbf{r}$ and spin $\mathbf{s}$ of all N electrons in the metal:

doi:10.1088/2053-2563/ab16c9ch2

$$H\Psi = \sum_{j=1}^{N} \left(-\frac{\hbar^2}{2m_e} \nabla_j^2 \Psi - Ze^2 \sum_R \frac{1}{|\mathbf{r}_j - \mathbf{R}|} \Psi \right)$$
$$+ \frac{1}{2} \sum_{j \neq k} \frac{e^2}{|\mathbf{r}_j - \mathbf{r}_k|} \Psi = E\Psi. \tag{2.2}$$

The contribution of the relatively massive nuclei is neglected with the help of the Born–Oppenheimer or adiabatic approximation, and as far as the electrons are concerned the nuclei remain static and they just provide classical potential. The potential energy term in equation (2.2):

$$U_{ion}(\mathbf{r}_j) = -Ze^2 \sum_R \frac{1}{|\mathbf{r}_j - \mathbf{R}|} \tag{2.3}$$

represents the attractive electrostatic potentials of the ions (nuclei + core electrons) fixed at the points $\mathbf{R}$ of the Bravais lattice associated with the periodic solid metal. On the other hand, the last term in equation (2.2) represents the interactions of the electrons with each other.

In equation (2.2), the first two terms can be represented by the sums over single particle operators. Hence, the solution of the Schrödinger equation would be simple if the electron–electron interaction represented by the last term in equation (2.2) is somehow neglected. The Schrödinger equation could then be separated by expressing the wavefunctions as:

$$\Psi(\mathbf{r}_1\mathbf{s}_1, \mathbf{r}_2\mathbf{s}_2, \ldots, \mathbf{r}_N\mathbf{s}_N) = \psi_1(\mathbf{r}_1\mathbf{s}_1)\psi_2(\mathbf{r}_2\mathbf{s}_2), \ldots, \psi_N(\mathbf{r}_N\mathbf{s}_N) \tag{2.4}$$

where ψ_j are a set of N orthonormal one-electron wave functions. This is prevented by the third term in equation (2.2), which depends on the coordinates of two electrons. It is this Coulomb interaction between the electrons that mainly gives rise to the computational difficulties in solving the N-particle Schrödinger equation.

We have discussed the importance of electron–electron interaction arising out of Coulomb interaction between electrons and the Pauli exclusion principle in chapter 1, section 1.7 in the context of the formation of the hydrogen molecule. There we have pointed out the similarity between the Heitler–London theory and the molecular orbital theory of hydrogen molecule formation with the tight binding model and nearly free electron model, respectively, of electron band theory of solids. Even in that relatively simple case, in reality a H_2 molecule is somewhere between these two extreme limits, and it is practical to start with the independent electron approximation, rather than the Heitler–London limit with strong electron correlation. This, however, changes when two protons are pulled apart to make further progress in the description of molecule formation. The larger the distance between the protons, the more suppression of the possible ionic-configurations is necessary in the ground state wave function of the hydrogen molecule. Hence electron correlation arising from Coulomb interaction becomes more and more important with the increasing bond length, until one ends up in the Heitler–London or strong electron correlation limit.

There is not much hope of solving equation (2.2) for large molecules and solids without further simplifications. However, the wavefunction expressed in equation (2.4) may still provide an approximate solution, and may describe some part of the electron–electron interaction. In this direction we first rewrite the Hamiltonian in equation (2.2) as:

$$H\Psi = \sum_{j=1}^{N}\left(-\frac{\hbar^2}{2m_e}\nabla_j^2\Psi - Ze^2\sum_{R}\frac{1}{|\mathbf{r}_j - \mathbf{R}|}\Psi\right)$$
$$+\frac{1}{2}\sum_{j\neq k}\frac{e^2}{|\mathbf{r}_j - \mathbf{r}_k|}\Psi = \sum_j H_j\Psi + \sum_{j\neq k}H_{jk}\Psi. \tag{2.5}$$

We now calculate the expectation value of energy E by incorporating the wave function expressed in equation (2.4) in the Schrödinger equation (2.5). The Hamiltonian H is now a sum of one-particle operators H_j and two-particle operators H_{jk}. As a result the matrix elements become products of integrals $\langle\psi_j|H_j|\psi_j\rangle$ or $\langle\psi_j\psi_k|H_{jk}|\psi_j\psi_k\rangle$ and integrals $\langle\psi_i|\psi_i\rangle$ $(i \neq j, k)$. Since ψ_j are normalized the latter becomes unity, and the expectation value of E can be expressed as:

$$E = \langle\Psi|H|\Psi\rangle = \sum_j\langle\psi_j|H_j|\psi_j\rangle + e^2\sum_{j\neq k}\left\langle\psi_j\psi_k\left|\frac{1}{|\mathbf{r}_j - \mathbf{r}_k|}\right|\psi_j\psi_k\right\rangle. \tag{2.6}$$

To make further progress one may use the variational principle, which states that a solution to the Schrödinger equation satisfying the Bloch condition with wave vector $\mathbf{k}$ and energy $E(\mathbf{k})$ will make equation (2.6) stationary. The best set of wavefunctions ψ_i for the ground state within the constraint of equation (2.4) will be represented by those ψ_j, which minimize E. One therefore needs to vary equation (2.6) with respect to ψ_j^* or ψ_j and equate the result to zero. To this end the variation is carried out after adding the normalization condition multiplied with Lagrange parameters E_j to equation (2.6):

$$\delta\left[E - \sum_j E_j(\langle\psi_j|\psi_j\rangle - 1)\right] = 0. \tag{2.7}$$

This leads to:

$$\langle\delta\psi_i|H_i|\psi_i\rangle + \sum_{j\neq i}\left\langle\delta\psi_i\psi_j\left|\frac{e^2}{|\mathbf{r}_j - \mathbf{r}_i|}\right|\psi_i\psi_j\right\rangle - E_i\langle\delta\psi_i|\psi_i\rangle$$
$$= \langle\delta\psi_i|H_i + \sum_{j\neq i}\left\langle\psi_j\left|\frac{e^2}{|\mathbf{r}_j - \mathbf{r}_i|}\right|\psi_j\right\rangle - E_i|\psi_i\rangle = 0. \tag{2.8}$$

Equation (2.8) is valid regardless of the variation $\delta\psi_i^*$, hence it implies that the ψ_i are defined by the equation:

$$\left[-\frac{\hbar^2}{2m_e}\nabla^2 + U_{ion}(\mathbf{r}) + e^2 \sum_{j(\neq i)} \int \frac{|\psi_j(\mathbf{r}')|^2}{|\mathbf{r} - \mathbf{r}'|}dr'\right]\psi_i(\mathbf{r}) = E_i\psi_i(\mathbf{r}). \qquad (2.9)$$

Here $U_{ion}(\mathbf{r})$ represented by equation (2.3) is the ionic potential seen by the electron. This set of equations (2.9) is known as the Hartree equations. This is a one-electron Schrödinger equation representing an electron (i) at position $\mathbf{r}$ in the potential field $U_{ion}(\mathbf{r})$ of the lattice ions and also under the influence of Coulomb potential arising from an average distribution of all other electrons ($j \neq i$) in the metal. The Hartree equations thus give the best approximation to the full many electron wave function that can be expressed in terms of a simple product of one-electron wave functions. The Coulomb interaction between the electrons is approximated by replacing the rest of the electrons by a smooth distribution of negative charge with charge density $n = -e\sum_j|\psi_j(\mathbf{r})|^2$.

Hartree equations are solved by *self-consistent field* approximation. In this method a form for $U_{el} = e^2 \sum_j \int dr'|\psi_j(\mathbf{r}')|^2\frac{1}{|\mathbf{r}-\mathbf{r}'|}$ is initially chosen, and on the basis of that equation (2.9) is solved. In the next step a new U_{el} is then computed from the resulting wave functions $\psi_j(\mathbf{r})$, and a new Schrödinger equation is solved. This iterative procedure is continued until further steps do not change the potential U_{el}. Even this approximation (2.4) of wavefunction for the exact N-electron Schrödinger equation still involves a significant amount of mathematical exercise with considerable numerical complexity.

The Hartree equations with self consistent field approximations, however, do not take account of certain important features of electron–electron interactions:

1. The wave function (equation (2.4)) is not consistent with the Pauli exclusion principle, according to which the sign of Ψ should change with respect to the interchange of any two of its arguments.
2. The electrons interact only with the field that is obtained by averaging over the positions of the remaining $N - 1$ electrons. It is a mean-field approximation to the electron–electron interaction, which takes into account only the electronic charge. This approximation neglects the important fact that the particular configuration of the rest of the $N - 1$ electrons affects the electron under consideration. In reality each electron is influenced by the motion of every other electron in the system; this leads to the important *correlation* property.

In order to have a remedy to the first problem, the following condition needs to be fulfilled for the trial wave function:

$$\Psi(\mathbf{r}_1\mathbf{s}_1, \ldots, \mathbf{r}_i\mathbf{s}_i, \ldots, \mathbf{r}_j\mathbf{s}_j, \ldots, \mathbf{r}_N\mathbf{s}_N) = -\Psi(\mathbf{r}_1\mathbf{s}_1, \ldots, \mathbf{r}_j\mathbf{s}_j, \ldots, \mathbf{r}_i\mathbf{s}_i, \ldots, \mathbf{r}_N\mathbf{s}_N). \quad (2.10)$$

This is achieved by the introduction of the *Slater determinant* as a replacement of the trial wave function (equation (2.4)) [1]. The Slater determinant is a linear combination of the product (equation (2.4)) and all other products arising from it by the permutation of the $\mathbf{r}_j\mathbf{s}_j$ among themselves, and then added together with pre-factors ±1. This would guarantee the requirement of antisymmetrization of wave function (equation (2.10)) as per the Pauli exclusion principle:

$$
\begin{aligned}
\Psi = {}& \psi_1(\mathbf{r}_1 \mathbf{s}_1)\psi_2(\mathbf{r}_2 \mathbf{s}_2) \cdots \psi_N(\mathbf{r}_N \mathbf{s}_N) \\
& - \psi_1(\mathbf{r}_2 \mathbf{s}_2)\psi_2(\mathbf{r}_1 \mathbf{s}_1) \cdots \psi_N(\mathbf{r}_N \mathbf{s}_N) + \cdots .
\end{aligned}
\tag{2.11}
$$

Equation (2.11) can be written more compactly as the determinant of an $N \times N$ matrix:

$$
\Psi(\mathbf{r}_1 \mathbf{s}_1, \, \mathbf{r}_2 \mathbf{s}_2, \, \ldots, \, \mathbf{r}_N \mathbf{s}_N) =
\begin{pmatrix}
\psi_1(\mathbf{r}_1 \mathbf{s}_1) & \psi_1(\mathbf{r}_2 \mathbf{s}_2) & \cdots & \psi_1(\mathbf{r}_N \mathbf{s}_N) \\
\psi_2(\mathbf{r}_1 \mathbf{s}_1) & \psi_2(\mathbf{r}_2 \mathbf{s}_2) & \cdots & \psi_2(\mathbf{r}_N \mathbf{s}_N) \\
\vdots & \vdots & \cdots & \vdots \\
\psi_N(\mathbf{r}_1 \mathbf{s}_1) & \psi_N(\mathbf{r}_2 \mathbf{s}_2) & \cdots & \psi_N(\mathbf{r}_N \mathbf{s}_N)
\end{pmatrix} .
\tag{2.12}
$$

With this wavefunction in equation (2.12) we calculate again the expectation value of energy $E = \langle \Psi | H | \Psi \rangle$ as:

$$
\begin{aligned}
E = {}& \left[-\frac{\hbar^2}{2m_e}\nabla^2 + U_{ion}(\mathbf{r}) \right]\psi(\mathbf{r}) + \sum_{j,k} \int dr\, dr' \frac{e^2}{|\mathbf{r} - \mathbf{r}'|} |\psi_j(\mathbf{r})|^2 |\psi_k(\mathbf{r}')|^2 \\
& - \sum_{j,k} \int dr\, dr' \frac{e^2}{|\mathbf{r} - \mathbf{r}'|} \delta_{s_j s_k} \psi_j^*(\mathbf{r})\psi_j(\mathbf{r}')\psi_k^*(\mathbf{r}')\psi_k(\mathbf{r}) .
\end{aligned}
\tag{2.13}
$$

The last term in the above equation (2.13) is negative and includes the product $\psi_j^*(\mathbf{r})\psi_j(\mathbf{r}')$ instead of the one-electron combination $|\psi_j(\mathbf{r})|^2$. Applying the variational principle as in the case of Hartree equations above, minimization of equation (2.13) with respect to ψ_j^* leads to a generalization of Hartree equations:

$$
\begin{aligned}
& -\frac{\hbar^2}{2m_e}\nabla^2\psi_i(\mathbf{r}) + U_{ion}(\mathbf{r})\psi_i(\mathbf{r}) + \sum_{j \neq i} \int \frac{e^2 |\psi_j(\mathbf{r}')|^2}{|\mathbf{r} - \mathbf{r}'|} dr' \psi_i(\mathbf{r}) \\
& - \sum_{j \neq i} \int e^2 \frac{\psi_j^*(\mathbf{r}')\psi_i(\mathbf{r}')}{|\mathbf{r} - \mathbf{r}'|} dr' \psi_j(\mathbf{r})\delta_{s_j s_i} = E_i\psi_i(\mathbf{r}) .
\end{aligned}
\tag{2.14}
$$

The above set of equations (2.14) is known as the *Hartree–Fock* equations. One can include the terms $i = j$ in both the summations of equation (2.14). They cancel out, and thus avoid the electron self energy.

The Hartree–Fock equations (2.14) are different from the Hartree equations (equation (2.9)) in that there is an additional term in the former. The Kronecker-delta $\delta_{s_j s_i}$ in this term takes a value of either one or zero, depending on whether the spin states at s_i and s_j are parallel or antiparallel to one another, respectively. Hence, this term is finite only for electrons with parallel spins. This term is known as the *exchange term*, and is written as:

$$\sum_j \int dr' \frac{e^2}{|\mathbf{r} - \mathbf{r}'|} \psi_j^*(\mathbf{r}') \psi_i(\mathbf{r}') \psi_j(\mathbf{r}) \delta_{s_j s_i} = -e \int dr' \frac{n_i^{HF}(\mathbf{r}, \mathbf{r}')}{|\mathbf{r} - \mathbf{r}'|} \psi_i(\mathbf{r}) \qquad (2.15)$$

where,

$$n_i^{HF}(\mathbf{r}, \mathbf{r}') = -e \sum_j \frac{\psi_j^*(\mathbf{r}') \psi_i(\mathbf{r}') \psi_i^*(\mathbf{r}) \psi_j(\mathbf{r})}{\psi_i^*(\mathbf{r}) \psi_i(\mathbf{r})}. \qquad (2.16)$$

This is in contrast with the third term in the Hartree equations (equation (2.9)), which is often referred to as the direct or Hartree term. Here, in place of charge density n_j defined by $n_j = -e|\psi_j(\mathbf{r})|^2$, we have the *exchange charge density n_i^{HF}*. This also represents a charge $-e$, which can be verified by integration over $\mathbf{r}'$. The crucial difference between n_j and n_i^{HF} lies in the fact that while n_j is distributed over the crystalline solid in the same way as the charge density of the $N - 1$ other electrons, n_i^{HF} depends on the position $\mathbf{r}$ of the particular electron considered. One can say that the motion of the electrons with the 'same spin or parallel spin' is correlated.

The basic difficulty in solving the Hartree–Fock equations lies in the fact that the interaction term in the Hartree–Fock equations depend on ψ_i, the very solutions one is trying to find out. Hence, to solve this equation one takes recourse to an iterative method. A trial wavefunction ψ_i is chosen for n_i^{HF} to start with and a better estimate of n_i^{HF} is obtained by solving the Hartree–Fock equations. This cycle is repeated until a self-consistent solution is found. There is another difficulty arising from the position (i) dependence of the interaction term, which means there is a different Hartree–Fock equation for each electron. We will see in section 2.6 that it is possible to get around this problem with Slater's approximation by taking an average over n_i^{HF}.

The Hartree–Fock wavefunction is a single determinant of one-electron orbitals and sometime it is possible that the best single determinant corresponding to minimum ground state energy breaks the symmetries of the Hamiltonian. This situation is referred to as unrestricted Hartree–Fock (UHF), and an example of that is a magnetically ordered ground state. Here time-reversal symmetry is broken, whereas all terms in the Hamiltonian remain symmetric under time reversal. In contrast, in the restricted Hartree–Fock the single determinant wavefunction retains the symmetries of the Hamiltonian even if a lower-energy solution is possible.

Summarizing the discussions above, at this point we can see that certain important parts of many-electron effects are still missing in the Hartree–Fock theory. The electron–electron interaction in both Hartree and Hartree–Fock equations is dealt with in such a way that each electron feels an average electric field created by the rest of the other electrons in the metal. In Hartree–Fock theory, however, due to the Pauli exclusion principle an electron displaces the other electrons with the same spin in its neighbourhood, and as a result their motion is correlated. It may be noted that the Pauli exclusion principle is taken into account in the Hartree–Fock theory only with the constraint of antisymmetry of the many-electron wave function, but not included explicitly in the Hamiltonian operator. This of course implies a sort of repulsive interaction between electrons with parallel spins. But it is prudent to note here that the Hartree–Fock equation exaggerates the role of the exchange interaction between electrons with parallel spins, since the Coulomb repulsion between the electrons is still completely ignored. An exact theory must include the important electron–electron correlation due to the Coulomb repulsion for all electrons. There must be a displacement for electrons irrespective of their spin orientation. This Coulomb repulsion is included only in an averaged way both in the Hartree and Hartree–Fock theories, and the correlation resulting from the Coulomb repulsion is missing in both theories. The difference between the Coulomb correlation and the exchange energy is known as the correlation energy. A precise quantitative definition of the correlation energy is the difference between the expectation values of the Hartree–Fock wavefunctions and the exact wave-functions [3].

2.2 Hartree–Fock theory of free electrons

The Hartree–Fock equations in general are quite difficult to deal with; one possible exception being the case of free electron gas. The Hartree–Fock equations can be solved exactly with the choice of ψ_i to be a set of orthonormal plane waves when the periodic potential is zero or constant. The electron gas is thought to be embedded in a background of constant positive charge. This model is often called the *jellium model*. The periodic lattice of the solid is replaced by the constant background, and the properties of the electron gas, in particular the *electron–electron interaction* becomes the main focus of interest.

The set of free-electron plane waves which gives a solution to the Hartree–Fock equation can be represented by [1]:

$$\psi_i(\mathbf{r}) = \left(\frac{e^{i\mathbf{k}_i \cdot \mathbf{r}}}{\sqrt{v_g}} \right) \times \text{spin-function} \tag{2.17}$$

where each wave vector less than k_F occurs twice in the Slater determinant, once for each spin orientation. The electronic charge density that determines U_{el} will be uniform. On the other hand the ions in a solid represented by a free-electron gas can be considered as a uniform distribution of positive charge with the same density as that of the electronic charge. As a result the potential of the ions will be canceled by the direct term i.e. $U_{ion} + U_{el} = 0$. In this situation it is only the *exchange term* that will remain. The exchange potential for the jellium model can be written as:

$$U^{ex} = -e \int \frac{n^{HF}(\mathbf{r} - \mathbf{r}')}{|(\mathbf{r} - \mathbf{r}')|} dr'. \tag{2.18}$$

Using equation (2.16) and taking ψ_i expressed as a plane wave in equation (2.17), U^{ex} can be rewritten as:

$$U^{ex} = -\frac{e^2}{2\pi^3} \int_0^{k_F} \int \frac{e^{i(\mathbf{k} - \mathbf{k}') \cdot (\mathbf{r} - \mathbf{r}')}}{|(\mathbf{r} - \mathbf{r}')|} dr' dk'. \tag{2.19}$$

We now represent the Coulomb interaction in terms of its Fourier transform [1]:

$$\frac{e^2}{|\mathbf{r} - \mathbf{r}'|} = 4\pi e^2 \frac{1}{(2\pi)^3} \int \frac{1}{q^2} e^{i\mathbf{q}\cdot(\mathbf{r}-\mathbf{r}')} dq. \tag{2.20}$$

Incorporating equation (2.20) in equation (2.19) leads to:

$$U^{ex} = -\frac{e^2 4\pi}{(2\pi)^3} \int_0^{k_F} \frac{1}{|(\mathbf{k} - \mathbf{k}')|^2} dk'. \tag{2.21}$$

After performing the integration in equation (2.21) we get:

$$U^{ex} = -\frac{2e^2 k_F}{\pi} F\left(\frac{k}{k_F}\right) \tag{2.22}$$

where,

$$F(x) = \frac{1}{2} + \frac{1 - x^2}{4x} \ln\left|\frac{1 + x}{1 - x}\right|. \tag{2.23}$$

The eigenvalues of the Hartree–Fock equation in the jellium model are expressed as:

$$E(k) = \langle\psi_k|H|\psi_k\rangle = \frac{\hbar^2 k^2}{2m_e} + \langle\psi_k|U^{ex}|\psi_k\rangle. \tag{2.24}$$

Now the expectation value of the exchange potential expressed in equations (2.21) and (2.22) is given by:

$$\langle\psi_k|U^{ex}|\psi_k\rangle = U^{ex}\langle\psi_k|\psi_k\rangle = -\frac{2e^2 k_F}{\pi} F\left(\frac{k}{k_F}\right). \tag{2.25}$$

Incorporating equation (2.25) into (2.24) one can write:

$$E(k) = \frac{\hbar^2 k^2}{2m_e} - \frac{2e^2}{\pi} k_F F\left(\frac{k}{k_F}\right). \tag{2.26}$$

This indicates that the plane waves can solve Hartree–Fock equations (2.14) and the energy of the one-electron state with wave vector $\mathbf{k}$ is given by equation (2.26). The energy of an electron is now given by $\frac{\hbar^2 k^2}{2m_e}$ plus a term representing the effects of electron–electron interaction. This additional term, depends on the wavevector $\mathbf{k}$ and causes a lowering of the single-electron energies. In order to calculate the contributions of these interactions to the total energy of the N-electron system, one needs to sum this correction over all $\mathbf{k} < \mathbf{k}_F$ taking also into account the two spin levels that are occupied for each $\mathbf{k}$. After that the summation needs to be divided by 2 because in summing the interaction energy of a given electron over all electrons, one is counting each electron pair twice. Ultimately the whole exercise leads to the expression for the total energy of the N-electron system [1]:

$$E_N = 2 \sum_{k<k_F} \frac{\hbar^2 k^2}{2m_e} - \frac{e^2 k_F}{\pi} \sum_{k<k_F} \left[1 + \frac{k_F^2 - k^2}{2kk_F} \ln\left| \frac{k_F + k}{k_F - k} \right| \right]. \qquad (2.27)$$

The dependence of the function $F(k/k_F)$ representing the exchange potential U^{ex} as a function of the wavevector $\mathbf{k}$ is presented in figure 2.1. The slope of this function diverges at $\mathbf{k} = \mathbf{k}_F$. However, the divergence is logarithmic in nature and not quite visible in figure 2.1. This leads to the anomalous feature namely the energy derivative becoming logarithmically infinite at $\mathbf{k} = \mathbf{k}_F$. This issue will be discussed below in section 2.5. In figure 2.2 we show the comparison of single-electron energy dispersion $\frac{\hbar^2 k^2}{2m_e}$ in the jellium model of free electrons or Hartree electrons and energy dispersion (equation (2.26)) for Hartree–Fock electrons. The energy dispersion of Hartree–Fock electrons is parabolic for small value of $\mathbf{k}$. While the energies are lower than Hartree electrons, the band width of the occupied states for the Hartree–Fock electrons is significantly larger.

2.3 Exchange hole, correlation hole and pair distribution function

Since the wave function within the Hartree–Fock theory here are plane waves, the relationship $\mathbf{p} = \hbar\mathbf{k}$ is valid. However, equation (2.26) clearly tells that the energy and momentum of the Hartree–Fock electron are not related by the classical

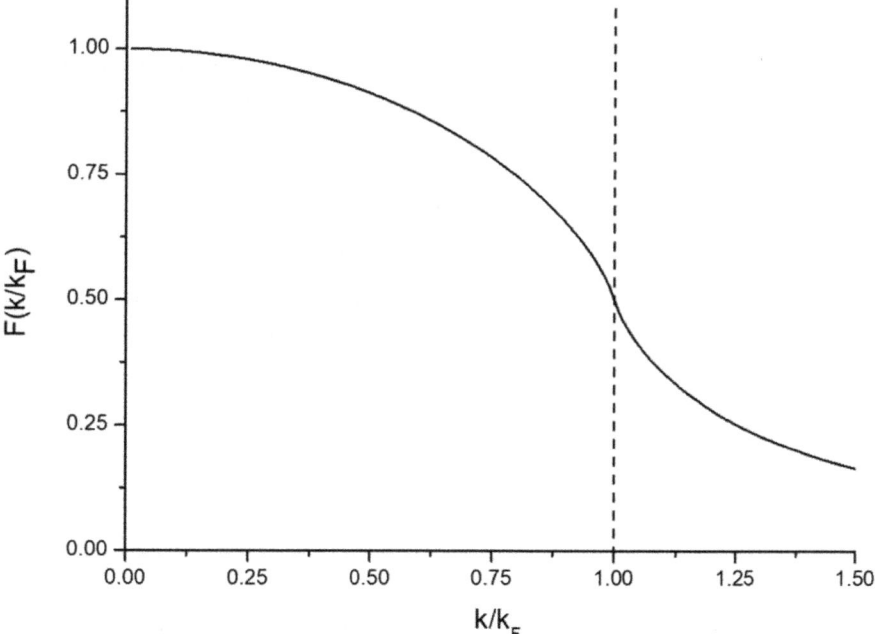

Figure 2.1. A plot showing the dependence of the function $F(k/k_F)$ (defined by equation (2.23)) on wave vector k. The slope of $F(k/k_F)$ function diverges logarithmically at $k = k_F$, but this is not quite visible even by changing the scale of the plot.

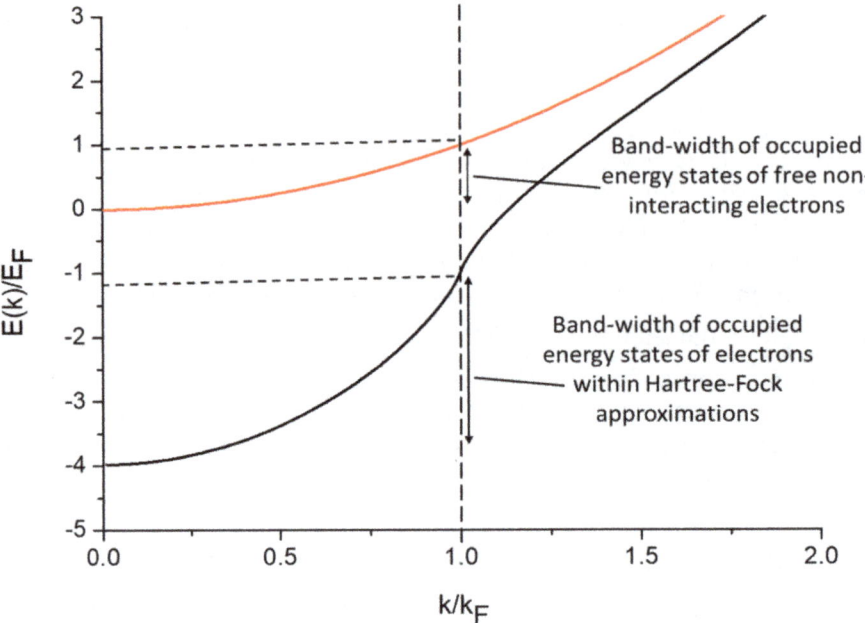

Figure 2.2. Comparison of single-electron energy dispersion in the jellium model of free electrons or Hartree electrons and energy dispersion for Hartree–Fock electrons. Energy $E(k)$ is expressed in terms of $\frac{\hbar^2 k_F^2}{2m_e}$.

equation $E = \frac{p^2}{2m_e}$. This can be understood in a better manner by considering the exchange charge density (see equation (2.16)), which can be rewritten by incorporating ψ_i as the plane wave (see equation (2.17)):

$$n_i^{HF}(\mathbf{r}, \mathbf{r}') = -\frac{e}{v_g} \sum_{k_i}^{N/2} e^{(i(\mathbf{k}_i - \mathbf{k}_j) \cdot (\mathbf{r} - \mathbf{r}'))}. \tag{2.28}$$

It is assumed here that the system is nonmagnetic, hence each state is occupied by two electrons. Since $n_i^{HF}(\mathbf{r}, \mathbf{r}')$ is different for each state, the meaning of equation (2.28) becomes more clear if an average is taken over all electrons:

$$[n_i^{HF}(\mathbf{r}, \mathbf{r}')]_{av} = \sum_i^N \frac{\psi_i^*(\mathbf{r}) \rho_i^{HF}(\mathbf{r}, \mathbf{r}') \psi_i(\mathbf{r})}{n(\mathbf{r})}$$

$$= -\frac{2e}{N v_g} \sum_{k_j}^{N/2} e^{-i\mathbf{k}_j \cdot (\mathbf{r} - \mathbf{r}')} \sum_{k_i}^{N/2} e^{i\mathbf{k}_i \cdot (\mathbf{r} - \mathbf{r}')}. \tag{2.29}$$

Replacing the summation by integration over the Fermi sphere and assuming that $N/2$ electrons in the sphere have the same spin we can obtain:

$$\sum_{k_i}^{N/2} e^{i\mathbf{k}_i \cdot (\mathbf{r}-\mathbf{r'})} = \frac{v_g}{2\pi^3} \int_0^{k_F} e^{i\mathbf{k}_i \cdot (\mathbf{r}-\mathbf{r'})} d^3k$$

$$= \frac{3}{2} N \frac{k_F r'' \cos(k_F r'') - \sin(k_F r'')}{(k_F r'')^3} \qquad (2.30)$$

where $\mathbf{r''} = \mathbf{r'} - \mathbf{r}$, and finally:

$$\left[n_i^{HF}(\mathbf{r}) \right]_{av} = \frac{9eN}{2} \frac{(k_F r \cos k_F r - \sin k_F r)^2}{(k_F r)^6}. \qquad (2.31)$$

The exchange charge density oscillates in a characteristic manner, and the oscillations are caused by the existence of a sharp Fermi surface. These oscillations are known as Friedel oscillations. For large value of $\mathbf{r}$, $[n_i^{HF}(\mathbf{r})]_{av}$ goes to zero as r^{-4}. On the other hand, for small value of $\mathbf{r}$ one can expand the sine and cosine terms in equation (2.31), and $[n_i^{HF}(\mathbf{r})]_{av}$ can be expressed as:

$$[n_i^{HF}(\mathbf{r} \to 0)]_{av} = \frac{9eN}{2} \frac{1}{(k_F r)^6} \left(k_F r - \frac{1}{2}(k_F r)^3 - k_F r + \frac{1}{6}(k_F r)^3 \right)^2$$

$$= -\frac{1}{2}eN. \qquad (2.32)$$

Thus the charge density actually seen by a Hartree–Fock electron is given by:

$$n - n^{HF} = \frac{eN}{v_g} \left[1 - \frac{9}{2} \frac{(k_F r \cos k_F r - \sin k_F r)^2}{(k_F r)^6} \right]. \qquad (2.33)$$

Equations (2.31) and (2.33) indicate that the concentration of electrons with the same spin is reduced in the vicinity of the electron considered, while the electrons with opposite spin are distributed uniformly. In other words the electron is surrounded by an 'exchange hole' with charge $+e$. If the Coulomb correlations beyond Hartree–Fock are taken into account, then we get a correlation hole $-\frac{1}{2}eN$ as $\mathbf{r} \to 0$ due to the presence of the electrons with opposite spin. Thus, the effective electronic charge density is actually reduced to zero in the vicinity of a given electron. In practice, the electron can be imagined to be accompanied by this exchange–correlation hole as it moves in the solid crystal. This implies a continuous rearrangement of the surrounding electrons, hence the free electron energy–momentum relation is no longer valid for a Hartree–Fock electron. The relation $E = \frac{\hbar^2 k^2}{2m_e}$, however, is said to be maintained, if one replaces the electron mass m_e by an *effective mass* m^*, which then from equation (2.27) is dependent on k and has a value greater than m_e. The effective mass of electron m^* is increased because the electron has to carry the 'exchange–correlation hole' with it.

The concept of exchange and correlation hole can be formally introduced in another way in terms of this important property of Hartree–Fock theory, namely the pair distribution function $g(\mathbf{r})$ [4]. This gives the probability of finding an electron at the position $\mathbf{r}$ with respect to an electron already present at $\mathbf{r} = 0$. There are two different pair distribution function $g(\mathbf{r})$ depending on the spin index of electron: $g_{\uparrow\uparrow} = g_{\downarrow\downarrow}$ and $g_{\uparrow\downarrow}$. The first spin index s in the expression $g_{ss'}$ represents the spin of the electron at $\mathbf{r} = 0$, while index s' stands for electron spin at $\mathbf{r} = \mathbf{r}$. Thus, $g_{\uparrow\downarrow}$ represents the probability of finding a spin-down electron at $\mathbf{r}$ with respect to a spin-up electron at $\mathbf{r} = 0$. The electrons in general are moving rapidly, and hence the pair distribution functions are averages for moving electrons.

We recall here that in the Hartree–Fock theory the N-electron wave function is represented by the Slater determinant and the square of this N-electron wave function gives the N-electron density matrix. Within Hartree–Fock approximations the pair distribution function is given by the two-particle density matrix [4]. This is estimated by integration of the N-electron density over $N - 2$ space coordinates and taking the similar expectation value over the $N - 2$ spin variables:

$$g(\mathbf{r}_1, \mathbf{r}_2) = v^2 \int dr_3, \ldots, dr_N |\Psi_{\lambda_1, \ldots, \lambda_N}(\mathbf{r}_1, \ldots, \mathbf{r}_N)|^2 \qquad (2.34)$$

where v is volume and λ_j are the quantum numbers describing the electron states. Assuming the one-electron wavefunctions ψ_λ to be orthogonal, this integration in the above equation (2.34) gives the sum over all possible pair wavefunctions:

$$g_{ss'} = \frac{v^2}{N(N-1)} \sum_{\lambda_i \lambda_j} \begin{vmatrix} \psi_{\lambda_i}(\mathbf{r}_1) & \psi_{\lambda_i}(\mathbf{r}_2) \\ \psi_{\lambda_j}(\mathbf{r}_1) & \psi_{\lambda_j}(\mathbf{r}_2) \end{vmatrix}^2. \qquad (2.35)$$

The sum over all $\lambda_i \lambda_j$ includes all occupied electron states, so each pair is summed twice.

We remind ourselves here that we are dealing with a homogeneous gas of N electrons spread over a large volume v with a uniform density $n_0 = N/v$. In such a homogeneous electron gas, the one-electron states must be represented by plane waves:

$$\psi_\lambda(\mathbf{r}) = \chi_S \frac{e^{i(\mathbf{k}\cdot\mathbf{r})}}{v^{1/2}} \qquad (2.36)$$

where χ_s represent spin functions. The spin averages are given by $\langle \chi_\uparrow \chi_\uparrow \rangle = \langle \chi_\downarrow \chi_\downarrow \rangle = 1$ and $\langle \chi_\uparrow \chi_\downarrow \rangle = \langle \chi_\downarrow \chi_\uparrow \rangle = 0$. Two pair distribution functions can be expressed now as:

$$g_{\uparrow\downarrow} = \frac{1}{n_0^2} \sum_{k_1, k_2} \left(\left| \frac{e^{i(\mathbf{k}_1 \cdot \mathbf{r}_1 + \mathbf{k}_2 \cdot \mathbf{r}_2)}}{v} \right|^2 + \left| \frac{e^{i(\mathbf{k}_1 \cdot \mathbf{r}_2 + \mathbf{k}_2 \cdot \mathbf{r}_1)}}{v} \right|^2 \right)$$

$$= \frac{2}{N^2} \left(\frac{N}{2} \right)^2 = \frac{1}{2} \qquad (2.37)$$

and,

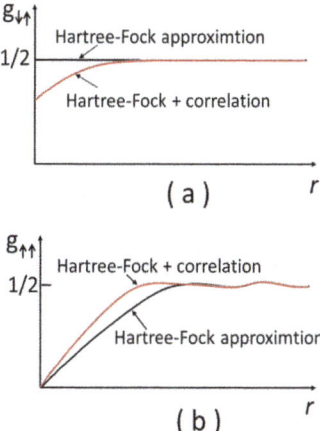

Figure 2.3. Pair distribution function $g(\mathbf{r})$ for electrons with: (a) anti-parallel spins and (b) parallel spins. The black curves are obtained under Hartree–Fock approximation, and the red curves include the effect of correlation.

$$g_{\uparrow\uparrow} = \frac{1}{n_0^2} \sum_{k_1,k_2} \left(\left| \frac{e^{i(\mathbf{k_1}\cdot\mathbf{r_1}+\mathbf{k_2}\cdot\mathbf{r_2})}}{v} \right|^2 - \left| \frac{e^{i(\mathbf{k_1}\cdot\mathbf{r_2}+\mathbf{k_2}\cdot\mathbf{r_1})}}{v} \right|^2 \right)$$

$$= \frac{2}{N^2} \sum_{k_1,k_2} (1 - e^{i(\mathbf{k_1}-\mathbf{k_2})\cdot(\mathbf{r_1}-\mathbf{r_2})}) \qquad (2.38)$$

$$= \frac{1}{2}[1 - \psi(\mathbf{r_1} - \mathbf{r_2})^2]$$

where,

$$\psi(\mathbf{r}) = \frac{2}{N} \sum_{k} e^{i\mathbf{k}\cdot\mathbf{r}}. \qquad (2.39)$$

Within the Hartree–Fock approximation the anti-parallel spin pair distribution function $g_{\uparrow\downarrow} = g_{\downarrow\uparrow} = \frac{1}{2}$ (see figure 2.3(a)). Two spin-functions are orthogonal, hence the cross term which arises from the determinant is zero. This implies an absence of correlation in the positions of anti-parallel spin. In the case of electrons with parallel spins, the cross term survives and the spin averages are unity. This results in the spatial dependence of the parallel spin pair distribution function $g_{\uparrow\uparrow} = g_{\downarrow\downarrow}$. Replacing the summation with the integration over the Fermi sphere and recalling the relation between the average electron density n_0 and Fermi wavevector k, $n_0 = 2 \int \frac{d^3k}{2\pi^3} n_k$, equation (2.39) can be rewritten as [4]:

$$\psi(r) = \frac{2}{n_0} \int \frac{d^3k}{2\pi^3} n_k e^{ik\cdot r} = \frac{3}{rk_F{}^3} \int_0^{k_F} k \, \sin(kr) dk$$

$$= \frac{3}{rk_F{}^3}[\sin(rk_F) - rk_F \cos(rk_F)]. \tag{2.40}$$

Combining equations (2.38) and (2.40) we see that the parallel spin pair distribution function goes to zero at $r = 0$ and approaches the value 1/2 at large distance (see figure 2.3(b)). This is of course an obvious consequence of the Pauli exclusion principle, which is built in many-electron Hartree–Fock wavefunctions.

The total pair distribution function for an electron is the sum of the results for electrons with parallel and anti-parallel spin configurations:

$$g(\mathbf{r}) = g_{\uparrow\uparrow} + g_{\uparrow\downarrow}. \tag{2.41}$$

The result is same for the central electron with spin-down configuration. The total pair distribution function goes to unity at large distance $\mathbf{r}$. This is the consequence of uniform density n_0 of electrons. The total electron charge density is $-en_0 g(\mathbf{r})$ and it gets reduced in the vicinity of the central electron since $g(\mathbf{r}) < 1$ near $\mathbf{r} \approx 0$. This depletion can be visualized as a hole in the electron charge density.

The pair distribution function is subject to the normalization integral:

$$n_0 = \int d^3r[g(\mathbf{r}) - 1] = -1. \tag{2.42}$$

According to this total charge missing from the hole is equivalent to the charge of one electron. One can say that the hole has a positive charge density since it implies absence of electrons, and the integral of this charge density is e.

The charge density of the homogeneous electron gas is uniform only on average. It is, however, not uniform for a particular electron since the other electrons cannot come near to this as they can to other points. Near the electron the positive background charge is not totally canceled by other electrons, because those electrons are not likely to be there in the vicinity. The electron is said to be interacting with its own hole charge.

In the Hartree–Fock approximation only the electrons with parallel spin configuration contribute to the hole, since electrons with anti-parallel spins are not affected i.e. $g_{\uparrow\downarrow} = 1/2$. It may, however, be possible to arrange $g(r)$ so that the potential energy is lower than that estimated within the Hartree–Fock approximation, while maintaining the sum rule expressed in equation (2.41) [4]. This can be achieved by allowing $g_{\uparrow\downarrow}$ to be less than 1/2 near the point $r = 0$ (see the red curve in figure 2.3(a)). Now the total hole charge must be unity. Hence, this change in $g_{\uparrow\downarrow}$ must be balanced by a faster increase in $g_{\uparrow\uparrow}$ with $\mathbf{r}$ near $\mathbf{r} = 0$ (the red curve in figure 2.3(a)). However, changes in $g_{\uparrow\uparrow}$ and $g_{\uparrow\downarrow}$ may cost some kinetic energy as well, so that one cannot just adjust $g_{\uparrow\uparrow}$ and $g_{\uparrow\downarrow}$ to maximize the potential energy alone [4]. The lowest energy state of the system will also have some correlation between the motion of electrons with anti-parallel spins. The correlation energy arises from this correlation between the motion of pairs of electrons.

2.4 Exchange energy and correlation energy in the electron gas

Within the Drude–Sommerfeld model the energy density of the free-electron gas is given by [1]:

$$\frac{E}{V} = \frac{1}{4\pi^3} \int_{k<k_F} dk \frac{\hbar^2 k^2}{2m_e} = \frac{1}{\pi^2} \frac{\hbar^2 k_F^5}{10 m_e}. \tag{2.43}$$

From equation (1.69), $N = \frac{k_F^3}{3\pi^2} V$. Hence, the contribution to the energy of the N-electron system E_N from the first term on the right hand side in equation (2.26) is $N \times \frac{3}{5} E_F$. Transforming the second term in equation (2.26) into an integral, this can be evaluated to give total energy of the N-electron system [1]:

$$E = N \left[\frac{3}{5} E_F - \frac{3}{4} \frac{e^2 k_F}{\pi} \right]. \tag{2.44}$$

The electron energy is often expressed in the literature as a function of the mean electron separation in the electron gas r_s. This is defined as [1]:

$$\frac{1}{n} = \frac{4\pi r_s^3 a_0^3}{3}; \quad r_s = \left(\frac{3}{4\pi n a_0^3} \right)^{\frac{1}{3}} \tag{2.45}$$

where a_0 is the Bohr radius and $n = N/V$. In most of the metals r_s/a_0 has a value between 2 and 6 [1], and large r_s means low electron concentration and vice versa. Equation (2.44) can now be written in terms of the Rydberg ($e^2/2a_0 = 1$ Ry = 13.6 eV) and the parameter (r_s/a_0) [1]:

$$\frac{E}{N} = \frac{e^2}{2a_0} \left[\frac{3}{5} (k_F a_0)^2 - \frac{3}{2\pi} (k_F a_0) \right] = \left[\frac{2.21}{(r_s/a_0)^2} - \frac{0.916}{(r_s/a_0)} \right] \text{Ry}. \tag{2.46}$$

The second term in equation (2.46) is quite comparable in size to the first term, since in metals r_s/a_0 is in the range 2–6. This clearly indicates that the electron–electron correlation cannot be overlooked in metals in any free electron estimate of the electron energy.

In the solid metals from the theoretical point of view the cohesive energy can be divided into two parts: (i) the energy of each electron as a particle in an interacting electron gas, and (ii) the energy of electrostatic attraction between the electron gas and the positively charged ions fixed in a periodic lattice of the solid. Within this premise let us discuss the case of alkali metals with one free electron per atom, considering that ions as point charges localized at the lattice sites of a body-centred cubic (BCC) lattice. The electrostatic energy per atom arising out of the ions and electron-gas in a BCC lattice is given by [1]:

$$U^{Coulomb} = -\frac{24.35}{(r_s/a_0)} \text{ eV/atom.} \tag{2.47}$$

The kinetic energy of the electrons in alkali metals can be approximately estimated from the ground state energy of free electron gas:

$$U^{Kinetic} = \frac{3}{5}E_F = \frac{30.1}{(r_s/a_0)^2} \text{ eV/atom.} \tag{2.48}$$

In the Hartree–Fock theory, the average change in the energy of an electron from the energy $\frac{\hbar^2 k^2}{2m_e}$ (obtained from the Drude–Sommerfeld free-electron model) due to the *exchange* is:

$$\langle E_{exchng} \rangle = U^{exchng} = -\frac{3}{4}\frac{e^2 k_F}{\pi} = -\frac{0.916}{(r_s/a_0)} \text{ Ry/atom}$$
$$= -\frac{12.5}{(r_s/a_0)} \text{ eV/atom.} \tag{2.49}$$

It may be noted that the exchange correction to the electron gas energy (equation (2.49)) is about half the size of the average electrostatic energy (equation (2.47)) and has density dependence. This underscores the importance of the electron–electron correlation in the cohesive energy of metals.

Combining the three contributions we get:

$$U = \frac{30.1}{(r_s/a_0)^2} \text{ eV/atom} - \frac{36.8}{(r_s/a_0)} \text{ eV/atom.} \tag{2.50}$$

Minimizing equation (2.50) with respect to r_s we get:

$$\frac{r_s}{a_0} = 1.6. \tag{2.51}$$

The experimental values of r_s/a_0 in alkali metals, however, range from 2 to 6 [1]. Apart from the lower estimation of r_s/a_0 the striking qualitative failure of equation (2.50) is the prediction of same r_s for all alkali metals. There are various possible reasons for this discrepancy. In this formulation the ions are being treated as point charges, while the real ion-cores have non-negligible radii. Hence the conduction electrons would be prevented from entering a non-negligible fraction of metallic volume occupied by the ion-cores. So even within this rather crude approximation the density of the electrons would be higher than what has been estimated and so would be kinetic energy. The exclusion from ion-core regions would also mean that the conduction electrons cannot get as close to the positive ion-cores as was assumed in the course of derivation of equation (2.50). This in turn would mean that the estimated electrostatic energy would be less negative. Finally it may be remembered that the Hartree–Fock exchange correction to the electron gas energy takes account of the correlation only between electrons with like spin configuration. It does not take account the Coulomb interaction between the electrons with opposite spins. In this respect the situation in transition metals with the presence of relatively narrow d-band becomes even more involved [5]. It may also be pointed out here that even the bond energies are significantly underestimated within the Hartree–Fock theory; some molecules like F_2, are not even bound.

The ground-state energy of a free electron gas has been calculated in both the high and the low-density limit, and it depends only on the inter electron spacing r_s [6]. The results of such calculations in terms of the improvement over the Hartree–Fock theory of free electron gas may be expressed as:

$$\frac{E}{N} = \left[\frac{2.21}{(r_s/a_0)^2} - \frac{0.916}{r_s/a_0} + E_C \right] \text{Ry.} \tag{2.52}$$

Here in this equation (2.52), E_C is called the *correlation energy*. In the limit of high electron densities i.e. $r_s/a_0 \ll 1$, the Coulomb interaction is a relatively small perturbation on the motion of the electrons. In this case the correlation energy may be expressed as a series of the following type [7]:

$$E_C = (A \ln(r_s/a_0) + C + D(r_s/a_0) \ln(r_s/a_0) + \cdots) \text{Ry.} \tag{2.53}$$

With the estimation of the coefficients A and C, equation (2.53) can be rewritten as [1]:

$$\begin{aligned}
\frac{E}{N} &= \left[\frac{2.21}{(r_s/a_0)^2} - \frac{0.916}{(r_s/a_0)} + E_C \right] \text{Ry} \\
&= \left[\frac{2.21}{(r_s/a_0)^2} - \frac{0.916}{(r_s/a_0)} + 0.0622 \ln(r_s/a_0) - 0.096 + \cdots \right] \text{Ry.}
\end{aligned} \tag{2.54}$$

On the other hand in the low-density limit the behavior of the electron gas is again simple. In this regime the Coulomb interaction exerts a dominating influence on the electrons. While the potential energy keeps the electrons apart, the kinetic energy for large r_s/a_0 (i.e. $r_s/a_0 \gg 10$) is insufficient to prevent the electrons becoming localized at fixed sites. The correlation energy can be expressed as a power series in $(1/r_s)^{\frac{1}{2}}$:

$$E_C = \left(\frac{U}{(r_s/a_0)} + \frac{V}{(r_s/a_0)^{\frac{3}{2}}} + \frac{W}{(r_s/a_0)^2} + \cdots \right) \text{Ry.} \tag{2.55}$$

The coefficients U, V and W had been estimated by Wigner [8]. The r_s value in metals lie in the region 2–6. That is essentially an intermediate density regime where the kinetic energy and the potential energy play roughly comparable roles in determining the electron behavior. In this regime there is no rigorous series expression for the correlation energy. Some detailed discussion on the electronic behavior at metallic densities can be found in the work of Bohm and Pines [9, 10]. Their correlation–energy calculation was based on an approximate interpolation between the contribution to the system energy arising from the long wavelength part and that arising from the short wavelength part of the Coulomb interaction, both limits being accurately calculated. The calculation of the correlation energy gave results that were in good agreement with Wigner's interpolation formula [8]:

$$E_C = -0.88/((r_s/a_0) + 7.8) \text{ Ry.} \tag{2.56}$$

Further refinement of the interpolation procedures by Noziers and Pines [6] for actual metallic densities gave an expression for correlation energy:

$$E_C \cong (-0.115 + 0.031 \ln(r_s/a_0)) \text{ Ry.} \tag{2.57}$$

Subsequently there have been many more works on the determination of correlation energy, but the discussion of such works is beyond the scope of the present book. Interested readers may refer to the early review by Löwden [11] and the book by Fulde [12]. We can summarize here saying that the correlation energy is a quantity representing the error in the energy calculated within the framework of the independent one electron model. It may also be noted that the correlation energy is not exactly the difference between the experimentally determined cohesive energy and the Hartree–Fock energy, since the experimentally determined energy also contains contributions from the relativistic terms and the zero-point vibrations of the nuclei (at the ionic sites), which are of course neglected in the Hartree–Fock formalism. It can be said that the cohesion in condensed matter is largely due to exchange and correlation. In fact the equilibrium bound state of minimum total energy arises from a balance between the kinetic energy and the exchange-correlation energy. The kinetic energy of the electrons tends to push atoms away from each other, whereas the exchange-correlation energy, tends to bring the atoms together.

2.5 Screening in the electron gas

There is another anomalous feature in equation (2.26) namely the derivative $\frac{\partial E}{\partial k}$ becomes logarithmically infinite at $k = k_F$. It is to be noted that the velocity of the electrons, which is an important parameter for the metallic properties, is given by $(\frac{1}{\hbar})\frac{\partial E}{\partial \mathbf{k}}|_{k=k_F}$. A singularity at $k = k_F$ in the one-electron energies also leads to an electronic specific heat at low temperature varying as $T/|\ln T|$, and not as T [1]. This singularity arises due to the divergence of the Fourier transform $\frac{4\pi e^2}{k^2}$ of the interaction $\frac{e^2}{r}$ at $k = 0$. This $k = 0$ divergence will be eliminated in a situation where the Coulomb interaction is replaced, say by an interaction of the form $e^2(e^{-k_0 r}/r)$. In this case the Fourier transform would be $4\pi e^2/(k^2 + k_0^2)$ and the unphysical singularity of the Hartree–Fock energies will be removed [1]. In real solids the potential appearing in the exchange term gets altered in just this way. This then takes account of the fields of electrons other than the two at $\mathbf{r}$ and $\mathbf{r}'$ under consideration, which rearrange themselves so as to partially cancel the fields the two electrons exert on one another [1]. This effect is known as *screening*.

Screening is one of the most important manifestations of electron–electron interactions, and this phenomenon as a general subject is discussed in detail in the book by Ashcroft and Mermin [1]. We will have a short discussion here on the screening in free electron gas, which is sometimes quite useful in the context of real metals. Considering the case of a positively charged particle placed rigidly at a given position in the electron gas, one assumes by an analogy with the theory of dielectric media that the potential ϕ^{ext} arising from the positively charged particle and ϕ the full physical potential (arising from both the positively charged particle and the cloud of screening electron it produces) are linearly related by an equation of the form:

$$\phi^{ext}(\mathbf{r}) = \int \epsilon(\mathbf{r}, \mathbf{r}')\phi(\mathbf{r}')dr'. \tag{2.58}$$

Here ϵ is the dielectric constant, which for a spatially uniform electron gas depends only on the separation between the points $\mathbf{r}$ and $\mathbf{r}'$, but not on their absolute position:

$$\epsilon(\mathbf{r}, \mathbf{r}') = \epsilon(\mathbf{r} - \mathbf{r}'). \tag{2.59}$$

Hence, equation (2.58) can be written as:

$$\phi^{ext}(\mathbf{r}) = \int \epsilon(\mathbf{r} - \mathbf{r}')\phi(\mathbf{r}')dr' \tag{2.60}$$

which in turn implies that the corresponding Fourier transforms satisfy:

$$\phi^{ext}(\mathbf{q}) = \epsilon(\mathbf{q})\phi(\mathbf{q}) \tag{2.61}$$

where $\epsilon(\mathbf{q})$ is the wave vector dependent dielectric constant of the metal. Equation (2.61) can be written in the form:

$$\phi(\mathbf{q}) = \frac{1}{\epsilon(\mathbf{q})}\phi^{ext}(\mathbf{q}). \tag{2.62}$$

The most natural quantity to calculate directly is the charge density n^{ind} induced in the electron gas by the total potential $\phi(\mathbf{r})$. Assuming a linear relation between n^{ind} and ϕ, their Fourier transform will satisfy a relation of the form:

$$n^{ind}(\mathbf{q}) = \chi(\mathbf{q})\phi(\mathbf{q}). \tag{2.63}$$

It can now be shown that [1]:

$$\frac{q^2}{4\pi}(\phi(\mathbf{q}) - \phi^{ext}(\mathbf{q})) = \chi(\mathbf{q})\phi(\mathbf{q}) \tag{2.64}$$

and in turn:

$$\epsilon(\mathbf{q}) = 1 - \frac{4\pi}{q^2}\chi(\mathbf{q}) = 1 - \frac{4\pi}{q^2}\frac{n^{ind}(\mathbf{q})}{\phi(\mathbf{q})}. \tag{2.65}$$

There are two widely used theories—Thomas–Fermi theory and Lindhard theory—to calculate $n^{ind}(\mathbf{r})$, which are both simplifications of a general Hartree calculation of the charge induced by the impurity. In the Thomas–Fermi method the total potential $\phi(\mathbf{r})$ is assumed to be a very slow varying function of $\mathbf{r}$. Using this method the total potential in a metal can be expressed as [1]:

$$\phi(\mathbf{q}) = \frac{1}{\epsilon(\mathbf{q})}\phi^{ext}(\mathbf{q}) = \frac{4\pi Q}{q^2 + k_0^2} \tag{2.66}$$

where k_0 is the Thomas–Fermi wave vector, and the external potential ϕ^{ext} is that of a point charge Q:

$$\phi^{ext}(\mathbf{r}) = \frac{Q}{r}, \ \phi^{ext}(\mathbf{q}) = \frac{4\pi Q}{q^2}. \tag{2.67}$$

The Fourier transform in (2.66) can be inverted to give:

$$\phi(\mathbf{r}) = \int \frac{dq}{(2\pi)^3} e^{i\mathbf{q}\cdot\mathbf{r}} \frac{4\pi Q}{q^2 + k_0^2} = \frac{Q}{r} e^{-k_0 r}. \tag{2.68}$$

We see that the total potential given by equation (2.68) is in Coulomb form with an exponential damping factor, which reduces the potential to a negligible magnitude at distances greater than the order of $\frac{1}{k_0}$. This is known in the literature as *screened Coulomb potential*. In free electron gas k_0 can be expressed as [1]:

$$k_0 = 0.815 k_F \left(\frac{r_s}{a_0}\right)^{\frac{1}{2}} = \frac{2.95}{(\frac{r_s}{a_0})^{\frac{1}{2}}} \mathring{A}^{-1}. \tag{2.69}$$

At metallic densities, r_s/a_0 is about 2–6. Hence k_0 is of the order of k_F, and the disturbances due to the external charge are screened within a distance similar to the inter electron spacing. Thus the electrons are highly effective in shielding external charges in metals.

In the Lindhard method, which is also known as the random phase approximation, one solves the Schrödinger equation by perturbation theory [1]:

$$-\frac{\hbar^2}{2m_e}\nabla^2\psi_i(\mathbf{r}) - e\phi(\mathbf{r})\psi_i(\mathbf{r}) = \epsilon_i\psi_i(\mathbf{r}) \tag{2.70}$$

with the requirement that the induced charge density n^{ind} is in linear order in ϕ. It is found that at large distances the screened potential of a point charge has a term that goes (at $T = 0$) as:

$$\phi(\mathbf{r}) \approx \frac{1}{r^3}\cos 2k_F r. \tag{2.71}$$

In this case the screening at large distances has more structure with a relatively weak decaying oscillatory term than the simple screened Coulomb potential obtained within the Thomas–Fermi theory.

Now coming back to the screening within the Hartree–Fock approximation, screening will also effect the interactions of two electrons with each other, since from the point of view of the remaining electrons in the electron gas in metal, these two can be considered as external charges. To take account of this, one way is to replace the electron–electron interaction that occurs in the exchange potential of

Hartree–Fock equations in the jellium model (see equation (2.21)) by its screened form, i.e. multiplying $\frac{1}{(\mathbf{k}-\mathbf{k}')^2}$ by the inverse dielectric constant $1/\epsilon(\mathbf{k}-\mathbf{k}')$. In the neighbourhood of $q = 0$, the screened interaction will approach e^2/k_0^2, and that will eliminate the singularity responsible for the anomalous divergence in the one electron velocity at $\mathbf{k} = \mathbf{k}_F$. The calculated velocity of electron for r_s values typical of metallic densities differs from its free electron value by only about 5%. The phenomenon of screening thus somewhat reduces the importance of electron–electron interactions in metallic solid.

2.6 Hartree–Fock–Slater method and density functional theory

It is quite clear from the discussions above that *exchange* and *correlation* are crucial for the existence of the solid materials, in fact they provide the glue in the formation of the solids starting from the atoms. At the same time it is quite clear that an estimation of these quantities are by no means simple in the case of real solids, particularly when both the kinetic and potential energies play comparable important roles. Within one-electron Hartree–Fock theory, which is very useful as a starting point, computation of HF orbitals in solid materials was an intractable problem in the 1940s. In 1951 Slater proposed a solution [13]—the Hartree–Fock–Slater (HFS) method—where one could simplify the Hartree–Fock equations in the presence of the periodic ionic potential of the solid, by replacing the exchange term in equation (2.14) by a local energy given by twice equation (2.49) with k_F evaluated at the local density. Slater proposed an equation where the effect of exchange was included by simply adding an additional potential $U^{exchange}(\mathbf{r}) = -2.95(a_0^3 n(\mathbf{r}))^{\frac{1}{3}}$ Ry to the Hartree term $U^{el}(\mathbf{r})$ (see Ashcroft and Mermin [1]). The Slater approximation was rather gross and ad hoc, but it was proposed at the time when the electronic computers were just becoming available for electronic structure calculations. Around that period, methods for solving the Schrödinger equation in solids, including the augmented plane-wave (APW) method [14] and Korringa–Kohn–Rostoker (KKR) method [15, 16] had also been developed. Together they turned out to be very useful and popular for studying various physical properties of solid materials. The original APW code and its early applications used the Slater potential, which was modified by an adjustable parameter α ($\alpha = 2/3$ gives the exchange energy of a free-electron gas); this approximation came to be known as the 'X-α method'. While the estimated bond energies within the HFS approximation were superior to Hartree–Fock bond energies, the errors were still quite large. Post HFS wave function theory and methods, now commonly termed as *ab initio* theory, is quite well developed. However, the computer-time scaling with system size N is several orders larger than for Hartree–Fock depending on the method, and the computational costs are quite high.

The ground state and other properties of many systems in an external field can be determined from the knowledge of electron density $n(\mathbf{r})$ alone. This is the basis of the well known *density functional theory*. The first intuitive approach in this direction was due to Thomas [17] and Fermi [18]. They studied atomic properties based purely on the electron density $n(\mathbf{r})$, with the assumption that the electrons form a gas satisfying Fermi–Dirac statistics and the electron–electron interaction energy was determined from the classical Coulomb potential. They adopted a local density approximation for the contribution to the kinetic energy, in which the contribution from point **r** is determined from the kinetic energy of a homogeneous electron gas with the local density $n(\mathbf{r})$. Dirac [19] introduced 'exchange' phenomenon into the 'Thomas atom' by recasting Hartree–Fock theory in terms of a 'density function'. Thomas–Fermi–Dirac (TFD) theory provided an approximate theory of electronic structure, which depended only on the total electronic density $n(\mathbf{r})$. TFD theory, however, failed qualitatively because it is unable to reproduce atomic shell structure self-consistently. TFD energies have errors of about 10% even with accurate input of densities from other sources, which is too large for computational purposes. Subsequently it was shown by Teller [20] that Thomas–Fermi theory could not bind molecules. For a more detailed discussion on the HFS method and early history of DFT the readers are referred to the excellent review articles by Jones [21] and Becke [22].

In a seminal paper in 1964 Hohenberg and Kohn [23] showed that all the information contained in a many electron wavefunction in principle can be obtained with the knowledge of electron density defined as:

$$
\begin{aligned}
n(\mathbf{r}) &= \langle \Psi | \sum_{i=1}^{N} \delta(\mathbf{r} - \mathbf{r}_i) | \Psi \rangle \\
&= N \int d r_1, r_2, \ldots, r_N \Psi^*(\mathbf{r}_1, \mathbf{r}_2, \ldots, \mathbf{r}_N) \delta(\mathbf{r} - \mathbf{r}_i) \Psi(\mathbf{r}_1, \mathbf{r}_2, \ldots, \mathbf{r}_N).
\end{aligned}
\tag{2.72}
$$

The relations proposed by Hohenberg and Kohn are stated in the form of two theorems [3]:

1. Theorem 1: For any system if the electron density of the ground state of a many electron system is known, the external potential U_{ext} in which the electrons reside can be determined uniquely up to an overall constant.
2. Theorem 2: The ground state energy of a many electron system can be defined in terms of a universal functional $E[n]$ of the electron density n. In any external potential $U_{ext}(\mathbf{r})$ the exact ground state energy of the system is the global minimum value of this functional $E[n]$. The exact ground state electron density $n(\mathbf{r}_0)$ is that electron density, which minimizes the energy functional.

In order to prove the theorem 1 of Hohenberg–Kohn, let us suppose that there are two different external potentials U_{ext}^1 and U_{ext}^2, which lead to the same electron density $n(r_0)$. Two Hamiltonians H_1 and H_2 corresponding to these external potentials will have different ground state wave functions Ψ_1 and Ψ_2. Assuming the ground state to be non-degenerate, ground state energy of H_1 is realized only by the wave functions Ψ_1; Ψ_2 is not the ground state wavefunction of H_1. Hence,

$$
E_1 = \langle \Psi_1 | H_1 | \Psi_1 \rangle < \langle \Psi_2 | H_1 | \Psi_2 \rangle.
\tag{2.73}
$$

The last term of the above equation can be written as:

$$\langle \Psi_2 | H_1 | \Psi_2 \rangle = \langle \Psi_2 | H_2 | \Psi_2 \rangle + \langle \Psi_2 | H_1 - H_2 | \Psi_2 \rangle. \tag{2.74}$$

Now two Hamiltonians with the same number of electrons can be different only in potential. Hence,

$$\langle \Psi_2 | H_1 - H_2 | \Psi_2 \rangle = \int n(\mathbf{r})(U_{ext}^1(\mathbf{r}) - U_{ext}^2(\mathbf{r}))dr. \tag{2.75}$$

Equation (2.74) can be rewritten using equation (2.75) as:

$$\langle \Psi_2 | H_1 | \Psi_2 \rangle = E_2 + \int n(\mathbf{r})(U_{ext}^1(\mathbf{r}) - U_{ext}^2(\mathbf{r}))dr. \tag{2.76}$$

So from equations (2.73) and (2.76) We can write:

$$E_1 < E_2 + \int n(\mathbf{r})(U_{ext}^1(\mathbf{r}) - U_{ext}^2(\mathbf{r}))dr. \tag{2.77}$$

We can find the same equation for E_2 by switching the superscripts 1 and 2:

$$E_2 < E_1 + \int n(r)(U_{ext}^2(\mathbf{r}) - U_{ext}^1(\mathbf{r}))dr. \tag{2.78}$$

Adding these two equations (2.77) and (2.78) We get:

$$E_1 + E_2 < E_1 + E_2. \tag{2.79}$$

This is indeed a contradiction, which clearly says that two external potentials differing by more than a constant cannot give rise to the same non-degenerate ground state electron density. This proves the Hohenberg–Kohn first theorem that ground state electron density can uniquely determine the external potential within a constant. Thus, the Hamiltonian is also uniquely determined except for a constant by the ground state electron density. Hence, the wavefunction of any state can be determined in principle by solving the Schrödinger equation with this Hamiltonian. The unique ground state wavefunction will have the lowest energy among all the possible solutions with the given electron density.

Before proving the second Hohenberg–Kohn theorem we first define the meaning of a functional of the density and restricted space of densities. The original work of Hohenberg–Kohn considered the electron densities $n(\mathbf{r})$ that are ground state densities of the electron Hamiltonian with some external potential U_{ext}. Such densities are called V-representable, and that defines a space of possible densities within which functionals of the density can be constructed [3]. If all properties are determined uniquely once $n(\mathbf{r})$ is specified, then each such property including the total energy functional can be expressed as a functional of $n(\mathbf{r})$:

$$\begin{aligned}
E[n] &= T[n] + E_{int}[n] + \int U_{ext}(\mathbf{r})n(\mathbf{r})d^3r + E_{nuc} \\
&= F[n] + \int U_{ext}(\mathbf{r})n(\mathbf{r})d^3r + E_{nuc}.
\end{aligned} \tag{2.80}$$

Here $T[n] + E_{int}[n]$ are the kinetic energy and potential energy of the interacting electron system, and E_{nuc} is the interaction energy of the nuclei. The functional $F[n] = T[n] + E_{int}[n]$ includes the kinetic and potential energy of the interacting electron system. The functional $F[n(\mathbf{r})]$ is universal in the sense that it is valid for any external potential U_{ext}.

Now in a system with the ground state density $n^1(\mathbf{r})$ corresponding to an external potential, the total energy functional is equal to the expectation value E_1 of the H_1 in the unique ground state with wavefunction Ψ_1:

$$E_1 = E[n^1] = \langle \Psi_1 | H_1 | \Psi_1 \rangle. \qquad (2.81)$$

A different electron density $n^2(\mathbf{r})$ will correspond to a different ground state with wavefunction Ψ_2. The energy E_2 of this state is greater than E_1, since

$$E_1 = \langle \Psi_1 | H_1 | \Psi_1 \rangle < \langle \Psi_2 | H_1 | \Psi_2 \rangle. \qquad (2.82)$$

Thus, the energy given by equation (2.81) in terms of the Hohenberg–Kohn functional corresponding to the correct ground state density $n^0(\mathbf{r})$ will be lower than the value of this energy for any other density $n(\mathbf{r})$. This proves the second theorem of Hohenberg–Kohn. If the functional $F[n(\mathbf{r})]$ in equation (2.80) is known, then minimization of the total energy of the system given by equation (2.81) by varying $n(\mathbf{r})$ will give the exact ground state energy and electron density.

Before we proceed further it may be worth having a short discussion on what is a functional and what is a functional derivative? A simple definition of functional is: 'function maps one number to another, while a functional assigns a number to a function' [24]. Let us consider as an example all functions $r(\theta)$, $0 \leqslant \theta \leqslant 2\pi$, which are periodic, i.e. $r(\theta) = r(\theta + 2\pi)$, which describe shapes of curves in two-dimensions. One can define the perimeter P as the length of the curve, and the area A as the area enclosed by it for all such curves. P and A are functionals of $r(\theta)$. We know for example that for an ellipse represented by the function $r(\theta) = \frac{1}{\sin^2\theta + 4\cos^2\theta}$ there is a single well-defined value of perimeter P and area A. One writes $P[r]$ and $A[r]$ to denote this functional dependence.

In order to differentiate functional, one first needs to define a functional derivative. Let us ask this question: if a small change is made in a function localized to one point, how will the value of a functional change due to this increase? To be more precise, if an infinitesimal change $\delta r(\theta) = \epsilon\delta(\theta - \theta_0)$ is made to the function $r(\theta)$, what will be the corresponding change say in A? This change can be expressed as:

$$A[r + \delta r] - A[r] = \frac{1}{2} \int_0^{2\pi} [(r + \epsilon\delta(\theta - \theta_0))^2 - r^2]d\theta$$
$$= \int_0^{2\pi} r\epsilon\delta(\theta - \theta_0)d\theta = \epsilon r(\theta_0). \qquad (2.83)$$

The functional derivative, denoted $\frac{\delta A}{\delta r(\theta)}$ is just the change in A divided by ϵ, or just $r(\theta)$ in this case. Since this is linear in the change in r, for any infinitesimal change in r, the general definition is:

$$A[r + \delta r] - A[r] = \int_0^{2\pi} \frac{\delta A}{\delta r(\theta)} \delta r(\theta) d\theta \tag{2.84}$$

where the functional derivative $\delta A/\delta r(\theta)$ is that function of θ which makes this formula exact for any small change in $r(\theta)$. Comparing with the case of usual derivative df/dx of a function $f(x)$, which reveals the change in f when x changes by a small amount i.e. $f(x + dx) - f(x) = (df/dx)dx + O(dx^2)$, the functional derivative tells how much a functional changes when its arguments change by a small amount, which in this case is a small function.

The next step to know is, how to maximize (or minimize) one functional subject to a constraint imposed by another functional. Suppose we want to know what will be the maximum area that can be enclosed inside a piece of loop of string of fixed length l. For this we have to maximize $A[r]$, subject to the constraint that $P[r] = l$. To solve such problems the method of Lagrange multipliers is employed, which involves construction of a new functional:

$$B(r) = A(r) - \mu P[r] \tag{2.85}$$

where μ is Lagrange multiplier, and which at this point is an unknown constant. This new functional $B[r]$ is extremized by equating its functional derivative to zero:

$$\delta B/\delta r = \delta A/\delta r - \mu \delta P/\delta r = r(\theta) - \mu = 0 \tag{2.86}$$

or,

$$r(\theta) = \mu = \text{constant.} \tag{2.87}$$

The above equation (2.87) tells that the optimum shape of the loop is a circle, and the radius of this circle can be obtained by putting the solution in the constraint, $P[r = \mu] = 2\pi\mu = l$. Thus, the largest area enclosed by a piece of string of length l is $A[r = \mu] = \frac{l^2}{4\pi}$.

Now coming back to the many electron system, a good approximation to the energy functional $E[n, U_{ext}]$ can be made by using the idea of decomposition introduced by Kohn and Sham [25]:

$$F[n] = T_s[n] + \frac{1}{2} \int dr n(\mathbf{r}) U_{Coulomb}(\mathbf{r}) + E_{XC}[n]. \tag{2.88}$$

Here, (i) T_s represents the kinetic energy that a system with density $n(\mathbf{r})$ would have in the absence of electron–electron interactions, (ii) $U_{Coulomb}$ is the classical Coulomb potential for electrons, which appeared earlier in Hartree equations, and (iii) E_{xc} represents the exchange–correlation (xc) energy. T_s is of comparable magnitude to the true kinetic energy T, and is treated here without approximation. Many of the deficiencies of the Thomas–Fermi–Dirac approach, such as the lack of a shell structure of atoms or the absence of chemical bonding in molecules and solids, are thus removed. Except the exchange–correlation energy E_{xc}, all other terms in equation (2.88) can be evaluated exactly, hence the approximations for this exchange–correlation energy term are crucial in the applications of the density functional theory. The variational principle applied to equation (2.88) yields [21]:

$$\frac{\delta E[n, U_{ext}]}{\delta n(r)} = \frac{\delta T_s}{\delta n(r)} + U_{ext}(\mathbf{r}) + U_{Coulomb}(\mathbf{r}) + \frac{\delta E_{XC}[n]}{\delta n(r)} = \mu \qquad (2.89)$$

where μ is the Lagrange multiplier associated with the requirement of constant particle number. Comparing this with the corresponding equation for a system with the same density in an external potential but without electron–electron interactions:

$$\frac{\delta E[n]}{\delta n(r)} = \frac{\delta T_s}{\delta n(r)} + U(\mathbf{r}) = \mu \qquad (2.90)$$

one can see that the mathematical problems are identical, if:

$$U(\mathbf{r}) = U_{ext}(\mathbf{r}) + U_{Coulomb}(\mathbf{r}) + \frac{\delta E_{XC}[n]}{\delta n(r)}. \qquad (2.91)$$

The solution of equation (2.90) can be obtained by solving the Schrödinger equation for noninteracting electrons:

$$\left[-\frac{\hbar^2}{2m_e} + U(\mathbf{r}) \right] \psi_i(\mathbf{r}) = \varepsilon_i \psi_i(\mathbf{r}) \qquad (2.92)$$

which yields:

$$n(\mathbf{r}) = \sum_{i=1}^{N} f_i |\psi_i(\mathbf{r})|^2. \qquad (2.93)$$

The functions ψ_i are the Kohn–Sham orbitals. The f_i are occupation numbers, which are noninteger at zero temperature when the orbitals are degenerate at the Fermi level and Fermi–Dirac occupancies at finite temperatures [21]. The condition on $U(\mathbf{r})$ given in equation (2.91) can be satisfied self-consistently. The set of equations (2.91)–(2.93) are the well known Kohn–Sham equations. The solution of this set of equations gives energy and density of the lowest state and all quantities derivable from them. In comparison with the Hartree–Fock potential the Kohn–Sham effective potential $U(\mathbf{r})$ is local in nature.

Summarizing the discussions above we can say that in the Kohn–Sham theory one assumes that the ground state density of the original interacting electron system is equal to that of some chosen non-interacting electron system. This leads to one-electron equations for the non-interacting electron system that can be considered as exactly solvable with all the difficult many-body terms incorporated into an exchange–correlation functional of the density. Thus the Kohn–Sham approach involves independent particles but an interacting density. The ground state density and energy of the original interacting system can be obtained by solving the one-electron Schrödinger equations with the accuracy limited only by the approximations involved in the exchange–correlation functional. The numerical advantages in the Kohn–Sham approach lies in the efficient methods for solving self-consistently single-particle Schrödinger-like equations with a local effective potential, and there is no restriction to small systems.

An important point in the density functional theory is the relationship between the interacting electron system, whose energy and density one is trying to find out, and the artificial non-interacting electron system for which one solve equations (2.92) and (2.93). This can be studied by considering the interaction $\lambda/|\mathbf{r} - \mathbf{r}_0|$ in the presence of an external potential U_{ext} such that the ground state of the Hamiltonian H_λ has density $n(\mathbf{r})$ for all λ. The Hamiltonian H_λ is expressed as [21]:

$$H_\lambda = T + U + \lambda(U_{ext} + U_{ee} - U) \tag{2.94}$$

where λ is a coupling constant which varies between 0 (non-interacting system) to 1 (physical system), U is the density dependent Kohn–Sham effective potential expressed in equation (2.91), and U_{ee} is electron–electron interaction. The exchange–correlation energy of the interacting electron system is then expressed in terms of an integral over the coupling constant λ [21]:

$$E_{xc} = \frac{1}{2} \int n(\mathbf{r})dr \int \frac{1}{|\mathbf{r} - \mathbf{r}'|} n_{xc}(\mathbf{r}, \mathbf{r}' - \mathbf{r})dr' \tag{2.95}$$

where,

$$n_{xc}(\mathbf{r}, \mathbf{r}' - \mathbf{r})dr' \equiv n(\mathbf{r}') \int_0^1 [g(\mathbf{r}, \mathbf{r}', \lambda) - 1]d\lambda. \tag{2.96}$$

Here the term exchange correlation hole density $n_{xc}(\mathbf{r}, \mathbf{r}' - \mathbf{r})$ is a quantity of great interest, which describes how the presence of an electron at the point $\mathbf{r}$ depletes the total density of the other electrons at the point $\mathbf{r}'$. The function $g(\mathbf{r}, \mathbf{r}', \lambda)$ represents the pair-correlation function of the system with density $n(\mathbf{r})$ and Coulomb interaction λU_{ee}. From equations (2.95) and (2.96) it is clear that E_{xc} is simply the energy resulting from the interaction between an electron and its exchange–correlation hole.

The isotropic nature of the Coulomb interaction U_{ee} has important consequences. Substitution of $\mathbf{r} - \mathbf{r}' = \mathbf{R}$ in equation (2.95) leads to:

$$E_{xc} = \frac{1}{2} \int n(\mathbf{r})dr \int_0^\infty dRR^2 \frac{1}{\mathbf{R}} \int d\Omega n_{xc}(\mathbf{r}, \mathbf{R}). \tag{2.97}$$

Equation (2.97) tells that the exchange–correlation energy depends only on the spherical average of $n_{xc}(\mathbf{r}, \mathbf{R})$. Thus, the approximations for E_{xc} that describe this average well will give an accurate value of the exchange–correlation energy. A sum-rule, requiring that the exchange–correlation hole contains one electron for all $\mathbf{r}$, arises from the definition of the pair-correlation function [21]:

$$\int n_{xc}(\mathbf{r}, \mathbf{r}' - \mathbf{r})dr' = -1. \tag{2.98}$$

This means that $n_{xc}(\mathbf{r}, \mathbf{r}' - \mathbf{r})$ can be considered as a normalized weight factor, and can be used to define locally the radius of the exchange–correlation hole for a particular value of $\mathbf{r}$:

$$\left\langle \frac{1}{\mathbf{R}} \right\rangle_r = - \int \frac{n_{xc}(\mathbf{r}, \mathbf{R})}{|\mathbf{R}|} dr. \tag{2.99}$$

Thus, E_{xc} can be expressed as [21]:

$$E_{xc} = -\frac{1}{2} \int n(\mathbf{r}) \left\langle \frac{1}{\mathbf{R}} \right\rangle_r dr. \tag{2.100}$$

E_{xc} is determined by the first moment of a function whose second moment is known exactly and known to depend only weakly on the details of n_{xc} provided the sum rule requirement defined by equation (2.98) is satisfied. In such situations approximations to E_{xc} can lead to good total energies even if the details of the exchange–correlation hole are described rather poorly.

The exact functional $E_{xc}[n]$ can be very complex, but significant progress has been made with remarkably simple approximations. In their original paper Kohn and Sham [25] suggested that solids may be considered as close to the limit of the homogeneous electron gas. The effects of exchange and correlation are local in character in this limit. Kohn and Sham [25] then introduced the local density approximation for the exchange–correlation functional in which $E_{xc}[n]$ is simply an integral over all space with the exchange–correlation energy per electron at each point assumed to be the same as in a homogeneous electron gas with that density:

$$E_{XC}^{LD} = \int dr n(\mathbf{r}) \varepsilon_{XC}[n(\mathbf{r})]. \tag{2.101}$$

Here ε_{XC} represents the exchange and correlation energy density of a homogeneous electron gas with density n. Local density approximation (LDA) is a good approximation if (i) the density is almost constant, (ii) at high densities, where the kinetic energy dominates the exchange and correlation terms. Subsequently, generalizations to spin-polarized systems, known as the local spin density approximation (LSDA) have been made, which can be written as [21]:

$$E_{XC}^{LSD} = \int dr n(\mathbf{r}) \varepsilon_{XC}[n_\uparrow(\mathbf{r}), n_\downarrow(\mathbf{r})] \tag{2.102}$$

where $n_\uparrow(\mathbf{r})$, $n_\downarrow(\mathbf{r})$ is the exchange and correlation energy per electron of a homogeneous, spin-polarized electron gas with spin-up and spin-down densities $n_\uparrow$ and $n_\downarrow$, respectively. In addition $X\alpha$ approximation has been used extensively in numerous calculations in the late 1960s and 1970s:

$$E_X^{X\alpha} = -\frac{3}{2} \alpha C \int dr n(\mathbf{r}) \left\{ [n_\uparrow(\mathbf{r})]^{\frac{4}{3}} + [n_\downarrow(\mathbf{r})]^{\frac{4}{3}} \right\} \tag{2.103}$$

where $C = 3(\frac{3}{4\pi})^{\frac{1}{3}}$.

LDA and LSDA are expected to work best for solids like nearly-free-electron metals, whose behaviour is close to a homogeneous electron gas. They are not good for very inhomogeneous cases like atoms and van der Waals solids where the electron density goes continuously to zero outside the atom. Another fault in LDA/LSDA is the presence of the spurious self-interaction term. This unphysical self-interaction term in the Hartree equations is exactly canceled out in the Hartree–Fock approximation by the non-local exchange interaction. This cancellation, however, is only approximate in the local approximation to exchange. The spurious self-interaction terms are negligible in the homogeneous electron gas, but quite large in confined systems such as atoms.

Many different types of approximations have been tried to bring the density functional theory in close correspondence with the experimental results. In the spirit of an expansion around homogeneous density, a dependence on gradients and higher derivatives of the electron density has been included in the functional, which now involves the magnitude of the gradient of the electron density $|\nabla n|$ as well as the value of electron density n at each point. Second-order gradient expansions lead to the improvement on LDA for very slowly varying electron densities. This approximation, however, is less accurate than LDA for realistic densities. This is because the second-order gradient expansion of the hole violates the exact hole constraints. However, several generalized forms or generalized gradient approximations (GGA) have been proposed:

$$E_{xc}^{GGA}[n] = \int dr n(\mathbf{r}) \epsilon_{xc}^{GGA}(n, \nabla n). \tag{2.104}$$

GGAs usually involve parameters, which are to be fixed by requirements to satisfy exact constraints and/or to match other calculations or experiments on a set of prototype systems for exchange and correlation [3].

Starting from the LDA and GGAs further ingredients have been added, for example, second-order gradients in meta-GGAs and the exact exchange energy in hybrid functional technique. The form of the integration for the exchange–correlation energy in equations (2.95) and (2.96), involving the coupling constant λ, is the basis for constructing of hybrid functional. They are a combination of orbital-dependent Hartree–Fock term and an explicit density functional. At $\lambda = 0$ the energy is just the Hartree–Fock exchange energy, whereas the potential part of the LDA or GGA functional is the most appropriate at full coupling $\lambda = 1$. The integral in equation (2.95) can be approximated by assuming a linear dependence on λ leading to the half-and-half form [3]:

$$E_{xc} = \frac{1}{2}[E_{xc}^{HF} + E_{xc}^{DFA}] \tag{2.105}$$

where DFA stands for an LDA or GGA functional.

There has been a very significant growth of density functional calculations since the early 1990s, which is partly due to its recognition as an 'approximate practical method' for calculations [21]. The interested readers are referred to this recent review article by Jones for an update in this field [21] and the book by Martin [3], for a pedagogical exposition on DFT. However, some deep concerns still remain [26, 27]. Moreover, the identification of the 'best functional' may be ambiguous and depends on the choice of statistical measure (mean error, mean absolute error, variance, etc) [21]. As far as the Mott insulators and Mott metal–insulator transition are concerned where potential energy dominates over the kinetic energy, LDA and GGA can give qualitatively incorrect information on these materials. The failure in the correct description of the physics in strongly correlated electron systems is generally attributed to the tendency in the exchange–correlation functional to excess delocalization of the valence electrons and thus over-stabilizing the metallic state. This tendency arises from the rather defective account of the exchange and correlation interactions in the exchange–correlation functional, which fails to cancel out electron self interaction contained in the Hartree term. In recent times, the study of the systems with strong electron correlation has been pursued with computational methods complementing DFT with model Hamiltonians [28]. One of the simplest approaches that were formulated to improve the description of the ground state of correlated electron systems, is modification of density functional calculations with the addition of an on-site Coulomb repulsion term U i.e. $LSDA + U$ scheme [21, 28]. The parameter U can be estimated within a density functional framework, but often it is obtained as a fit to experiment.

2.7 Fermi liquid theory in metals

We have seen in chapter 1 that various properties of metals including heat capacity can be explained by treating the conduction electrons in the metals as a gas of non-interacting electrons. However, it remained a puzzle for several years: why this independent electron approximation should be so successful for a system of electrons that is clearly interacting? If the electron–electron interaction is turned on gradually in a metal represented by non-interacting electrons, then two kinds of effects are expected [1]:

1. There will be a modification in energy of each one-electron level. This kind of effect has been discussed above to some extent within the framework of Hartree–Fock approximation and its further refinements.

2. There will be scattering of electrons in and out of the one electron levels, the energy levels will no longer be stationary. The strength of the scattering will determine whether that is serious enough to invalidate the independent electron picture. If the scattering rate is low enough, one could introduce a relaxation time and treat the electron–electron scattering in the same manner as the other scattering mechanisms are treated in the transport theories within the relaxation time approximations.

To address these issues Landau introduced a theory [29] for interacting spin-1/2 fermion systems to investigate the stability of Fermi-gas against perturbations. The basic ideas behind Landau's theory are the principle of adiabaticity and Pauli exclusion principle. It was argued that starting from ground state of a non-interacting fermion system if the interaction is turned on slowly, the system will end up with the ground state of a Fermi liquid. The interactions conserve the total particle number, spin and momentum. It can be said formally that if the full Hamiltonian possesses the same symmetries of some unperturbed Hamiltonian, its ground states can be obtained by perturbative transformation of the simple ground state. In a similar way the low-lying excitations of the non-interacting Fermi gas can be mapped to the low-lying excitations of the Fermi liquid. The low lying excitations of the Fermi liquid are called Landau quasiparticles, which have a finite lifetime τ arising due to interactions with other quasi particles.

Particles can only be added in the non-interacting system only with momentum $\mathbf{p}(\hbar k) > \mathbf{p}_F$, and so this gives quasiparticle excitation with $\mathbf{p} > \mathbf{p}_F$. It may be remembered here that the Fermi energy E_F is not changed by interactions. No particles can be added to the non-interacting system with $\mathbf{p} < \mathbf{p}_F$, but a particle can be removed from the Fermi sea to form an excited state of the $N - 1$ particle system. Switching on the interaction here gives a quasi-hole state with momentum $-\mathbf{p}$.

Since Coulomb interaction (even in the screened-Coulomb form) is quite strong, it is expected that the electron–electron scattering rate (which is inversely proportional to τ) would be quite high. However, the Pauli exclusion principle reduces the scattering rate very significantly in many cases, and this reduction occurs when the electronic configuration of the many electron system differs only marginally from its thermal equilibrium configuration [1]. To elaborate this let us consider the collision between two particles with momenta $\mathbf{p}_1$ and $\mathbf{p}_2$ and leaving with the final momenta $\mathbf{p}_3$ and $\mathbf{p}_4$. The momentum and energy has to be conserved in such a process, and this leads to:

$$\mathbf{p}_1 + \mathbf{p}_2 = \mathbf{p}_3 + \mathbf{p}_4 \tag{2.106}$$

and

$$E_1 + E_2 = E_3 + E_4. \tag{2.107}$$

The scattering rate can be estimated qualitatively using the Fermi golden rule:

$$\Gamma = \frac{2\pi}{\hbar} \sum_f |\langle f|H_{interaction}|i\rangle|^2 \delta(E_f - E_i). \tag{2.108}$$

The scattering rate is proportional to the number of available final states f. Let us consider the specific case of filled Fermi sphere and one extra electron with energy $E_1 > E_F$, and estimate the allowed final states when this electron collides with any particle inside the Fermi sphere with energy $E_2 < E_F$. Since the Pauli exclusion principle forbids all final states inside the Fermi sphere, the final state must have two electrons outside the Fermi sphere i.e. $E_3 > E_F$ and $E_4 > E_F$. If the energy of the initial particle is close to the Fermi energy, then the number of final states become very small and both E_3 and E_4 needs to remain within an energy shell of thickness proportional to $E_1 - E_F$. This implies that the final states are limited by a factor proportional to $(E_1 - E_F)^2$. The power of two arises since out of the three energies E_2, E_3 and E_4, one is fixed by the conservation of energy while leaving the other two as free parameters. If E_1 tends to E_F, then $E_2 = E_3 = E_4 = E_F$, then the Fermi surface is the only phase space available for scattering, which is an area of zero in phase space. Thus, the Pauli exclusion principle reduces the number of scattering channels that are compatible with energy and momentum conservation. With the assumption that the interacting electron system has a Fermi surface, the excitations near the Fermi surface are Landau quasi-particles, just like the electron and hole excitations in the case of the non-interacting electron system, i.e. Fermi gas.

The concept of quasi-particles helps to simplify somewhat the collective interactions in a many particle system. A charge particle moving through a gas of similarly charged particle will repel the other particles from its trajectory. This event can be formally described within a framework without invoking charge particle interactions. Instead the charge particle is accompanied in its trajectory by a compensating cloud of charges with opposite sign. The interaction with other charges is effectively replaced by the inertia of charge-cloud, which the charge particle carries with it. Hence the system of interacting particle is effectively replaced by a non-interacting system of quasiparticles with a heavier effective mass. In the Fermi liquid, the quasiparticle decays through excitations of electron–hole pairs across the Fermi energy E_F, and within an energy range $|E_1 - E_F|$ about the Fermi surface the scattering rate between electrons will be of the order $(E_1 - E_F)^2$. Thus there is a one-to-one correspondence between the energetically low-lying one-electron excitations of the Fermi gas and the Fermi liquid. For the thermodynamical and electrical transport properties in metals only low energy excitations are relevant. This is because the Fermi energy in metals is usually much larger than thermal energies of interest. The residual interaction between Landau quasi-particles are described in terms of a few Landau parameters [1], and there is no qualitative difference between a Fermi gas and a Fermi liquid; both represents a metallic state. While the excitation spectrum remains qualitatively similar, there is, however, a shift in energies in Fermi liquid. Let us again study the specific case of a filled Fermi sphere plus one electron with momentum $\mathbf{p}$, with $\mathbf{p} > \mathbf{p}_F$. If the energy of the particle is not far from the Fermi energy i.e. $(\mathbf{p} - \mathbf{p}_F) \ll \mathbf{p}_F$, the excited state of the Fermi gas corresponds to the excitation energy:

$$E_P - E_F = \frac{p^2}{2m_e} - E_F \equiv \frac{p_F}{m_e}(p - p_F). \tag{2.109}$$

In the presence of weak interactions the excitation energy will still be linear in $(\mathbf{p} - \mathbf{p}_F)$, but the coefficient will no longer be $\frac{p_F}{m_e}$. The new excitation energy is customarily expressed as:

$$E_P - E_F = \frac{p_F}{m^*}(p - p_F) \tag{2.110}$$

where m^* is defined as effective mass of the quasi particles. It may be noted that the Fermi momentum $\mathbf{p}_F$ is not changed. There will be a new density of state corresponding to the new dispersion relation of quasi-particles, which is expressed as:

$$g(E_F) = \frac{m^* p_F}{\pi^2 \hbar^3}. \tag{2.111}$$

This new density of states correspond to the effective mass m^* of quasi-particles, and not to the bare electron mass m_e in an ideal Fermi gas. This provides a relation of the effective mass m^* in Fermi liquid theory to the experimentally determined electronic contributions to the specific heat. Indeed the experimental studies have indicated the increase in effective mass of quasi-particles in a certain class of metallic systems with significant electron–electron correlation, i.e. heavy fermion systems [30]. For example the electronic coefficient of specific heat in the intermetallic compound $CeCu_6$ is about 1600 mJ mol^{-1} K^{-2} in comparison to a few mJ mol^{-1} K^{-2} in pure Cu. This indicates the effective mass m^* of quasi-particles in $CeCu_6$ to be 1000 times heavier than bare electron mass.

The important question now is, under which conditions the concept of a Fermi surface is still valid within the solely phenomenological Fermi-liquid theory [31]. While the perturbation theory supports the existence of Fermi surface, subject to the convergence of the series expansion, one may also argue only on physical grounds that the Fermi-liquid picture provides a self-consistent description of the influence of electron–electron interactions in metals in the following way [31]. In metallic crystals the electrons are highly mobile, and thus their mutual interaction is well screened. This in turn allows application of Fermi-liquid theory for the description of the physical properties of metal. The screened electron–electron interaction does not cause any qualitative change in the metallic properties. This chain of argument obviously breaks down at the correlation driven metal–insulator transition, and the problem of such transition i.e. Mott transition (the core subject matter of this book) is closely related to the question of the stability of Fermi-liquid theory [31]. For a more detailed information on the Fermi-liquid theory the readers are referred to the book by Pines and Nozières [32].

2.8 Slater insulator

Slater [13] in the early 1950s suggested that an insulating state could arise in a solid by taking into account electron–electron interaction within the unrestricted Hartree–Fock approximation (see section 2.1) of many electron systems. Slater's view of the insulating behavior due to electron–electron interactions was based on the features of Peierls insulators (see chapter 1, section 1.6). Instead of presenting a detailed theoretical exposition of the Slater insulator, we will rather focus here on a physical picture to understand the phenomenon. Here the narratives will closely follow the presentation in the book by Gebhard [31]. A theoretical treatment of Slater insulator will be presented in chapter 3, section 3.7.

In chapter 1, section 1.6 of this book, we discussed the Peierls transition in a solid to an insulating state and the consequent charge density wave. In Peierls insulators, the electron band structure of the solid is modified by the formation of a superlattice structure due to electron–lattice interactions. In a close analogy to the mechanism giving rise to the Peierls insulating state, the electron–electron interactions may also generate a periodic modulation of their spatial charge and/or spin distribution. In contrast to the Peierls insulating state, the Slater insulator is stabilized by the electron–electron interaction instead of the electron–lattice interaction.

The basic principle behind the Slater insulating state can be clearly described in the case of s electrons on a bipartite lattice [13, 31]. This bipartite lattice is a lattice that could be separated into two interpenetrating sublattices (A and B), so that the nearest neighbour of any site are members of the opposite sublattice. Further, it is assumed for simplicity that the lattice is half filled such that there is on average one electron per site. The spin-up and spin-down electrons repel mutually, hence these electrons will prefer to arrange themselves on alternating lattice sites to minimize their Coulomb repulsion; they will form a self-stabilizing spin-density wave with a wave vector Q commensurate with the lattice. The electrons in the Slater insulating state tend to avoid each other, thus there will be a gain in potential energy as compared to a translationally invariant state. This gain however, is balanced by an increase in kinetic energy because of the restricted motion of the electrons. Similar to the situation in Peierls insulator (see section 1.6), the lattice unit cell here is doubled, while the first Brillouin zone is reduced to half in size. At the boundaries of the new magnetic Brillouin zone, the bands will split because of the Q-periodic potential, and the energy of the occupied levels is lowered [31]. The band splitting will cause a gap for charge excitations at half band-filling, and the system turns into an insulator. Thus the onset of a long-range antiferromagnetic order is a necessary condition for the existence of Slater insulating state. In other words, in Slater's picture antiferromagnetism comes first, and the insulating behaviour second. However, even in the early 1950s (when Slater proposed his theory) there were experimental results available on the transition metal oxides and fluorides showing that those materials continued to be in the insulating state even in the paramagnetic regime well above the antiferromagnetic transition temperature [26]. Eventually it became quite clear that the Slater insulator is a self-consistently stabilized band-type insulator, where the insulating state arises out of the electron exchange effect due to the Pauli exclusion principle. This turns out to be an over-simplified description for the real solids with significant electron–electron correlation [31].

Further reading

1. Ashcroft N W and Mermin N D 1976 *Solid State Physics* (Philadelphia, PA: Saunders College).
2. Madelung O 2004 *Introduction to Solid State Theory* (Berlin: Springer).
3. Martin R M 2008 *Electronic Structure: Basic Theory and Practical Methods* (Cambridge: Cambridge University Press).
4. Mahan G D 2000 *Many Particle Physics* (New York: Plenum).
5. Fulde P 1995 *Electron Correlations in Molecules and Solids* (Berlin: Springer).
6. Gebhard F 1998 *Mott Metal–insulator Transition* (Berlin: Springer).
7. Pines D and Nozières P 1989 *The Theory of Quantum Liquids* (New York: Benjamin).

Self-assessment questions and exercises

1. Hartree equations for metals include the Coulomb interaction between conduction electrons in an approximate manner. Which important aspects of electron–electron interaction, however, are missing in this approximation?

2. What is the essential qualitative difference between Hartree equations and Hartree–Fock equations?

3. Hartree–Fock equations take into account electron–electron correlation to some extent. However an important part of electron–electron interactions is still missing. What is this important part?

4. What is the qualitative difference between Hartree charge density $\rho = -e \sum_j |\psi_j(\mathbf{r})|^2$ and the Hartree–Fock exchange charge density $\rho_i^{HF}(\mathbf{r}, \mathbf{r}') = -e \sum_j \frac{\psi_j^*(\mathbf{r}')\psi_i(\mathbf{r}')\psi_i^*(\mathbf{r})\psi_j(\mathbf{r})}{\psi_i^*(\mathbf{r})\psi_i(\mathbf{r})}$? Verify that Hartree–Fock charge density $\rho_i^{HF}(\mathbf{r}, \mathbf{r}')$ actually represents a charge $-e$.

5. What is the difference between restricted and unrestricted Hartree–Fock theory? Give an example of the application of unrestricted Hartree–Fock theory.

6. Show that the exchange term ((2.18) and (2.19)) of Hartree–Fock equations for a free electron gas can be expressed as $U^{ex} = -\frac{e^2}{2\pi^3} \int_0^{k_F} \int \frac{e^{i(k-k')(\mathbf{r}-\mathbf{r}')}}{|(\mathbf{r}-\mathbf{r}')|} d\mathbf{r}' d\mathbf{k}'$. Assume that the solution for Hartree–Fock equations for free electron gas is in the form of plane waves.

7. Show that the charge density seen by a Hartree–Fock electron is given by $\frac{eN}{v_g}\left[1 - \frac{9}{2}\frac{(k_Fr \cos k_Fr - \sin k_Fr)^2}{(k_Fr)^6}\right]$.

8. Show that the energy–momentum relation for a free-electron is no longer valid for a Hartree–Fock electron in the electron gas model.

9. While applying the Hartree–Fock theory for homogeneous electron gas, the charge density of the electron gas is considered to be uniform. But for a particular Hartree–Fock electron, is the charge density really uniform? If not, why?

10. What is the reason behind the failure of the Hartree–Fock theory to properly estimate the cohesive energy of metals?

11. There is an anomalous feature in the expression for one-electron energies given by equation (2.26), which would affect the electron velocity and heat capacity of metals. What is this anomalous feature and how does it arise? Why does this anomalous feature ultimately not become effective in real metals?

12. What are exchange energy and correlation energy in a crystalline solid, and what role do they play in determining the material's properties? In which condition is a material known as a strong correlated electron system?

13. Estimate the ground state energy of a two-electron system using Hartree–Fock theory.

14. In which condition is local density approximation (LDA) really a good approximation within the framework of density functional theory (DFT)? Can DFT with LDA be applied, say, in studying the properties of solid argon or neon?
15. The electron self-interaction term is a problem in DFT with LDA or LSDA, but not in Hartree–Fock theory. Why?
16. What are the basic principles of Landau Fermi liquid theory? What are Landau quasi-particles?
17. What is the influence of the Pauli exclusion principle on the electron–electron scattering rate in the Fermi liquid?

References

[1] Ashcroft N W and Mermin N D 1976 *Solid State Physics* (Philadelphia, PA: Saunders College)
[2] Madelung O 2004 *Introduction to Solid State Theory* (Berlin: Springer)
[3] Martin R M 2008 *Electronic Structure: Basic Theory and Practical Methods* (Cambridge: Cambridge University Press)
[4] Mahan G D 2000 *Many Particle Physics* (New York: Plenum)
[5] Friedel J and Sayers C M 1977 *J. Phys.* **38** 697
[6] Nozières P and Pines D 1958 *Phys. Rev.* **111** 442
[7] Gell-Mann M and Brueckner K 1957 *Phys. Rev.* **106** 364
[8] Wigner E P 1938 *Trans. Faraday Soc.* **34** 678
[9] Bohm D and Pines D 1952 *Phys. Rev.* **82** 625
Bohm D and Pines D 1952 *Phys. Rev.* **85** 338
Bohm D and Pines D 1953 *Phys. Rev.* **92** 609
[10] Pines D 1953 *Phys. Rev.* **92** 626
[11] Löwden P 1958 *Adv. Chem. Phys.* **2** 207
[12] Fulde P 1995 *Electron Correlations in Molecules and Solids* (Berlin: Springer)
[13] Slater J C 1951 *Phys. Rev.* **81** 385
Slater J C 1951 *Phys. Rev.* **82** 538
Slater J C 1953 *Phys. Rev.* **91** 528
[14] Slater J C 1937 *Phys. Rev.* **51** 846
[15] Korringa J 1947 *Physica* **13** 392
[16] Kohn W and Rostoker N 1954 *Phys. Rev.* **94** 1111
[17] Thomas L H 1927 *Proc. Cambridge Philos. Soc.* **23** 542
[18] Fermi E 1927 *Rend. Accad. Lincei* **6** 602
[19] Dirac P A M 1930 *Proc. Cambridge Philos. Soc.* **26** 376
[20] Teller E 1962 *Rev. Mod. Phys.* **34** 627
[21] Jones R O 2015 *Rev. Mod. Phys.* **87** 8977
[22] Becke A D 2014 *J. Chem. Phys.* **140** 18A301
[23] Hohenberg P and Kohn W 1964 *Phys. Rev.* **136** B864
[24] Burke K 2003 *ABC of DFT* http://master-mcn.u-strasbg.fr/wp-content/uploads/2015/09/DFT.pdf
[25] Kohn W and Sham L J 1965 *Phys. Rev.* **140** A1133
[26] Anderson P W 2011 *More and Different: Notes from a Thoughtful Curmudgeon* (Singapore: World Scientific)

[27] Burke K 2012 *J. Chem. Phys.* **136** 150901

[28] Capelle K and Leiria Campo V Jr 2013 *Phys. Rep.* **528** 91

[29] Landau L D 1957 *Sov. Phys. JETP* **3** 920
Landau L D 1957 *Sov. Phys. JETP* **8** 70

[30] Stewart G R 1984 *Rev. Mod. Phys.* **66** 755

[31] Gebhard F 1998 *Mott Metal-insulator Transition* (Berlin: Springer)

[32] Pines D and Nozières P 1989 *The Theory of Quantum Liquids* (New York: Benjamin)

Chapter 3

Mott insulators and related phenomena: a basic introduction

3.1 Localized framework of solid state

In this section we will recapitulate some of the topics discussed in the previous chapters, which will be helpful to understand and appreciate the important role of electron–electron interactions in the solid materials. The narrative in this section will closely follow the presentation in the book by Madelung [1].

The properties of crystalline solids can be described theoretically from two opposite points of view. The first approach is the delocalized description of solids in terms of electrons and quasi-particles. This has been exemplified in the form of electron band theory in chapters 1 and 2. We have seen how the translational invariance of the lattice in a crystalline solid leads to the description of electrons in terms of plane waves modulated by a function with the lattice periodicity, i.e. Bloch functions as solutions of the one-electron Schrödinger equation. The probability of finding a Bloch electron at a particular point within a Wigner–Seitz cell is the same for all equivalent positions in other cells. The delocalization of the electron states in the electron band theory is expressed by the designation of 'extended states'. This viewpoint encapsulates the important concept of elementary excitations. A characteristic feature of all these excitations is that a wave vector of definite value is associated with each state in their energy spectrum, whereas the location of the state itself remains undefined. To this end the electrons in a solid are described in terms of Bloch functions, which are plane waves modulated by the lattice periodicity.

In the second approach, the solid is thought to be composed of individual atoms with characteristic properties, and the description starts from localized states, i.e. the states of atoms embedded at the individual lattice sites. The various phenomena occurring in the solid are considered as local processes taking place at individual atoms, but influenced by the fact that the atoms are fixed at the periodic lattice sites in a solid. The interaction between these atoms will enable the local excitations to propagate through the solid. We had some discussion on this tight-binding approach in chapter 1, section 1.6 and observed that the Bloch function can be expressed in terms of atomic orbitals ϕ_n (equation (1.125)):

$$\psi_{k,n} = \frac{1}{\sqrt{N}} \sum_R e^{i\mathbf{k}\cdot\mathbf{R}} \phi_n(\mathbf{r} - \mathbf{R}). \tag{3.1}$$

However, we also saw that the function defined in equation (3.1) is not a solution of the Schrödinger equation of the crystal potential, and we needed to make a linear combination of the atomic orbitals with coefficients to be determined to satisfy the Schrödinger equation.

We have seen in chapter 1, section 1.4 that the Bloch functions are periodic in **k**-space, and hence the Bloch functions in the nth energy band can be expressed as a Fourier series:

$$\psi_{k,n} = \frac{1}{\sqrt{N}} \sum_R e^{i\mathbf{k}\cdot\mathbf{R}} a_n(\mathbf{r} - \mathbf{R}). \tag{3.2}$$

These new functions a_n are called *Wannier functions* and are expressed as:

$$a_n = \frac{1}{\sqrt{N}} \sum_k e^{(i\mathbf{k}\cdot(\mathbf{R}-\mathbf{r}))} u_n(\mathbf{k}, \mathbf{r}). \tag{3.3}$$

The summation covers all **k**-vectors in a Brillouin zone. Since $u_n(\mathbf{k}, \mathbf{r})$ has the periodicity of the lattice, the a_n are dependent only on the difference $\mathbf{r} - \mathbf{R}$. Hence, every Wannier function is centred on the mid-point of a Wigner–Seitz cell. They are localized around individual lattice sites. A comparison between the Bloch function and Wannier function is shown in figure 3.1. Wannier functions can be used for characterizing the electronic properties. The arrangement of the atoms in a solid of given structure becomes very important, and the interpretation of solid state phenomena is then based on the collective properties of the atoms in a solid. Within this localized framework, the electron moves from one localized state at one atom into another localized state of another atom at a different lattice site, and this electron transfer process involves interaction with all the charged particles in its vicinity. This is in contrast with the electron band model, where the collective properties of the lattice are included in the properties of itinerant electron, and further interactions in the lattice are described by electron transfer from one delocalized state to another with a change in the electron energy and wave vector.

These local and non-local frameworks to study crystalline solid are actually equivalent in the case of a full energy band—the valence band of an insulator. This band does not contribute to electrical conductivity, since the contributions from two electrons with equal energy but opposite wave vector cancel exactly in a fully occupied band. In a local approach there will be no transfer of electron from one atom to the next, because the corresponding states of the neighbouring atoms are also occupied. The electron band model approach in terms of delocalized excitations and their interaction is a suitable choice in many cases, for example in the study of

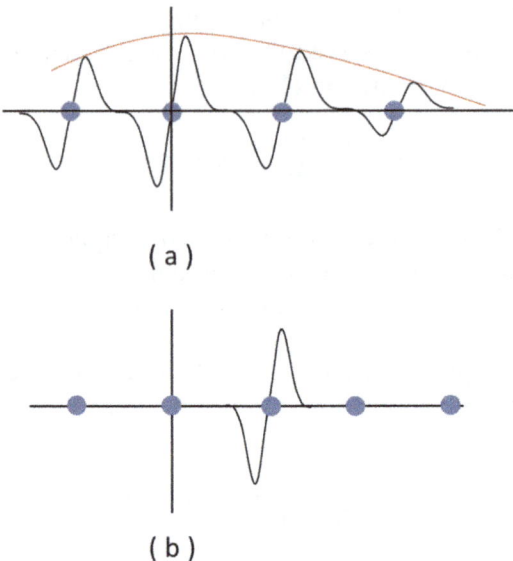

(a)

(b)

Figure 3.1. Schematic representations for comparison between the Bloch function and Wannier function. (a) The Bloch function is a plane wave modulated by a function with the periodicity of the lattice. (b) The Wannier function is localized around an individual lattice site.

optical excitation of an electron out of a fully occupied band by the absorption of a photon of given energy and wave vector. On the other hand, local description is suitable in the study of chemical bonds in a solid, which is concerned with the mutual interaction of valence electrons of all atoms in the lattice. The readers are referred to the book by Madelung [1] for a detailed discussion on the methodology to study chemical bonds within the framework of localized states. The local description becomes particularly useful when the electron–electron interaction becomes important in defining one-electron states in a perfect infinite lattice, where the electron band model approximation fails. This local approach will be followed extensively in the sections below to study the electron correlation effects in the crystalline solid.

3.2 Interacting electron gas: Wigner crystallization and Mott insulating state

We have seen in chapter 2 that the mean energy of an electron within the Hartree–Fock approximation in the jellium model can be given by the equation (2.46), which we reproduce here:

$$\frac{E}{N} = \frac{e^2}{2a_0}\left[\frac{3}{5}(k_F a_0)^2 - \frac{3}{2\pi}(k_F a_0)\right] = \left[\frac{2.21}{(r_s/a_0)^2} - \frac{0.916}{(r_s/a_0)}\right] \text{Ry.} \qquad (3.4)$$

The first term in the right-hand side(s) represents kinetic energy and the second term is the potential energy of the system after taking into account of the

Hartree–Fock exchange correlation between the electrons of same spins. With reduction in the electron density (i.e. increasing the mean electron separation r_s) the kinetic energy contribution decreases faster in strength relative to the potential energy. We can see from equation (3.4) that at large r_s, i.e. low electron density, the kinetic energy scales with density as r_s^2 and the potential energy as r_s, and hence the potential energy dominates the kinetic energy. In this situation as $r_s \to \infty$ electrons in the lowest energy state tend to organize in such a way that they are as far away from each other as possible. In other words, the electrons become organized in a crystal like periodic array known as a *Wigner lattice*. This also suggests the possibility of the localization of the electrons through correlation effects. The most stable crystal structure in three-dimension for a pure $1/r$ interaction is a body-centred cubic (bcc) lattice, and for two-dimension it is a hexagonal lattice. On the other hand at high electron density i.e. at small r_s, the kinetic energy will dominate and the system goes to a non-interacting Fermi gas state $r_s \to 0$. This suggests that somewhere at the intermediate electron density, the crystalline order of Wigner lattice is lost, and there is a phase transition. Thus, the formation of a Wigner crystal is considered to be the purest example of a 'Mott transition' driven entirely by the electron–electron interaction.

At very low temperatures, the electron spins can become ordered in a Wigner crystal phase due to the exchange of electrons on different lattice sites [2]. A three-dimensional bcc lattice can be represented by two interpenetrating sub-lattices. Here the pair exchange dominates and gives rise to an antiferromagnetic order. Such interpenetrating sub-lattices are not possible in a two-dimensional hexagonal lattice, and in addition because of strong electron correlation, exchange involving three or more spins becomes important. These effects frustrate conventional antiferromagnetic ordering and gives rise to the possibility of the lack of spin order in a Wigner crystal in two-dimension even at very low temperatures. Such a spin state is called a spin liquid [2]. We will see some examples of the spin-liquid state in chapter 4 and chapter 6.

The crucial role of electron interactions in solid was brought forward clearly in an early paper of Nevill Mott in 1949 [3]. The interesting outcome of this interaction can be studied through a thought experiment, by slowly bringing together a large number of hydrogen atoms or sodium atoms or any array of mono-valent metallic atoms. Each atom at the lattice sites contribute one electron, which then can move in the lattice formed by the nuclei or ionic-cores and it is subjected to the Coulomb repulsion by other electrons. The possibility of the electrons moving to the other lattice sites will depend on the distance a between the atoms forming the lattice (see figure 3.2) i.e. the overlap between the outermost atomic orbitals. The hopping of the electrons from site to site will tend to delocalize the electrons over the whole lattice, and that will lead to a kinetic energy gain. However, the presence of two electrons at the same lattice site, which happens during the process of electrons hopping from site to site, will cause intrasite Coulomb repulsion between such electrons. This indicates that there will be an energy cost because of this Coulomb repulsion. Now what happens if the interatomic separation a is increased from its equilibrium value to a new larger value? According to the electron band theory the valence band will be narrower since the overlap of atomic orbitals will be smaller (see the lower panel of figure 3.2). There

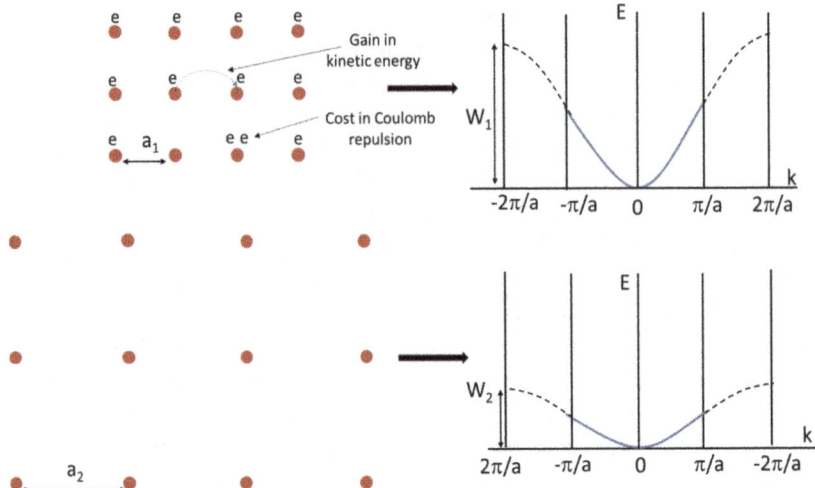

Figure 3.2. Schematic representation of the crystal lattice of a monovalent solid with half filled band. Hopping of the electrons between the lattice sites leads to the gain in kinetic energy. Increasing the lattice constant from a_1 to a_2 narrows the bandwidth from W_1 to W_2, while retaining the metallic property. Double occupancy of electrons at a lattice site is subject to Coulomb repulsion, and thus costs in energy.

will be a corresponding increase in the ratio between the Coulomb repulsion energy and kinetic energy. The system will always be metal since the band is half-filled, and this metallic state is realized for small values of a. On the other hand, for very large interatomic distances the system will consist of neutral atoms fixed to the lattice sites, and the system is no more conducting and behaves like an insulator. At such large distances the overlap between the atomic orbitals is small, and thus gives a very small kinetic energy gain associated with the hopping of the electrons among the lattice sites. The intrasite Coulomb repulsion becomes dominant in this situation leading to the localization of single electron at each lattice. This happens above a critical value of interatomic separation, and the system becomes a Mott insulator, which cannot be described within the realm of electron band theory. In contrast to the band insulators, which results from the lack of available electron states at low energies, the repulsion among the electrons plays the central role in a Mott insulator.

More formally, we can say that within the framework of electron band theory of solids (chapter 1, section 1.5), a band will form as a consequence of the overlap of wave functions on each atom, and the solid will be a metal. However, the formal equivalence between a set of localized orbitals $\phi_n(\mathbf{r} - \mathbf{R})$ and a set of Bloch functions $\psi_k(\mathbf{r})$ as suggested in the tight binding approximation (chapter 1, section 1.6) is valid only if one neglects the Pauli exclusion principle and electron–electron interaction. If there is only one electron per atom, the determinantal wave function of all the states $\psi_k(\mathbf{r})$ in the half-filled band is not the same as the determinant of all the localized atomic orbitals $\phi_n(\mathbf{r} - \mathbf{R})$. There will be tendency to have two electrons on some atoms, and leave some atoms deficient in electron and in turn with a distribution of charge that would have large positive contributions to the total Coulomb energy of the system. A set of localized states may then be a better trial wave function for such

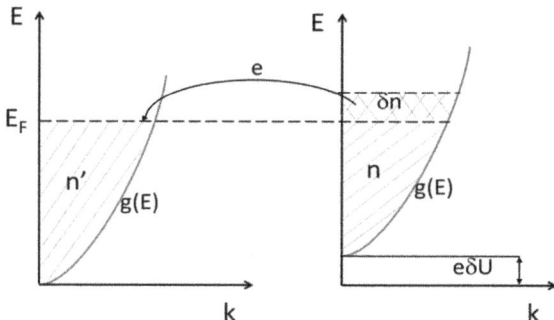

Figure 3.3. Schematic presentation showing the effect of switching on a perturbation potential δU on the properties of Fermi free electron gas.

a system than the set of delocalized band states. The former type of localized wave function represents an insulator, whereas the later type a metal.

If one takes an electron out of one of the array of localized states in such a monovalent atom system, this will leave behind an empty ion i.e. a positive charge. The electron will be attracted back to this ion, and may even form a bound state there, so that it will not really be free to move through the solid and carry an electric current. However, if there are already a number of such electrons present, then those will tend to screen the charge of the ion. We have already seen in chapter 2, how the perturbation takes place in the homogeneous electron distribution when an electrical charge is introduced in a metal, which in turn screens the electric field of the charge. In chapter 2, section 2.5 we have seen that within the Thomas–Fermi approximation, the total potential in the presence of such screening takes the form of a *screened Coulomb potential*. We shall study this aspect here in some more detail.

We assume that perturbation potential δU to be small (so that $|e\delta U| \ll E_F$), and it is expected that there will be a local raising of the electron density of states $g(E)$ by an amount $e\delta U$ (figure 3.3), where e is electronic charge. Imagining the situation that the perturbation potential is switched on from zero, then some electrons will leave this perturbing region so that the Fermi level E_F remain constant throughout the solid. This condition of homogeneity arises because $E_F = \mu$ (chemical potential) is a thermodynamic function of state. The change in the electron density n for not too large a perturbation potential can be expressed in terms of the electron density of states at the Fermi level [4]:

$$\delta n(\mathbf{r}) = g(E_F)|e|\delta U(\mathbf{r}). \tag{3.5}$$

It can be assumed that $\delta U(\mathbf{r})$ arises essentially by the induced charge, except in the near proximity of the perturbing charge. Thus $\delta n(\mathbf{r})$ is related to δU via the Poisson equation:

$$\nabla^2(\delta U) = \frac{e}{\epsilon}\delta n = \frac{e^2}{\epsilon}g(E_F)\delta U \tag{3.6}$$

where ϵ is the dielectric constant.

With $\lambda^2 = e^2 g(E_F)/\epsilon$ the differential equation (3.6) will have a nontrivial solution:

$$\delta U(r) = \frac{\alpha}{r} e^{-\lambda r} \tag{3.7}$$

which gives the expected Coulomb potential for $\lambda \to 0$, where the screening effect disappears. The quantity $r_{TF} = 1/\lambda$ is called the Thomas–Fermi screening length:

$$r_{TF} = [e^2 g(E_F)/\epsilon]^{-\frac{1}{2}}. \tag{3.8}$$

In the case of free-electron gas model we have from equation (1.71) $E_F = \frac{\hbar^2}{2m_e}(3\pi^2 n)^{\frac{2}{3}}$ and from equation (1.76) $g(E_F) = \frac{1}{2\pi^2}\left(\frac{2m_e}{\hbar^2}\right)^{\frac{2}{3}} E^{\frac{1}{2}}$. From these two relations:

$$g(E_F) = \frac{1}{2\pi^2} \frac{2m_e}{\hbar^2} (3\pi^2 n)^{\frac{1}{3}}. \tag{3.9}$$

Equations (3.8) and (3.9) lead to:

$$\frac{1}{r_{TF}^2} = \frac{4}{\pi} (3\pi^2)^{\frac{1}{3}} \frac{n^{\frac{1}{3}}}{a_0} \tag{3.10}$$

and

$$\frac{1}{r_{TF}} \simeq 2 \frac{n^{\frac{1}{6}}}{a_0^{\frac{1}{2}}}; \quad r_{TF} \simeq 0.5 \left(\frac{n}{a_0^3}\right)^{-\frac{1}{6}} \tag{3.11}$$

where a_0 is the Bohr radius.

It is clear from equation (3.11) that above a critical electron density n_c the screening length r_{TF} becomes so small that the electron can no longer stay in a bound state. This situation leads to the metallic behaviour. On the other hand, below this critical concentration the potential well of the screened field spreads quite enough so that a bound state becomes possible. The electron can now be localized in a covalent or ionic bond. Such localized electron states leads to insulating property in a solid where the highest occupied states form localized bonds. It is possible to make a simple estimate of when a bound electron state is possible (in a screened Coulomb potential) by assuming that the screening length must be much greater than Bohr radius a_0:

$$r_{TF}^2 \simeq \frac{1}{4} \frac{a_0}{n^{\frac{1}{3}}} \gg a_0^2 \tag{3.12}$$

or,

$$n^{-\frac{1}{3}} \gg 4a_0. \tag{3.13}$$

This estimate, which was originally proposed by Mott [5], says that a solid will not show metallic behaviour, rather it becomes an insulator when the average electron separation $n^{-\frac{1}{3}}$ becomes much greater than four times the Bohr radius.

It may be noted that these two pictures of electron localization presented above are actually based on conceptual experiments. They reveal the possible influence of correlation even within the framework of the jellium model and one-electron approximation. The features of electron localization are common in these frameworks but with some essential differences. In a crystal lattice the attractive potential provided by the nuclei tend to localize the electrons near each nucleus, so that a metal to insulator transition can take place at a much larger electron density and temperature.

The actual Wigner lattice or the Wigner crystal is a charge-density wave state, and it is an insulator since it is pinned by impurities. The Wigner lattice is the state with minimum potential energy, and the kinetic energy comes from the zero-point motion of the electrons in around the lattice sites. A similar electronic charge-density wave forms also at the Verwey transition. The essential difference between the Wigner crystal and Verwey insulator lies in that the later type of insulating state happens at higher electron concentration regime. The wave length of the charge-density wave in the Verway insulators is of the order of the lattice constant, and it is frequently commensurate with the underlying ion lattice. A transition into such an electronic charge-density wave was first observed by Verwey in magnetite Fe_3O_4 samples around 120 K [6]. A detail discussion of the experimental results on the Verwey insulator/transition will be presented in chapter 6. The Wigner–Verwey transitions take place due to the Coulomb repulsion between electrons on different lattice sites. This is in contrast to the Coulomb repulsion at the same lattice site, which together with the Hund's-rule exchange integrals is responsible for producing the magnetic moments. A sufficient large intra-site Coulomb interaction leads to the Mott insulating state.

3.3 Mott insulator: towards a formal definition

Nevill Mott himself provided a definition for a Mott insulator as a solid that would be a metal if no magnetic moments were formed in the solid. He wrote '… we describe a Mott insulator as an antiferromagnetic material that would be a metal if no moments were formed' [5]. It is to be noted here that it is the formation of the magnetic moment that is of primary importance and not the antiferromagnetic order. Mott emphasized this aspect by saying that 'in the insulating state, the gap is given by …, and it depends on the existence of moments and not on whether or not they are ordered' [5]. In a Mott insulator, the magnetic moments form because the electrons try to avoid each other to minimize their Coulomb repulsion. At this stage a formal definition of the Mott insulator may be introduced [7]: 'For a Mott insulator the electron–electron interaction leads to the occurrence of (relative) local moments. The gap in the excitation spectrum for charge excitations may arise either from the long-range order of the pre-formed moments (Mott–Heisenberg insulator) or by a quantum phase transition induced by charge and/or spin correlations (Mott–Hubbard insulator).'

3.4 Pair distribution function and magnetic moments

It is quite clear that the repulsive Coulomb interaction between electrons in a solid tends to keep the electrons away from each other. This tendency of electron localization, however, is in conflict with the kinetic energy of electrons, which tends to spread out the electrons over the whole crystal. In order to assess the strength of the electron–electron interactions in comparison to their kinetic energy, one needs to have some criterion to get a measure the relative strength of these kinetic and potential energies.

To determine the strength of the electron–electron interaction one needs to know the probability distribution to find a σ' electron at r_1 when there is already a σ electron at $\mathbf{r}$. (Here σ and σ' represent spin-up and spin-down electron states.) This information is contained in the pair distribution function, which we have discussed in detail in chapter 2, section 2.3. The pair distribution function gives the relative probability of finding an electron at $\mathbf{r}'$ given that one is there at $\mathbf{r}$. We have learnt that there are two different contributions to the pair correlation function and summarize the main results here to make the present discussion a self-contained one.

1. **Effect of exchange**: The Pauli exclusion principle ensures that two electrons of parallel spin cannot occupy the same energy state. As a result, the pair distribution function for electrons with parallel spin will be non-zero even in the absence of the electron–electron interaction. In the homogeneous electron system i.e. Fermi-gas ground state, the pair distribution function for parallel spins can be expressed as:

$$g_{\uparrow\uparrow} = \frac{1}{2}[1 - \psi(\mathbf{r} - \mathbf{r}')^2] \tag{3.14}$$

where,

$$\psi(\mathbf{r}) = \frac{3}{(rk_F)^3}[\sin(rk_F) - rk_F \cos(rk_F)]. \tag{3.15}$$

The Pauli exclusion principle reduces the probability on the scale of the reciprocal Fermi wave number $1/k_F$ to find an electron in the vicinity of another electron of same spin; it creates an 'exchange hole'.

Now considering the Coulomb repulsion between the electrons as a perturbation on the Fermi gas, since the electrons with same spin sign are already kept apart from each other by the Pauli exclusion principle, there will be a reduction in their mutual Coulomb interactions. The electron wave functions remain unchanged to the lowest order in perturbation theory such that the electron–electron interaction among electrons with same spin sign is reduced in comparison to the interaction between electrons of opposite spin sign [7]. The resultant energy gain is termed as 'exchange interaction', which leads to the parallel orientation of the electron spins in open shells in atoms i.e. first Hund's rule, and provides an explanation of ferromagnetism in transition metal solids.

2. **Effect of correlation**: The pair distribution function $g_{\uparrow\downarrow}(\mathbf{r}, \mathbf{r}')$ between electrons of antiparallel spin encapsulates the contributions beyond the exchange effect [7]. It represents the true many-body effects. The systems with a non-vanishing pair distribution function between electrons of anti-parallel spins is termed 'correlated electron systems' [7]. We have seen in chapter 2, section 2.3 that the interaction also changes the pair distribution function between electrons of the same spin sign. Apart from the 'exchange hole', an electron in general will also be surrounded by a 'correlation hole', since the repulsive coulomb interaction will also prevent electrons of different spin coming close to each other. A system will be termed as 'strongly correlated' if the pair distribution functions for electrons of parallel spin and electrons of antiparallel spin are quite comparable in size.

The pair correlation function, however, is not easily accessible either experimentally or theoretically, and it is instructive to look for a less complete but relatively more (experimentally) accessible probe for investigating the physical properties [7]. For this purpose local magnetic moment in the z-direction (assuming the z-axis as the quantization axis) can be a useful parameter. This is expressed as:

$$[M^z(R)]^2 = \langle \hat{S}^z(R)\hat{S}^z(R) \rangle \tag{3.16}$$

where $\hat{S}^z(R) = [\hat{n}\!\uparrow(R) - \hat{n}\!\downarrow(R)]/2$. Here n is the electron density and $\hat{n}$ is the corresponding number operator in second quantized form (see section 3.6 below and Appendix C).

The instantaneous magnetic moment can be subdivided into three parts [7]

$$[M^z(R)]^2 = [M^z(R)]^2_{FG} + [M^z(R)]^2_{LRO} + [M^z(R)]^2_{C} \tag{3.17}$$

where,

$$[M^z(R)]^2_{FG} = \frac{1}{8}\langle \hat{n}(R) \rangle (2 - \langle \hat{n}(R) \rangle) \tag{3.18}$$

$$[M^z(R)]^2_{LRO} = \frac{1}{2}\langle \hat{S}^z(R) \rangle^2 \tag{3.19}$$

$$[M^z(R)]^2_{C} = -\frac{1}{2}g_{\uparrow\downarrow}(R, R). \tag{3.20}$$

The first term in the equation (3.17) is termed the Fermi-gas contribution. This arises from the Pauli exclusion principle, and is independent of the electron–electron interactions. The second term has non-zero value, only when there is long range order in the spin-system. This happens in the case of ferromagnetism or antiferromagnetism. The third term represents the correlation contribution to the magnetic moment.

It is useful to define the relative local magnetic moment, which is expressed as [7]:

$$[M^z(R)]^2_{Rel} = [M^z(R)]^2 - [M^z(R)]^2_{FG} = [M^z(R)]^2_{LRO} + [M^z(R)]^2_C. \qquad (3.21)$$

This relative local magnetic moment, which arises inherently because of the electron–electron interaction is absent in the Hartree–Fock–Slater effective one-electron theories. But, as emphasized by Mott, this is very important for the physics of correlated electron systems.

The local moments $[M^z(R)]^2$ as defined in equation (3.16), will behave as free spins in experiments with short time scales or (equivalently) high energy scales. The static spin susceptibility in solids for example, follows a Curie law with a Curie constant proportional to $[M^z(R)]^2$ only when the temperature is large enough compared to all other energy scales. The relative local moments (equation (3.21)) will be present at finite frequencies for all non-zero interaction strengths, and the effects of local moments need to be included in a proper description of the metal–insulator transition where the low-energy Fermi-liquid picture inevitably breaks down.

3.5 Mott–Hubbard insulator

To carry the discussion we will start with an arrangement of monovalent atoms like hydrogen (H) in a cubic lattice with a lattice constant a as we have done earlier in section 3.2. The hydrogen atom at each lattice point carries one valence electron (s electron) with it. If the monovalent solid had been a metal we would have said that the valence band of the metal is half filled. (While sodium, a monovalent solid, is a metal, there is still debate whether metallic behaviour has been observed in hydrogen even at very high pressure.) The valence electrons in such metals will of course be delocalized and move freely in the crystalline solid. If the width of this s-band is reduced, this will effectively increase the lattice constant ultimately to an extent so that there will be practically no interaction between the lattice atoms. In this situation the band will turn into a discrete electronic level of isolated atoms. Metallic conduction will no longer be possible in this limit of the isolated atoms, although within the band model approximation there is still a half-filled band; the electron band model approximation thus breaks down in the limiting case of very narrow band. The tendency of localization of the electrons on the lattice sites will imply an increasing correlation between the electrons, and hence for the system with narrow bands the electron correlation effects will certainly be very important.

We will now try to understand the Mott insulating state from a tight-binding lattice model with N electrons distributed over $L = N$ lattice sites. This is the case of half filling with one electron at each lattice site. In such a tight-binding model, conduction takes place by the movement of electrons through the lattice from one site to another (see figure 3.4). These electrons reduce their kinetic energies by hopping from one lattice site to the other. To deal with the discrete lattice in the tight-binding model, where the particle nature of electron is more prevalent, it is customary to work within the framework of second quantization (see appendix C). To this end one works in a basis where the electron creation operator c_i^+ creates a

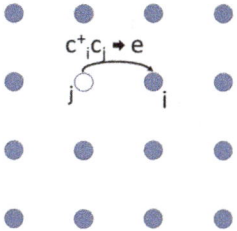

Figure 3.4. Schematic diagram showing the operator $c_i^+c_j$ causing an electron to hop from site j to site i. This process saves the cost of kinetic energy.

particle at a particular lattice site labeled by i and electron annihilation operator c_j annihilates a particle at a particular lattice site labeled by j. When an electron hops between points j and i, the kinetic energy saving in this process is termed as t_{ij}. It is clear that t_{ij} will depend on the overlap of atomic orbitals. The Hamiltonian therefore is a sum over pairs of lattice sites, and it essentially represents a sum over all processes in which an electron hops between lattice sites:

$$H = \sum_{ij}\left(-t_{ij}c_i^+c_j\right). \tag{3.22}$$

The operator $c_i^+c_j$ causes an electron to hop from the site j to the site i (see figure 3.4), and each term in the Hamiltonian in this sum represents processes in which a particle at site j is annihilated and then created again at site i. We now consider for simplicity the particular case of $t_{ij} = t$ for nearest neighbours and $t = 0$ otherwise. Accordingly the Hamiltonian is rewritten as:

$$H = -t\sum_{i\tau} c_i^+c_{i+\tau}. \tag{3.23}$$

Here the sum over τ involves the nearest neighbours. This tight-binding Hamiltonian can be diagonalized by making a change of basis with the help of Fourier transform [8]:

$$c_i = \frac{1}{\sqrt{v}}\sum_k e^{i\mathbf{k}\cdot\mathbf{r}_i}c_k \tag{3.24}$$

and

$$c_i^+ = \frac{1}{\sqrt{v}}\sum_q e^{-i\mathbf{q}\cdot\mathbf{r}_i}c_q^+. \tag{3.25}$$

With the help of equations (3.24) and (3.25) the Hamiltonian in equation (3.23) can be expressed as:

$$H = -\frac{t}{v}\sum_{i\tau}\sum_{kq} e^{-i(\mathbf{q}-\mathbf{k})\cdot\mathbf{r}_i - i\mathbf{k}\cdot\mathbf{r}_\tau}\, c_q^+c_k. \tag{3.26}$$

With the knowledge $\frac{1}{v}\sum_i e^{-i(\mathbf{q}-\mathbf{k})\cdot\mathbf{r}_i} = \delta_{qk}$ and summing over $\mathbf{q}$ the Hamiltonian can be written as [8]:

$$H = -t \sum_{\tau} \sum_{k} e^{i\mathbf{k}\cdot\mathbf{r}_{\tau}} c_k^+ c_k. \tag{3.27}$$

This Hamiltonian is diagonal in momentum and this can be rewritten as:

$$H = \sum_{k} E_k c_k^+ c_k \tag{3.28}$$

where E_k is the energy dispersion and is expressed as:

$$E_k = -t \sum_{\tau} e^{i\mathbf{k}\cdot\mathbf{r}_{\tau}}. \tag{3.29}$$

In the case of a two-dimensional square lattice, the sum over τ involves the vectors $(a, 0)$, $(-a, 0)$, $(0, a)$ and $(0, -a)$. This gives an energy dispersion relation:

$$E_k = -2t(\cos k_x a + \cos k_y a). \tag{3.30}$$

In general this cosine dispersion relation in the tight-binding approximation for independent electrons can be expressed as:

$$E_k = -2t \sum_{l=1}^{d} \cos k_l a \tag{3.31}$$

with a bandwidth $W = 2Zt$, where $Z = 2d$ is the number of nearest neighbours.

Pauli's exclusion principle will demand that the electron can hop from one site to another only if the other site is occupied by an electron with opposite spin or it remains unoccupied. During the process of hopping from one lattice site to the other the electron will also experience Coulomb repulsion from the electron with opposite spin direction, which is already occupying the new site (see figure 3.5). To set up this potential energy part of the Hamiltonian, one starts with the assumption that the effective electron–electron interaction will be local i.e. at each lattice site only [7]. It is also assumed that the electrons dominantly sit on the lattice sites and not between. This intra-atomic potential, defined as the difference between ionization energy and electron affinity, is called the 'Hubbard' U.

$$\begin{aligned} U &= E(H \to H^+) - |E(H \to H^-)| \\ &= E(H^-) + E(H^+) - 2E(H). \end{aligned} \tag{3.32}$$

The calculation of the Hubbard U, thus, involves processes where an electron is transferred to infinity (ionization) or another electron is added from infinity (electron affinity).

The overlap between the atomic orbitals will be small at large distances ($a \gg a_0$ (Bohr radius)) between the atoms. This will indicate to the situation where $W \ll U$.

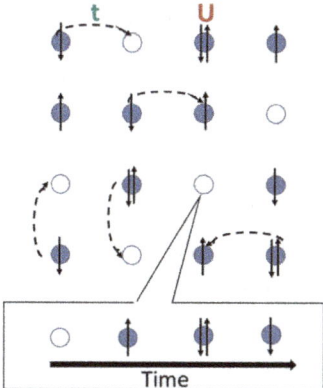

Figure 3.5. Schematic diagram representing a crystalline lattice with interacting electrons within the framework of Hubbard model. The circles represent the rigid (within the adiabatic approximations) lattice sites. The electrons (represented by up and down arrows) move from one lattice site to the next with a hopping amplitude t. Pauli's exclusion principle ensures that only two electrons with opposite spins can occupy a lattice site. These electrons encounter an Coulomb interaction potential U. A lattice site can either be unoccupied, singly occupied by electrons with up or down spins, or doubly occupied by electrons with opposite spins.

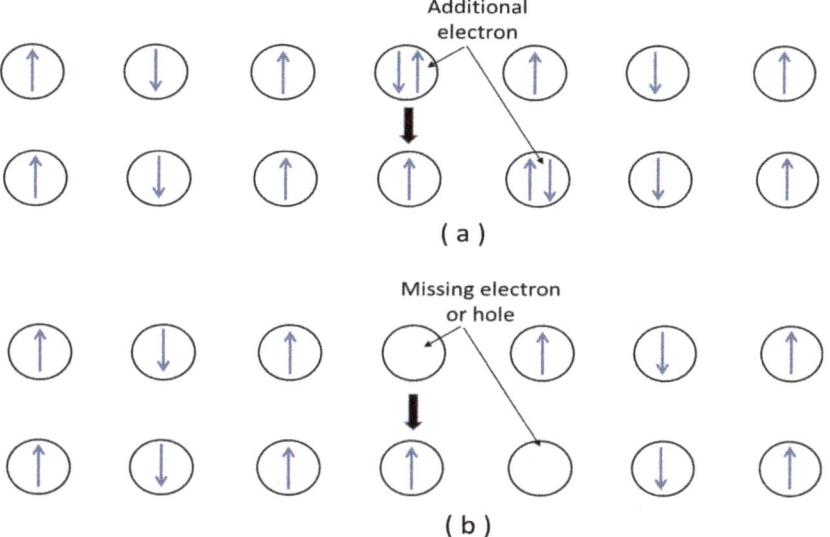

Figure 3.6. Schematic diagram showing the motion of (a) an electron in the upper Hubbard band, (b) a hole in the lower Hubbard band. The circles represent the atomic sites. Electrons are represented by up and down arrows, and the hole by the absence of an electron i.e. arrow at the atomic site.

All the lattice sites will be singly occupied in the ground state, since one is at half band-filling. If an additional electron is now added to the system, this new electron will sit in one of the atoms as shown in figure 3.6(a). The energy necessary to add this extra charge or electron to the ground state is $\mu^+(N) = E_0(N + 1) - E_0(N)$. Adding electrons to the atoms at the lattice site in the periodic lattice leaves it in an excited configuration. The internal degrees of freedom like orbital angular momentum and spin in the remaining atoms in the system will scatter these excitations with energy U. These states with a wave number k can be described by a many-electron wavefunction as in equation (3.2), and they will propagate through the lattice incoherently and broaden to form an energy band of width W_1, which is a function of (U/t). This energy is given by $\mu^+(N) = U - W_1/2$, and this band of charge excitations is known as 'upper Hubbard band' [7]. This band is not a one-electron band, and it actually describes the spectrum of charge excitations for an extra electron added to the ground state of the half-filled electron system. On the other hand, if an electron is removed from an atomic site at half band-filling (see figure 3.6(b)), the new many-electron wavefunction will represent the movement of this lack of electron or hole. Now, the energy required to remove an electron is $\mu^-(N) = E_0(N) - E_0(N - 1)$, and this is given by $\mu^-(N) = \frac{1}{2}W_2$, where W_2 is a function of (U/t). The related energy spectrum for the removal of an electron from the half-filled ground state forms the 'lower Hubbard band' of width W_2 [7]. Hubbard bands, thus, describe propagating empty and doubly occupied sites in a half-filled lattice. For $(W_1 + W_2)/2 \ll U$ at half band filling the chemical potential $\mu(N)$ is not expected to be continuous. A gap of width $\Delta\mu(N) = (\mu^+(N) - \mu^-(N)) \approx U - (W_1 + W_2)/2$ appears in the charge excitations spectrum and the system becomes an insulator.

We will try to explain the meaning of the parameter U and the origin of Hubbard bands in an even more simplistic way [1]. Let us consider a solid of infinite lattice constant. The electron energy of this solid can be represented by $E = N_1 t_0 + N_2(2t_0 + U)$. Here N_1 is the number of lattice points occupied by one electron and N_2 is the number of lattice points occupied by two electrons. Here t_0 is the energy needed to bind an electron on an isolated atom, and $t_0 + U$ is the energy needed to attach a second electron of opposite spin. Therefore U represents the Coulomb interaction energy of two electrons located at the same atom on a lattice site. When one electron is accommodated at each lattice site i.e. $N_1 = N$ and $N_2 = 0$, then in the ground state N electrons have the energy t_0. There will be a strict localization of electrons in this case. We will now study the ground state of a system with a finite lattice constant. Further, the electrons at the neighbouring sites have opposite spin directions i.e. antiferromagnetic ground state. If one more electron with a given spin direction is now introduced in this lattice, it can be accommodated at one of the $N/2$ atoms having an electron with opposite spin. The rest of the $N/2$ sites are forbidden by Pauli's exclusion principle since those are occupied by electrons with same spin direction. The energy of this added electron at an isolated lattice site is $t_0 + U$. However, in a lattice with finite lattice constant there will be a finite interaction between all the $N/2$ sites, which can take this added electron. Hence, this electron energy will now split into a band centred around $t_0 + U$. With the same argument the energy of the removal of an electron t_0 will split into a corresponding band. There exists a band gap, as long as the bandwidth remains smaller than the separation $(t_0 + U) - t_0 = U$. The band gap will disappear at a critical lattice constant, which determines the amount of energy splitting, thus leading to a transition from the localized electron to band electron picture [1]. Figure 3.7 presents a schematic representation of a Mott–Hubbard insulator.

A reduction in the distance between the H atoms will increase the overlap in their atomic wave functions, which in turn will lead to an enhancement of the charge screening and the tendency of electron delocalization. The combined bandwidth of the upper and lower Hubbard band $W_1 + W_2$ will also increase. The two bands will finally overlap for $a \approx a_0$ causing the gap for charge excitation to vanish. The band theory will qualitatively apply for $W \gg U$, predicting a single half-filled band, and the system will behave as a paramagnetic metal.

A metal–insulator transition is expected in the intermediate regime around a critical value $U_C \approx W$. It is pointed out here again that upper and lower Hubbard bands do not represent single-electron states, and the metal–insulator transition is not merely the result of a simple band crossing. We will discuss this in more detail in chapter 5.

Based on the various points discussed above, a formal definition of Mott–Hubbard insulator can be presented at this stage [7]: 'For a Mott–Hubbard insulator the electron–electron interaction leads to the formation of a gap in the spectrum for single charge excitations. The correlations force a quantum phase transition from a correlated metal to a paramagnetic Mott–Hubbard insulator, in which the local magnetic moments do not display long-range order.'

It may please be noted that the metal–insulator transition within the realm of above definition is understood to be a quantum phase transition, which is strictly limited to zero temperature and so are the definitions of a metal and an insulator. On one hand, the electrons tend to move over the entire solid crystal to optimize their

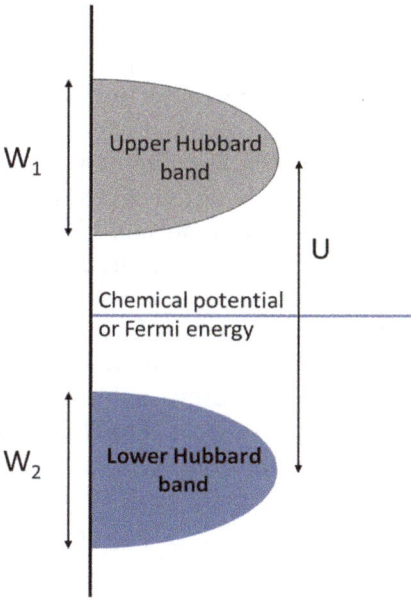

Figure 3.7. Schematic representation of a Mott–Hubbard insulator.

kinetic energy, and on the other hand the various interactions between the electrons tend to localize them for minimizing their potential energy. This competition between the kinetic and potential energies of electrons in a solid influences the electrical conduction of the concerned solid at zero temperature; the ground state can display dominantly either the itinerant (metallic) or localized (insulating) properties of electrons. An externally applied pressure can shift this energy balance to induce an insulator to metal transition. The gap for charge excitations into extended states is robust, and this gap does not vanish in the presence of other excitations [7].

A true quantum phase transition does not exist at any finite temperature. However, at finite temperatures both metals and insulators display a finite electrical conductivity at $T > 0$, and one can experimentally distinguish between a good and bad conductor. So a drastic change in the electrical conductivity may be observed even at finite temperatures, when the concerned system goes through the quantum phase transition value for the external parameters (say, pressure) that control the relative strength U/W of the kinetic and potential energy of the electrons. One may still speak of a Mott–Hubbard insulator in practical terms in the finite temperature regime, which is small compared to the single-electron gap i.e. $k_B T \ll \Delta\mu$.

3.6 Theoretical approaches in Mott physics

We shall now try to put the rather phenomenological picture of the Mott insulating state presented so far (in the sections above) in a more formal theoretical framework. We have discussed in chapter 1 that the energy scale for the global stability of the solid is about 1–10 eV per atom. This remains the relevant energy scale for the studies of Mott insulators. In addition, the Born–Oppenheimer or adiabatic approximation will remain valid for separating the motion of the electrons from the ionic motions at temperatures well below the condensation energy of the solids, and the solid will be considered a perfect periodic lattice of immobile ions.

The most widely used model to understand Mott physics is the Hubbard model [9, 10]. This model is represented by a Hamiltonian consisting of two terms: (i) the kinetic energy t defined by the hopping of the electrons between the lattice sites in a solid and (ii) the electrostatic Coulomb repulsion U between electrons at the lattice sites (see figure 3.5). Thus the, Hubbard model studies the interesting physics of the competition between the optimization of kinetic energy, which favours delocalization of electrons, and potential energy, which favours electrons to become localized at the lattice sites. The model Hubbard Hamiltonian is based on a tight-binding Hamiltonian representing the kinetic energy, and a two-body potential energy term representing the electron–electron interactions. The Hubbard Hamiltonian is expressed in the framework of second quantization as [8]:

$$ H = \sum_{ij}(-t_{ij})c_i^+c_j + \frac{1}{2}\sum_{ijkl} c_i^+c_j^+U_{ijkl}c_kc_l. \tag{3.33} $$

Now, the electrons carry a spin σ, which can point up or down. Accordingly the Hamiltonian is rewritten as:

$$H = \sum_{ij\sigma}(-t_{ij})c_{i\sigma}^+ c_{j\sigma} + \frac{1}{2}\sum_{ijkl\sigma\sigma'} c_{i\sigma}^+ c_{j\sigma'}^+ U_{ijkl\sigma\sigma'} c_{k\sigma'} c_{l\sigma}. \tag{3.34}$$

To proceed further it is assumed that there is only one band at the Fermi energy E_F and all other bands are energetically far from the Fermi energy. This is termed the single band Hubbard model. With only a single band at the Fermi energy the intra-atomic Coulomb interaction is the same for all the lattice sites. It is further assumed that during the process of hopping the electrons do not undergo any spin-flip. The interaction between electrons, however, is due to the Coulomb repulsion between their charges. Hence, the electrons with antiparallel spins will interact with each other just as much as electrons with the parallel spins. In the Hubbard model the Coulomb interaction is considered to be significant only between two electrons at the same lattice site, and these electrons interact via a constant potential energy U. The Pauli exclusion principle ensures that the spins coming into the same site must possess different spin directions (see figure 3.5), and this is where the influence of electron spin comes into the picture. The Hamiltonian of this relatively simplified Hubbard model is expressed as:

$$H = \sum_{ij\sigma}(-t_{ij})c_{i\sigma}^+ c_{j\sigma} + U\sum_i n_{i\uparrow}n_{i\downarrow} \tag{3.35}$$

where $n_i = c_i^+ c_i$ are number operators.

Let us now consider a simple two lattice sites Hubbard model with nearest-neighbour hopping only. If there is only one spin-up electron, that single electron can reside on the first site and is described by a configuration $|\uparrow, 0\rangle$ or on the second site with a configuration $|0,\uparrow\rangle$. A general state for this electron can be written as the superposition $|\psi\rangle = a\,|\uparrow,0\rangle + b\,|0,\uparrow\rangle$, where a and b are complex numbers. Since there is only one electron there is no potential term and the Hubbard Hamiltonian can be written as:

$$H = \begin{pmatrix} 0 & -t \\ -t & 0 \end{pmatrix}. \tag{3.36}$$

Solving this Hamiltonian equation we can obtain a ground state $|\psi\rangle = \frac{1}{\sqrt{2}}(|\uparrow, 0\rangle + |0, \uparrow\rangle)$ with energy $E = -t$ and an excited state $|\psi\rangle = \frac{1}{\sqrt{2}}(|\uparrow, 0\rangle - |0, \uparrow\rangle)$ with energy $E = t$.

If there are two electrons in the system with different spins, one can write a general state as $|\psi\rangle = a\,|\uparrow\downarrow, 0\rangle + b\,|\uparrow, \downarrow\rangle + c\,|\downarrow, \uparrow\rangle + d\,|0, \uparrow\downarrow\rangle$. In this case the Hubbard Hamiltonian is written as [8]:

$$H = \begin{pmatrix} U & -t & -t & 0 \\ -t & 0 & 0 & -t \\ -t & 0 & 0 & -t \\ 0 & -t & -t & U \end{pmatrix}. \tag{3.37}$$

This Hamiltonian can be diagonalized to get a ground state energy of $E = \frac{U}{2} - \frac{1}{2}(U^2 + 16t^2)^{1/2}$. This ground state is represented by a wave function $|\psi\rangle = N(|\uparrow\downarrow, 0\rangle + W|\uparrow, \downarrow\rangle + W|\downarrow, \uparrow\rangle + |0, \uparrow\downarrow\rangle)$, where N is some normalization constant and $W = \frac{U}{4t} + \frac{1}{4t}(U^2 + 16t^2)^{1/2}$. Thus, the ground state for two electrons is always a singlet. For $U = 0$, the ground state energy is $-2t$, whereas for large-interaction i.e. $U/t \gg 1$ the ground state energy is $-4t^2/U$.

Summarizing the above discussions it can be said that the Hubbard Hamiltonian primarily contains the necessary terms to understand the basic properties of correlated electron systems. Here the kinetic energy term reflects the itinerant features of electrons namely the metallic conductivity, and the potential or interaction operator influences the correlated motion of electrons, and in the limiting case of their localization leading to an insulating state. In spite of its simplifications (somewhat unrealistic) the Hubbard model provides a basic framework to understand metal to Mott insulator transition. In this qualitative sense the Hubbard model serves as the standard model of correlated electron systems [7]. To facilitate further discussions it might be useful to suitably rewrite the Hubbard Hamiltonian as $H = \hat{T} + U\hat{D}$, where $\hat{D} = \sum_i n_{i\uparrow}n_{i\downarrow}$ is the number of double occupancies, and also to introduce some explicit and implicit variables of the Hubbard model [7]:

1. The interaction strength U/t or U/W, where W is the one-electron band width with dispersion relation $E(k)$.
2. The dimension d ($d = 1, 2, 3, \ldots$) and the lattice structure.
3. The hopping matrix elements t_{ij}, with a value of $-t$ if ij represent nearest neighbours in the lattice or zero otherwise.
4. The density of σ electrons ($\sigma = \uparrow, \downarrow$), $n_\sigma = N_\sigma/L$, where $L = N_N$ denotes the number of ion-sites and $N = N_\uparrow + N_\downarrow$ is the total number of electrons.
5. The temperature $k_B T/W$.

In the one band Hubbard model the interaction term includes a dominant local part i.e. the Hubbard interaction $U\hat{D}$. The Hubbard model can be treated exactly in some limiting cases:

Fermi gas limit: The complete solution of the Hubbard Hamiltonian is possible for vanishing interactions i.e. $U = 0$. The remaining kinetic energy operator, $H(U = 0) = \hat{T}$ is diagonal in momentum space, and the model describes properties of a free Fermi gas, in particular an ideal metal.

Atomic limit: In this limiting case, $H(t = 0) = U\hat{D}$ is diagonal in position space, and the model can be solved relatively easily. There can be lattice site occupancies with two electrons (double occupancy), one electron (single occupancy with spin $\uparrow$ or spin $\downarrow$), and zero electron i.e. hole. Only single electron occupancies and holes are present in the ground state for $N \leqslant L$, whereas for $N > L$ only double and single occupancies are allowed. The number of double occupancies can classify the excited states, and a calculation involving the free energy density in the grand canonical ensemble in the presence of an external magnetic field H_0 can lead to the average density of doubly occupied sites, $\bar{d} = \langle \hat{D} \rangle / L$, which obeys a relation [7]:

$$e^{-\frac{U}{k_B T}} = \frac{\bar{d}(1 - n - \bar{d})}{(n\uparrow - \bar{d})(n\downarrow - \bar{d})}. \tag{3.38}$$

This is a law of mass action between doubly occupied sites and holes on one hand and singly occupied sites on the other hand, which is controlled by the ratio of the interaction strength to the temperature [7]. All the excited states and ground state are highly degenerate with respect to the spin and charge degrees of freedom. The lattice sites do not communicate with each other, and as a result the system is an insulator.

Infinite-U limit: This limit of strong interactions $U \gg W$ is different from the atomic limit, in that while in the case of atomic limit there is no communication between the lattice sites, a finite coupling is still possible in the limit $U \rightarrow \infty$. On the other hand, if U is quite small it is expected that the electron band structure is only slightly perturbed from the case of non-interacting electron gas, and the system remains a metal. Thus, there appears to be a critical value of U, where the system is expected to undergo a metal–insulator transition.

The Hubbard model in one-dimension was solved exactly by Lieb and Wu [11]. The result suggests that except for the case when there is one electron per lattice site i.e. half-filling, the solution of the Hubbard model for all U/W at $T = 0$ is a metal with the Fermi surface the same as that for $U = 0$. At 'half filling' the model predicts the existence of the Mott insulator for any value of U. Thus, there will be no Mott transition as a function of U/W. Rather one can say that the critical value of U in one-dimension is $U \rightarrow 0^+$. However, the Mott insulator can be made metallic by either adding or removing a small number of electrons from the Mott insulator. On the other hand in infinite dimensions, the critical value of U is approximately equal to the electron bandwidth and there will be a nontrivial metal–insulator transition [12].

To the lowest order in perturbation theory in the parameter W/U around the limit of infinitely strong interactions, and for $n \leqslant 1$, one obtains in the subspace of no double occupancies (i.e. $\hat{D} = 0$) [7]:

$$H = \sum_{i \neq j, \sigma} t_{ij}(1 - n_{i,-\sigma})c_{i,\sigma}^+ c_{j,\sigma}(1 - n_{j,-\sigma}). \tag{3.39}$$

This model in equation (3.39) is termed the $\hat{T}$-model or 'infinite-U Hubbard model'. For the half filling case ($n_\uparrow + n_\downarrow = n = 1$) and $U = \infty$ this model leads to an insulating ground state. In general the exact solution of this $\hat{T}$-model is not possible [7]. However, there are two exceptions. The first one is a chain in one-dimension, and the second one a single hole in arbitrary dimensions $N = L - 1$. In the first case the electrons do not hop over each other and the system is 2^N-fold spin-degenerate. Since the spin of an electron plays no role, the spectrum of this system is equivalent to a fully ferromagnetically polarized system. In the second case the state with total spin $S = N/2$ is a possibly degenerate ground state of the system. This is known as the Nagaoka state and the system is fully ferromagnetically polarized [7].

The Hubbard model (and its various extensions) in the context of strong electron correlation is a complete subject by itself, and further detailed discussion on it is beyond the scope of this book. Interested readers may like to read the books by Martin, Ceperley and Reining [2], Gebhard [7], Fulde [13] and Freerick [14], and the review article by Imada *et al* [15] for an excellent exposition on this subject. We will now continue with this minimal introduction of the Hubbard model, which will be helpful in the subsequent discussions in the present book.

In the next section and chapter 4 we will see that the Hubbard model at half band filling and strong interactions i.e. large U describes the antiferromagnetic state in magnetic insulators. In fact the super-exchange mechanism in magnetic insulators formulated by P W Anderson [16, 17] can actually be correlated to the Hubbard Hamiltonian.

In summary, the Hubbard model presents a picture of competition in a crystalline solid, between electron delocalization effects coming from the optimization of the kinetic energy and localization effects coming from the potential energy. In the limit where the intrasite Coulomb interaction U becomes very large, no lattice site in the solid can be doubly occupied by electrons. In this situation the solid becomes an insulator if there is on average one electron per lattice site (called a half-filled band). On the other extreme, if U is very small, the electron band structure is only perturbed slightly from the one obtained from the independent electron approximation, and the solid will be a metal. Hence it is natural to expect that there will be a critical value of U, where the solid will undergo a metal–insulator transition. Across this crossover regime, had there been no electron–electron interactions there would be a single band formed from the overlap of the atomic orbitals in the solid, and the band would be full when two electrons (one with spin-up and the other with spin-down) occupied each lattice site. However, now the two electrons sitting on the same site would feel a large Coulomb repulsion. Mott argued that this electron–electron interaction would split the band in two: (i) a lower band formed by the electrons occupying empty sites, and (ii) an upper band formed by electrons that occupied a site already taken by another electron. When there is one electron available per lattice site (half-filled band) the lower band would be full, and the solid will be an insulator.

3.7 Mott–Heisenberg insulator

In the discussion above on the Mott–Hubbard insulator, the possibility of exchange interaction between localized spins has been totally ignored. We now ask the question, in the absence of an external field, which way the spins will align. There are two obvious possibilities for a square or cubic lattice; the spins would either be aligned with their neighbors or they would be anti-aligned. This means there would be either ferromagnetic or antiferromagnetic state. Let us consider the possibility of an antiferromagnetic state (denoted by $|X_0\rangle$) as shown in figure 3.8(a). This state is an eigenstate with zero energy in the absence of any intersite hopping, since there is precisely one electron at each site. It is known that tunneling of electrons to neighboring sites is allowed even for $U \gg W$ and the electron can make a 'virtual' hop to a neighboring site, as shown in figure 3.8(b). So the hopping from one lattice

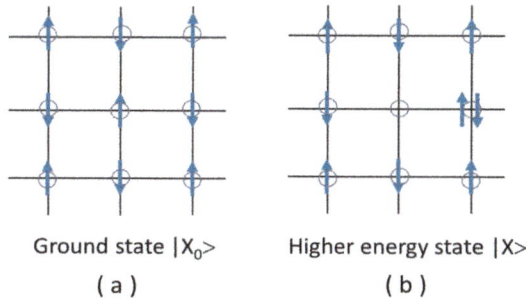

Ground state $|X_0\rangle$ Higher energy state $|X\rangle$

(a)   (b)

Figure 3.8. Schematic representation of the spin configurations in the Mott–Heisenberg antiferromagnetic state. (a) The proposed antiferromagnetic ground state $|X_0\rangle$ in the limit of $U \gg W$; (b) a higher-energy state $|X\rangle$ obtained by electron hopping from one site onto a neighboring site.

site to the next with an amplitude $-t$ can be added perturbatively to the system. The state $|X\rangle$ in figure 3.8(b) is a higher energy state, which will have an energy U in the absence of hopping because of the double electron occupancy. Now, using second-order perturbation theory one can obtain:

$$E = E_{X_0} + \sum_{X_0} \frac{|\langle X_0 | H_{hoppong} | X \rangle|^2}{E_{X_0} - E_X} = E_{X_0} - \frac{Nzt^2}{U}. \tag{3.40}$$

In the above equation the sum is over all $|X\rangle$ states, which can be reached through a single hop from the state $|X_0\rangle$. The number of such terms are counted to be Nz, where z is the number of nearest neighbors and N is the number of lattice sites. Now, the spin state of electrons is conserved during the process of hopping and during a hop the electron spins cannot flip. So, if the spins were all aligned to start with, no virtual hop to the state $|X\rangle$ can take place, because it will violate the Pauli exclusion principle. The double occupancy of electrons with parallel spin is strictly prohibited. So it is concluded that the antiferromagnetic state has a lower energy in a Mott insulator in the limit of large U in comparison to the ferromagnetic state. Thus, the formal perturbation theory in second order around the half-filled ground state with all electrons localized at the lattice sites, provides an antiferromagnetic coupling $J \sim t^2/U$ between the spin-1/2 electrons on neighboring lattice sites [7].

We can summarize the above discussion in the following way. One can think of an electron to be confined at each site by the interaction with its neighbors. The electron cannot make any excursions to neighboring sites in the ferromagnetic case because of the Pauli exclusion principle. In the antiferromagnetic case the excursions are allowed for the electrons, although the energy is higher when the electron hops into neighboring sites. However, in this process electron wavefunction spreads out, which always lowers the energy of the electron.

The energy scales for charge excitations ($\delta\mu \approx U - W$) and spin excitations ($J \sim t^2/U$) are well separated for large interactions $U \gg W$, and one is in the Mott–Hubbard insulating regime. However, in the temperature regime around or below that of the exchange interaction, the spin energy scale cannot be ignored. A thermodynamic phase transition takes place from the Mott–Hubbard insulating state to antiferromagnetic Mott–Heisenberg insulating state at a characteristic temperature $T = T_N$, the Néel temperature. The local magnetic moment, however, does not change appreciably in the transition through the Néel temperature [7]. The antiferromagnetic phase is replaced by a small region of Nagaoka type ferromagnetism for very large values of U [18, 19].

The Mott–Hubbard gap is absent for small interactions below the quantum phase transition at $U_c \approx W$. However, correlations for all $U > 0$ lead to the formation of magnetic moments, and these moments may undergo a long-range magnetic order at the Néel temperature via a thermodynamic phase transition from a correlated paramagnetic metallic state to an antiferromagnetic Mott–Heisenberg insulating state.

The various points discussed above lead to the following formal definition of Mott–Heisenberg insulator [7]: 'The Mott–Heisenberg insulating state is the result of a thermodynamic phase transition in which the pre-formed local (magnetic) moments (antiferromagnetically) order below the critical (Néel) temperature.'

In the present context we shall now revisit the Slater antiferromagnetic insulator introduced earlier in chapter 1, section 1.8 and try to distinguish the same from the 'Mott–Heisenberg' antiferromagnetic insulator. We shall work within the framework of unrestricted Hartree–Fock approximation allowing the best Hartree–Fock wavefunction corresponding to minimum ground state energy to break the symmetries of the Hamiltonian. Assuming that we are in small U limit, we replace the Coulomb interaction term in the Hubbard Hamiltonian (equation (3.35)) by mean-field Coulomb energy as in the Hartree–Fock approximation. This transformed the two-particle Coulomb interaction into an effective single-particle interaction [20].

$$H_U^{HF} = -U \sum_i [n_{i\uparrow}\langle n_{i\downarrow}\rangle + \langle n_{i\uparrow}\rangle n_{i\downarrow} - \langle n_{i\uparrow}\rangle\langle n_{i\downarrow}\rangle]. \tag{3.41}$$

If one is looking for a ferromagnetic solution, then $\langle n_{i\sigma}\rangle = n\sigma = (n/2) + \sigma m$, where $m = (n_\uparrow - n_\downarrow)/2$ and $n = (n_\uparrow + n_\downarrow)$, and equation (3.41) can be suitably rewritten as:

$$H_U^{HF} = U \sum_i \left(-2mS_i^z + m^2 + \frac{n^2}{4}\right) \tag{3.42}$$

where S_i^z is the z component of the spin operator. We now write the Hamiltonian in $\mathbf{k}$ space taking the Bloch states (equation (3.2)) as the one-electron basis. The spin operator S_i^z is then rewritten as [20]:

$$
\begin{aligned}
S_i^z &= \frac{1}{N}\sum_{kk'} e^{i((\mathbf{k}-\mathbf{k}')\cdot\mathbf{R}_i)} \frac{1}{2}\sum_\sigma \sigma c_{k\sigma}^+ c_{\kappa'\sigma} \\
&= \frac{1}{N}\sum_{kk'} e^{i((\mathbf{k}-\mathbf{k}')\cdot\mathbf{R}_i)} S_z(\mathbf{k}, \mathbf{k}')
\end{aligned}
\tag{3.43}
$$

where $\mathbf{R}$ is the lattice vector in $\mathbf{k}$-space.

The mean-field potential energy term H^{HF} possess the same periodicity as the lattice and, hence it does not couple states with different $\mathbf{k}$ vectors. This implies that the Hamiltonian will contain contributions only from $S_z(\mathbf{k}, \mathbf{k})$, and it can be expressed as:

$$H = \sum_k \sum_\sigma E_k n_{k\sigma} + U \sum_k \left[-2mS_i^z(k, k) + m^2 + \frac{n^2}{4} \right].$$ (3.44)

The HF correction due to the Coulomb interaction U will split the bands with opposite spin, and lead to new one-electron eigenvalues [20]:

$$E_{k\sigma} = E_k + \frac{U}{2} - \sigma U m$$ (3.45)

where the chemical potential $\mu = U/2$.

In the limit of small t/U and at half filling the system is a ferromagnetic insulator and $m = 1/2$. The total energy of the ground state is then expressed as:

$$E_{Ferro} = \frac{1}{N} \sum_k (E_k - \mu) = \frac{1}{N} \sum_k \left(E_k - \frac{U}{2} \right) = -\frac{U}{2}.$$ (3.46)

We now replace the same periodic lattice with two sublattices A and B with opposite spins, which will allow a two-sublattice antiferromagnetic state. The Bloch electron states of the original lattice is accordingly replaced:

$$\Psi_{k\sigma}(\mathbf{r}) = \frac{1}{\sqrt{2}} \left[\Psi_{k\sigma}^A(\mathbf{r}) + \Psi_{k\sigma}^B(\mathbf{r}) \right]$$ (3.47)

or,

$$\Psi_{k\sigma}^\alpha(\mathbf{r}) = \frac{1}{\sqrt{N_\alpha}} \sum_{i_\alpha} e^{i\mathbf{k} \cdot \mathbf{R}_i^\alpha} a_{i_\alpha \sigma}(\mathbf{r})$$ (3.48)

where α stands for A and B and $a_{i_\alpha \sigma}(r)$ are Wannier functions.

We choose the two Bloch functions Ψ_k and $\Psi_{Q_2 \sigma}$ as one-electron basis, where $Q_2 = (\pi/a, \pi/a, 0)$ is the vector associated with the antiferromagnetic instability. The Coulomb interaction in the Hartree–Fock approximation is then expressed as:

$$H_U^{HF} = U \sum_{i \in A} \left(-2mS_i^z + m^2 + \frac{n^2}{4} \right) + U \sum_{i \in B} \left(2mS_i^z + m^2 + \frac{n^2}{4} \right).$$ (3.49)

This interaction leads to the coupling of Bloch electron states with $\mathbf{k}$ vectors, which are made equivalent by the folding of the Brillouin zone. The Hartree–Fock Hamiltonian thus takes the form:

$$H = \sum_k \sum_\sigma E_k n_{k\sigma} + \sum_k \sum_\sigma E_{k+Q_2} n_{k+Q_2 \sigma}$$
$$+ U \sum_k \left[-2mS_i^z(k + Q_2, k) + 2m^2 + 2\frac{n^2}{4} \right].$$ (3.50)

With the restriction of the sum over k to the Brillouin zone of the antiferromagnetic lattice, this Hamiltonian leads to the two-fold degenerate eigenvalues:

$$E_{k\pm} - \mu = \frac{1}{2}(E_k + E_{k+Q_2}) \pm \frac{1}{2}\sqrt{(E_k - E_{k+Q_2})^2 + 4(mU)^2} \qquad (3.51)$$

where μ is the chemical potential. A gap will open up at the point of the original Brillouin zone, where the energy bands E_k and E_{k+Q_2} cross. The Fermi level also crosses the bands at this point for half filling and for $mU = 0$. Thus, the system is an insulator for any finite value of mU. In the limit of small t/U and assuming $m = 1/2$, one can expand the energy eigenvalues in the powers of E_k/U. Thus, one can find for the occupied states

$$E_k - \mu \sim -\frac{1}{2}U - \frac{E_k^2}{U} = -\frac{1}{2}U - \frac{4t^2}{U}\left(\frac{E_k}{2t}\right)^2. \qquad (3.52)$$

The total energy of the ground-state for the antiferromagnetic supercell is then expressed as $2E_{Antiferro}$ with:

$$E_{Antiferro} = -\frac{1}{2}U - \frac{4t^2}{U}\frac{1}{N}\sum_k\left(\frac{E_k}{2t}\right)^2 \sim -\frac{1}{2}U - \frac{4t^2}{U}. \qquad (3.53)$$

From equations (3.46) and (3.53) we can see that the antiferromagnetic insulating state has a lower energy than the ferromagnetic state and the energy difference per pair of spins between ferromagnetic and antiferromagnetic state is of the order of $\frac{4t^2}{U}$. This is quite similar to the result obtained earlier from the Hubbard model using second order perturbation theory. However, it is quite clear that in the Slater antiferromagnetic insulator obtained within unrestricted Hartree–Fock approximations is a self-consistently stabilized electron-band type insulator, where the insulating state arises out of the electron exchange effect due to the Pauli exclusion principle. The onset of a long-range antiferromagnetic order is a necessary condition for the existence of the Slater insulating state, or in other words the antiferromagnetism comes first, and then the insulating behaviour. In addition the local magnetic moments are absent in the Slater insulator in the temperature regime above the antiferromagnetic transition temperature or Néel temperature. In contrast. the antiferromagnetic Mott insulating state is clearly driven by electron correlation and at large U the Hubbard model reduces to an effective Heisenberg model with localized moments from the outset. The transition to the paramagnetic state from the Mott–Heisenberg antiferromagnetic insulator can either be to a correlated metallic state or to the Mott–Hubbard insulating state. Thus, the concept of the 'Mott–Heisenberg insulator' provides a clear distinction between the ideas of Hartree–Fock–Slater self-consistent single-electron theory and the Mott physics of many-electron correlations. Experimental studies involving x-ray absorption spectroscopy (XAS), resonant elastic x-ray scattering (REXS) and optical conductivity have been used successfully to distinguish between Mott versus Slater-type metal–insulator transition. We shall discuss some such experimental results in chapter 6.

3.8 Multi-band Mott insulator

In the discussion so far on the Mott insulating state, the starting point was a single band of s electrons with variable distance between their energy orbitals. Such s electrons in most real systems alway form broad bands under ambient conditions. In this situation the mutual Coulomb interaction remains well screened, and hence the correlation effects in the s band solids are usually almost negligible. In the case of transition metal atoms like Fe, Co, Ni, Cu, V etc the physics is expected to be more interesting because of the additional presence of d electrons. The bandwidth of d-electrons in transition metals under ambient conditions is significantly smaller than that of s bands in alkali metals, and the correlation effects in such narrow d-bands are expected to be more important.

The identification of the Hubbard parameter U for transition metals is relatively more difficult, as there are other interactions not included in the framework based on a single band Hubbard U [7]. The d levels are degenerate and there are: (i) Hund's rule couplings present in the atom tending to maximize the total spin magnetic moment; (ii) crystal field effects in the solid tending to lift the five-fold degeneracy of the atomic d levels. In addition, the local Hubbard U is considerably smaller than that estimated for s-band electrons from the difference between the ionization energy and the electron affinity (see equation (3.32)). The later calculation of the Hubbard U involved processes where an electron was transferred to infinity or an electron was added from infinity. In contrast in a d-band material one may just deposit electrons on an energetically close s-level [7]:

$$
\begin{aligned}
U &= E(3d^{n-1}4s^2) - E(3d^n4s^1) - |E(3d^{n+1}4s^0) - E(3d^n4s^1)| \\
&= E(3d^{n-1}4s^2) - E(3d^{n+1}4s^0) - 2E(3d^n4s^1).
\end{aligned}
\tag{3.54}
$$

This mechanism considerably reduces the Hubbard U. The appearance of a local charge disturbance in a solid is partially screened by surrounding electrons. An estimate of the screening effect for transition metals shows that $U \approx 2$ eV is at most half the bandwidth [7]. In comparison to alkali metals the bandwidth W is much smaller in transition metals, but the Coulomb interaction is too well screened to give rise to an insulating ground state in the transition elements. The pronounced magnetic properties of the transition metals, however, indicates that electron correlations indeed play an important role in these metals.

In order to reduce the screening effect by electrons, the $4s$ electrons (which are in wide bands) in the transition metals are needed to be bound chemically. This condition is satisfied in various transition metal-oxide compounds by oxidation. This can be visualized within the framework of band structure as a hybridization of the $2p$ levels of the oxygen atoms with the $4s$ bands of the transition-metal atoms. The mainly oxygen-like bonding bands lie below the Fermi energy and are occupied, whereas the mostly $4s$-like antibonding bands are shifted above the Fermi energy and are unoccupied [7]. As a result the $4s$ bands become much less effective in screening Coulomb interaction of the $3d$ electrons. It has been observed experimentally that most of the transition-metal oxides are indeed insulators rather than metals, the prominent examples being manganese oxide (MnO) and nickel oxide (NiO). We will discuss the experimental results on Mott insulating states in more detail in chapters 4 and 6.

Electronic band structure calculations predicts that all transition metal oxides would be metallic, since the 3d bands are partially filled. The experimentally observed insulating ground state in transition-metal oxides, which is in contradiction with such predictions, provides a strong evidence of the correlation effects in these materials [7]. At this stage it is also worth noting that some transition metal oxides like calcium oxide (CaO) and titanium oxide (TiO) are still metallic. While it is now quite clear that electronic band structure calculations cannot provide a complete explanation of the electronic properties of transition metal oxide compounds, it may still provide some results, which may be useful in estimating the importance of screening in such compounds. For example, in CaO and TiO the antibonding 4s bands and 3d bands are predicted to be overlapping strongly and cross the Fermi energy [7]. The band structure calculation will indicate that the Hubbard interaction is screened well enough in CaO and TiO so that the metallic behavior is retained even after taking electron correlations into account. On the other hand, the 4s bands are split off the 3d bands for the transition-metal oxides from Mn through Cu. In these compounds, the Coulomb interaction in the narrow 3d bands becomes quite important and leads to the Mott-insulating state.

3.9 Charge transfer insulator

The band structure calculations for transition metal oxides mentioned above does not take into account of the Coulomb interactions, which may lead to splitting of the 3d bands into upper and lower Hubbard bands. The Mott–Hubbard insulator is expected in the systems where the oxygen p-band lies under the lower Hubbard band, as shown in figure 3.9(a). However, the oxygen p bands in some of these compounds could energetically be placed between the two Hubbard bands; this effect is not incorporated in standard band structure calculations [7]. In such a case the energy for a charge excitation is determined by the charge-transfer energy Δ, rather than the Hubbard U [21]. The transport of an electron from a strongly correlated transition-metal d level in one atom to another transition-metal atom requires the Hubbard energy:

$$U = E(d^{n+1}) + E(d^{n-1}) - 2E(d^n). \tag{3.55}$$

On the other hand, to shift the same d electron to an oxygen p level one will need the charge transfer energy:

$$\Delta = E(d^{n-1}) - E(d^n) + E(p^{n'+1}) - E(p^{n'}). \tag{3.56}$$

In the situation where $U > \Delta$, the energy gap for charge excitations will be determined by the charge-transfer energy Δ, and not by the on-site interaction Hubbard U. The insulator is then termed as a charge-transfer insulator, instead of a (multi-band) Mott–Hubbard insulator. Figure 3.9(b) shows a schematic representation of charge-transfer insulator. We will discuss the current experimental situations

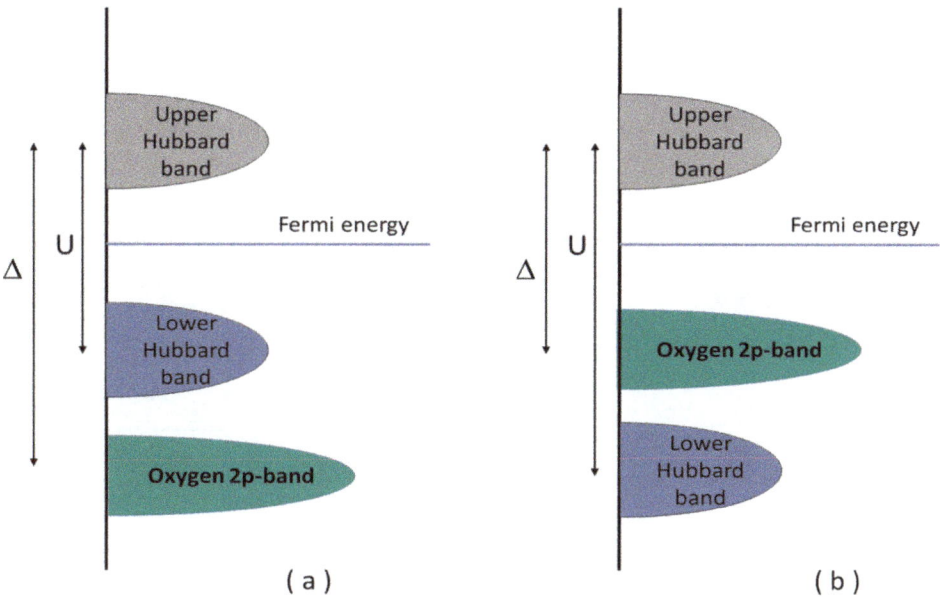

Figure 3.9. Schematic representation of the (a) Mott–Hubbard insulator and (b) charge-transfer insulator possible in transition metal oxides.

in connection with Mott–Hubbard insulators versus charge-transfer insulators in some detail in chapter 4.

3.10 Comparison between band-insulator, Mott-insulator and charge transfer insulator

A band insulating state arises due to the interaction of electrons with the periodically placed ionic-potential in the solid. In conventional electron band insulators and semiconductors all bands (valence band) lying below the Fermi energy are filled at $T = 0$, and the band above (conduction band) separated by the energy gap is empty. If an energy corresponding to the band gap is provided to the solid, electrons are excited from the valence band to the conduction band and are delocalized, hence can contribute to the electrical conductivity of the material. Simultaneously an empty state (or hole) is created in the valence band, and the hole as well as the electron placed in the valence band can move freely through the solid.

In contrast, in a Mott–Hubbard insulator, the insulating state arises because the hopping of electron from one site to another is inhibited by intra- and inter-orbital Coulomb repulsions. When a charge carrier is excited across the Mott–Hubbard gap, both electrons and holes are supposed to contribute to the electrical conductivity of the solid. However, the effective mass of the excited electrons and holes are larger in such correlated systems because the excited state is localized too, hence the contribution of the electron to electrical conductivity of the solid is inhibited.

A band insulator and a Mott insulator can be differentiated through spectral weight transfer, a phenomenon ubiquitous to Mott-insulators. The spectral weight is

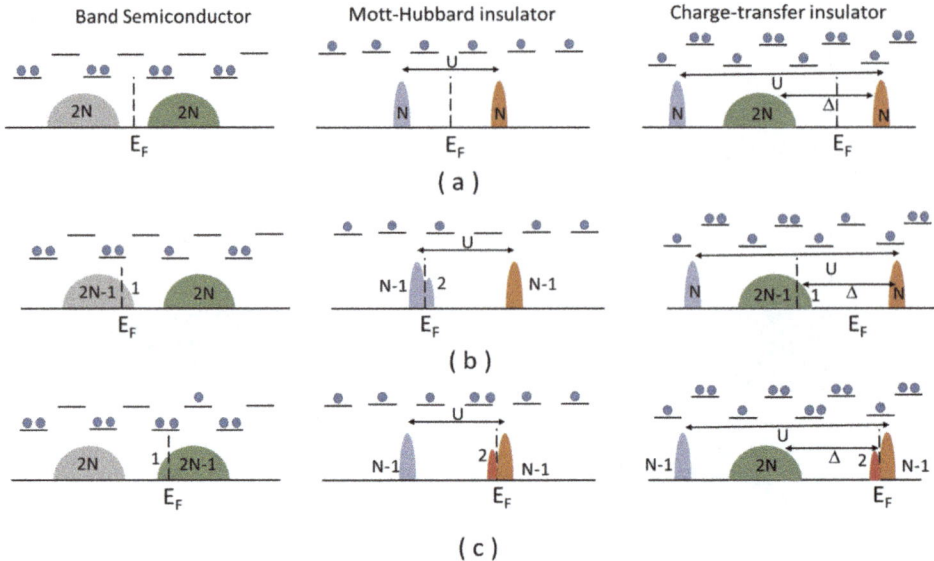

Figure 3.10. A schematic drawing of the electron-removal and electron-addition spectra for a band semi-conductor (left), a Mott–Hubbard insulator in the localized limit (middle) and a charge transfer insulator in the localized limit (right). (a) Undoped (half filling), (b) one-hole doped, and (c) one-electron doped. The bars just above the figures represent the sites and the dots represent the electrons. (Adapted figure with permission from [22]. Copyright 1993 by the American Physical Society.)

given by the spectral function $A(k, w)$, which for an electron in a crystal is given by $A(k, w) = -2 \operatorname{Im} G(k, w)$, where $G(k, w)$ is the retarded Green function describing the motion of the electron (see appendix D). A detailed comparison of the spectral weight transfer in band insulator (semi conductor), Mott–Hubbard insulator and charge transfer insulator has been provided by Meinders *et al* [22]. We will reproduce below the physical arguments, and the interested readers can see the original paper by Meinders *et al* [22] for the mathematical details. The spectral weight transfer can be studied experimentally by total photoelectron and inverse photoelectron spectroscopy.

We start with a band insulator (undoped semiconductor) with a filled valence band and an empty conduction band separated by an energy gap E_G, which can be described within the framework of independent electrons. The total electron removal and addition spectrum for such a band insulator is shown in the left of figure 3.10(a). If there are a total number of N lattice sites in the solid, then there will be $2N$ occupied states and $2N$ unoccupied states, separated by the energy gap E_G. In the event of doping the semiconductor with the addition of one hole, the chemical potential will shift into the former occupied band, provided one can neglect the impurity potential of the dopant. The total electron removal spectral weight will be $2N - 1$ (just the number of electrons in the ground state) and the total electron addition spectral weight will be $2N + 1$ (total number of holes in the ground state) [22]. The electron addition spectrum may be divided into two parts, a high energy-scale part (the conduction band) and a low energy-scale part, which is the unoccupied part of the valence band (see figure 3.10(a)). The low energy spectral weight equals 1 [22]. The same arguments follow for an electron-doped semiconductor. The total spectrum is just given by a repositioning of the Fermi energy (chemical potential) and the low energy spectral weight grows with the amount of doping, while at the same time the spectral weight of the high-energy band is not changed. There is no redistribution of spectral intensities upon doping a simple band semiconductor [22].

The middle part of figure 3.10(b) shows schematically the total photoelectron and inverse photoelectron spectrum of a Mott–Hubbard insulator at half filling. The total electron-removal spectral weight is equal to the number of occupied levels, while the total electron-addition spectral weight is equal to the number of unoccupied levels [22]. Each therefore has an intensity equal to N. There will be $N - 1$ singly occupied sites on doping the Mott–Hubbard insulator with one hole, so the total electron removal spectral weight will be $N - 1$. On the other hand, for electron addition there are $N - 1$ ways for adding the electron to a site that was already occupied. The intensity of the upper Hubbard band will be $N - 1$, and not N. For the empty site which is left, there are two ways of adding an electron (spin up, spin down), both belonging to the lower Hubbard band. There are now $N - 1$ electron removal states near the Fermi-level, two electron addition states near the Fermi level and $N - 1$ electron addition states in the upper Hubbard band. The same arguments follow for the case of electron doping. A doping concentration x thus yields a low energy spectral weight of $2x$ and a high energy spectral weight of $1 - x$; Nx states transferred from high to low energy [22].

The electron removal and addition spectrum for a charge transfer insulator in the localized limit (without any hybridization among the correlated bands and the in-between un-correlated band) is shown in the extreme right of figure 3.10(a). In the case of doping a charge transfer solid with electrons, the situation is similar to the Mott–Hubbard case with a transfer of spectral weight from high to low energy. On the other hand, with hole doping the situation is quite similar to that of the semiconductor without any transfer of spectral weight. Thus there is a fundamental asymmetry between hole and electron doping for a charge transfer solid in the localized limit [22].

3.11 Mott–Anderson insulator

In many real systems, both lattice disorder and the electron–electron interactions may have comparable strengths. In such cases, the mechanisms of electron localization proposed by Mott (discussed above in this chapter) and Anderson (discussed in section 1.11) cannot be treated separately; they will influence each other. To this end one may start with a Mott–Hubbard insulator in the strongly localized limit. Each lattice site will have two energy levels, $E_0 = 0$ and $E_1 = U$ in the absence of any lattice disorder. Each lattice site is singly occupied when the system is half filled, and the level E_1 will remain empty. There will be one local magnetic moment at each site, and a gap of U to charge excitations.

When disorder is introduced to the lattice, each of the energy levels is shifted by a randomly fluctuating lattice potential $-W/2 < E_i < W/2$. Normal metals do not display an Anderson metal-to-insulator transition since their bandwidth W is large in comparison to the fluctuations of the lattice potential. For the half filling case the chemical potential is $\mu = E_F = U/2$. There will be no change in the system in the case of $W < U$, since all the excited levels $E_1'(i) = U + E_i$ remain empty. For strong disorder, the Anderson transition will be significantly modified by the Coulomb interaction between the charge carriers and the charged impurities, and among themselves. Those lattice sites with $E_i > U/2$ have the level $E_0(i) = 0 + E_i > \mu$, which will remain empty. On the other hand, those sites with $E_i < -U/2$ have the excited level $E_1'(i) = 0 + E_i < \mu$ and will be doubly occupied. So for $W > U$ a fraction of the lattice sites are either doubly occupied or empty. The Mott gap is now closed, but a fraction of the sites will continue to have localized magnetic moments. This state is an inhomogeneous mixture of a Mott and an Anderson insulator. It may be worth pointing out here that apart from being a perfect insulator, another important characteristic property of the Mott insulator is its incompressibility. The charge compressibility measures the change in average electronic density due to changes in chemical potential μ. The compressibility is generally non-zero in the Anderson insulator, because of the incoherent states localized around impurities, which contribute to the compressibility.

3.12 Relativistic Mott insulator

Our discussion has been so far mostly keeping the $3d$-electron systems in mind, where the typical energy scale of Coulomb electron repulsion energy U ranges from 4 to 7 eV and kinetic energy or hopping integral t ranges from 1–2 eV [15]. So large U ($U > t$ or W) could easily drive many of such systems into a Mott-insulating state. Moving down the periodic table from the $3d$ to $4d$ to the $5d$ series, the d orbitals become more extended, which would tend to increase the bandwidth W and decrease the Coulomb electron repulsion U. Indeed the typical W values in $5d$ transition metal systems range from 3 to 4 eV, while the U value lies in the range of 0.5–2 eV. In such systems Coulomb repulsion does not have the necessary strength to open a Mott–Hubbard gap (since $U/W < 1$), and most $5d$ transition metal compounds are expected to be metallic within both the simple Mott–Hubbard picture and conventional one-electron band theory.

In recent times much interest has been generated in the non-trivial physics arising out of strong spin–orbit coupling, especially with the theoretical proposal of topological insulators in early 2000 and the subsequent experimental findings (see [23, 24]). Spin–orbit coupling (SOC) is a relativistic effect providing an interaction between the electron spin and electron angular momentum in atoms. In a solid, in usual discussion on electrons, SOC is at best considered as small perturbation. SOC may not, however, be small perturbations in solids with heavy elements, and indeed it increases in proportional to the fourth power of atomic number leading to interesting qualitative effects. One such example being of course the topological insulators, where the interest so far has been in the class of solids with heavy s and p

electron elements such as Bi, Pb, Sb, Hg, and Te. These materials are now already well known with their topologically protected Dirac-like surface states, and currently are under intense investigations for further interesting phenomena.

Lying between two limits, one of strong electron correlation (i.e. Mott insulators) and the other of strong spin–orbit coupling (i.e. topological insulators), the 5d-electron systems are in an intermediate interaction regime showing an interesting interplay between electron correlation and spin–orbit coupling. As we have mentioned above that the d orbitals in 5d-compounds become more extended and tend to reduce the Coulomb electronic repulsion U and hence diminish electron correlation effects. At the same time the SOC in these materials increases quite significantly. This leads to enhanced splittings between otherwise degenerate or almost degenerate orbitals and bands, and thus reduction of the kinetic energy in many cases. The last effect can offset the reduction in Coulomb electron repulsion energy U, and allow correlation physics back to the forefront. This combination of strong spin–orbit coupling and electron correlation has opened up a new field of *relativistic Mott-insulators* [25–27].

The first identified relativistic Mott insulator is an iridium-based 5d transition metal compound Sr_2IrO_4 [25]. The insulating mechanism in such relativistic materials originates from the splitting of t_{2g} states into SOC-induced $J_{eff} = L_{eff} + S$ states. Without consideration of the SOC as in the case of 3d transition metal compounds the d-orbital states split into e_g and t_{2g} states under octahedral crystal field [28]. In this situation 5d electrons in Ir^{4+} ions of relativistic Mott-insulator Sr_2IrO_4 would partially fill the t_{2g} states. This results in a metallic phase as shown in figure 3.11(a). However, under the influence of strong SOC the t_{2g} states split into effective total angular momentum $J_{eff} = 1/2$ and 3/2 states. The energy separation between these states is ζ_{SO} (see figure 3.11(b)). The $J_{eff} = 3/2$ states are completely filled, but $J_{eff} = 1/2$ remain half-filled. So a metallic phase is still expected. To explain the insulating behavior of 5d Mott-insulators, the Coulomb electron repulsion energy U needs to be taken into account. The effect of the

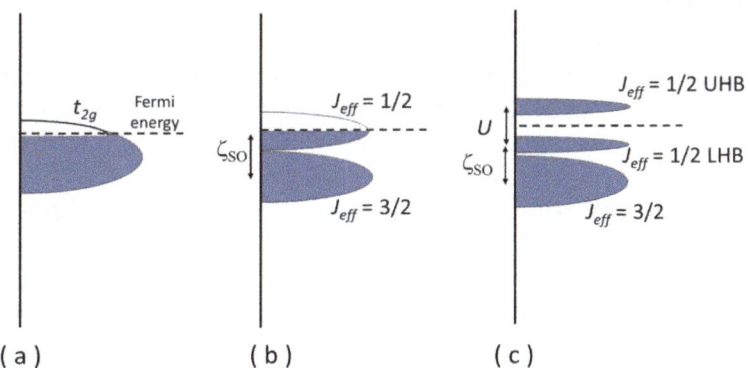

Figure 3.11. Schematic diagrams showing electron energy for the $5d^5$ configuration of the prototype relativistic Mott-insulator Sr_2IrO_4 (a) with only crystal field, (b) with crystal field and spin–orbit coupling, and (c) with a combination of crystal field, spin–orbit coupling and Coulomb electron repulsion energy U. (Adapted from [27] with the permission of John Wiley & Sons.)

Coulomb interaction U gets enhanced due to the formation of this narrow $J_{eff} = 1/2$ band, and this opens up a Mott gap even with a relatively small value of U (see figure 3.11(c)).

3.13 Excitonic insulator

We recall here that in a Bloch–Wilson or band insulator, an energy gap between the lowest conduction band and the highest valence band arises due to the interaction between electrons and the periodic ion potential. A band insulating state is possible only in solids with an even number of valence electrons per unit cell. However, semi metals with a small overlap between the conduction and the valence band are susceptible to a metal to band-insulator transition by the application of an external pressure. Semi-metallic ytterbium indeed shows such signature of metal to band insulator transition under external pressure. The pressure increases the mixing of the $4f$ and $5d$ shells, and ultimately leads to the opening of a hybridization gap in this isostructural band crossing transition [7]. Another interesting phenomenon may sometime occur at very low temperatures in the vicinity of the critical pressure, when the electron–ion interaction no longer dominates all other interactions. When the gap is relatively small, electron–hole pairs may spontaneously form and bind into excitons since their binding energy overcomes the single-electron gap. This is known as excitonic insulator, and it persists to a region for which the single electron band theory would predict a metallic behavior [7]. This transition from a metallic state to an excitonic insulator is often referred to as the 'Mott transition', since it was first predicted by Mott [29]. The driving force behind this metal–insulator transition is the closing of the band gap, and actual metal-band insulator transition is modified by the electron–electron interaction. This is, however, quite different from the Mott insulating state (discussed in the earlier sections) arising due to pure electron–electron interaction.

Further reading

1. Mott N F 1994 *Metal Insulator Transition* (London: Taylor and Francis).
2. Madelung O 1978 *Introduction to Solid State Theory* (Berlin: Springer).
3. Martin R M, Reining L and Ceperley D M 2016 *Interacting Electrons: Theory and Computational Approaches* (Cambridge: Cambridge University Press).
4. Gebhard F 1998 *Mott Metal–insulator Transition* (Berlin: Springer).
5. Fulde P 1995 *Electron Correlations in Molecules and Solids* (Berlin: Springer).
6. Freerick J K 2006 *Transport in Multi Layered Nanostructures: The Dynamical Mean-field Theory Approach* (London: Imperial College Press).
7. Lancaster T and Blundell S J 2014 *Quantum Field Theory for the Gifted Amateur* (Oxford: Oxford University Press).
8. Imada M, Fujimori A and Tokura Y 1998 *Rev. Mod. Phys.* **70** 1039.

Self-assessment questions and exercises

1. Both the electron band insulator and Mott-insulator are insulating in nature. Do the electrons as charge carriers behave in the same manner in both these classes of materials? If not, what is the essential difference?

2. Estimate critical electron density in a correlated electron gas to enable it to become a Mott insulator.

3. Among the transition metal monoxides, CaO and TiO are metallic in nature. Based on the arguments of electron kinetic and potential energies, can it be understood why these $3d$-oxide systems are not Mott insulators and behave in a very different manner?

4. What is the essential difference between a Wigner–Verwey insulator and canonical Mott–Hubbard insulator in crystalline solids?

5. Consider a single site Hubbard model, which is a collection of independent sites i.e. hopping amplitude $t = 0$. Write down the different possibilities of the occupation of the lattice site. Show that the average occupation is given by $\langle n \rangle = 2 \times \frac{e^{\beta\mu} + e^{2\beta\mu - \beta U}}{1 + 2e^{\beta\mu} + e^{2\beta\mu - \beta U}}$. Here μ is the chemical potential and U is the electron interaction energy.

6. In continuation of the previous exercise, it is quite instructive to plot the average occupation $\langle n \rangle$ as a function of chemical potential μ for various temperatures $T = 2, 0.5$ and 0.25, while keeping U fixed with $U = 4$. At $T = 0.25$, check what happens to the average occupancy when μ varies in the range extending from $\mu = -U/2$ and $\mu = +U/2$. How will you interpret the results physically?

7. What is the essential difference between a Mott–Heisenberg insulator and a Slater insulator?

8. How does one differentiate between an electron band insulator, a Mott–Hubbard insulator and a charge transfer insulator, experimentally as well as theoretically?

9. The d-electron band $5d$-oxide system is relatively wide in comparison to $3d$-oxide system, hence electron correlation is not expected to play a dominant role. In spite of this, some of the $5d$-oxide systems have now been identified as Mott-insulators. What is the mechanism of the Mott-insulating behaviour in such oxide systems?

References

[1] Madelung O 1978 *Introduction to Solid State Theory* (Berlin: Springer)
[2] Martin R M, Reining L and Ceperley D M 2016 *Interacting Electrons: Theory and Computational Approaches* (Cambridge: Cambridge University Press)
[3] Mott N F 1949 *Proc. Phys. Soc.* A **62** 416
[4] Ibach H and Lüth H 2003 *Solid State Physics* 3rd edn (Berlin: Springer)
[5] Mott N F 1994 *Metal Insulator Transition* (London: Taylor and Francis)
[6] Verwey E J W 1939 *Nature* **144** 327
[7] Gebhard F 1998 *Mott Metal-insulator Transition* (Berlin: Springer)

[8] Lancaster T and Blundell S J 2014 *Quantum Field Theory for the Gifted Amateur* (Oxford: Oxford University Press)

[9] Hubbard J 1963 *Proc. R. Soc. Lond. Ser.* A **276** 238

[10] Hubbard J 1964 *Proc. R. Soc. Lond. Ser.* A **281** 401

[11] Lieb E and Wu F C 1968 *Phys. Rev. Lett.* **20** 1445

[12] Bulla R 1999 *Phys. Rev. Lett.* **83** 136

[13] Fulde P 1995 *Electron Correlations in Molecules and Solids* (Berlin: Springer)

[14] Freerick J K 2006 *Transport in Multi Layered Nanostructures: The Dynamical Mean-field Theory Approach* (London: Imperial College Press)

[15] Imada M, Fujimori A and Tokura Y 1998 *Rev. Mod. Phys.* **70** 1039

[16] Anderson P W 1959 *Phys. Rev.* **115** 2

[17] Anderson P W 1963 *Magnetism* ed G T Rado and H Suhl (New York: Academic)

[18] Nagaoka Y 1966 *Phys. Rev.* **147** 392

[19] Obermeier T, Pruschke T and Keller J 1997 *Phys. Rev.* B **56** R8479

[20] Pavarini E 2015 *Many-Body Physics: From Kondo to Hubbard Modeling and Simulation* vol 5 ed E Pavarini, E Koch and P Coleman (Julich: Forschungszentrum Julich)

[21] Zaanen J, Sawatzky G A and Allen J W 1985 *Phys. Rev. Lett.* **55** 418

[22] Meinders M B J, Eskes H and Sawatzky G A 1993 *Phys. Rev.* B **48** 3916

[23] Hasan M Z and Kane C L 2010 *Rev. Mod. Phys.* **82** 3045

[24] Qi X-L and Zhang S-C 2011 *Rev. Mod. Phys.* **83** 1057

[25] Kim B J *et al* 2008 *Phys. Rev. Lett.* **101** 076402

[26] Witczak-Krempa W, Chen G, Kim Y B and Balents L 2014 *Annu. Rev. Condens. Matter Phys.* **5** 57

[27] Kim S Y, Lee M-C, Han G, Kratochvilova M, Yun S, Moon S J, Sohn C, Park J-G, Kim C and Noh T W 2018 *Adv. Mater.* **30** e1704777

[28] Khomskii D 2014 *Transition Metal Compounds* (Cambridge: Cambridge University Press)

[29] Mott N F 1984 *Rep. Prog. Phys.* **47** 909

IOP Publishing

Mott Insulators
Physics and applications
Sindhunil Barman Roy

Chapter 4

Mott physics and magnetic insulators

Classically, magnetic moment μ arises from an electrical current density $\mathbf{j}(r)$:

$$\mu = \frac{1}{2} \int \mathbf{r} \times \mathbf{j} d^3 r. \tag{4.1}$$

Such classical moments interact via dipole–dipole interaction. This mechanism, however, cannot explain magnetism of real materials, because according to the Bohr–van Leuween theorem charges cannot flow in the classical systems in thermodynamic equilibrium [1]. In quantum mechanics, however, an electron in state $\psi(\mathbf{r})$ can be correlated to a momentum current contribution:

$$\mathbf{j}(r) = \frac{-ie\hbar}{2m_e} (\psi^*(\mathbf{r})\nabla\psi(\mathbf{r}) - \psi(\mathbf{r})\nabla\psi^*(\mathbf{r})). \tag{4.2}$$

This current density can be non-vanishing for a complex wavefunction ψ and gives rise to a magnetic moment, which is proportional to the expectation value of the angular momentum $\mathbf{L}$:

$$\mu_L = -\frac{e\hbar}{2m_e}\langle \mathbf{L} \rangle = -\mu_B \langle \mathbf{L} \rangle. \tag{4.3}$$

The constant of proportionality μ_B is termed as Bohr magneton, and an atomic orbital has a magnetic moment $\mu = -\mu_B m \hat{z}$ proportional to the magnetic quantum number m. In addition, the electron spin $\mathbf{S}$ carries a magnetic moment:

$$\mu_s = -\mu_B g_e \langle \mathbf{S} \rangle. \tag{4.4}$$

Thus, the atomic moments are of the order of Bohr magneton number μ_B, and for two such atomic moments at a distance of 1 Å the magnetostatic energy is of the order of 0.05 meV. Such energy corresponds to a temperature of less than 1 K, which is in contrast to the magnetic ordering temperature above 500 K observed in many materials. Hence, magnetism in real materials must arise from an interaction, which is very different from the magnetostatic interaction of dipoles. We will see below that the Pauli exclusion principle in combination with the Coulomb repulsion between electrons and the hopping of electrons between lattice sites in a solid are responsible for the interaction of magnetic moments in solid materials.

The microscopic theory of magnetism is quantum mechanical in origin. In well known elemental ferromagnetic solids like iron (Fe), nickel (Ni), and cobalt (Co), the microscopic magnetic moments arise from the localized $3d$ electrons of Fe, Co, and Ni atoms or ions at the periodic lattice sites. These microscopic magnetic moments become aligned in the same direction i.e. orders ferromagnetically below the ferromagnetic transition temperature or Curie temperature. In 1928 Heisenberg introduced a quantum theory of ferromagnetism [2], where he proposed the presence of a ferromagnetic exchange interaction in the form of $-2J_{ij}\mathbf{S}_i \cdot \mathbf{S}_j$. Here $(J > 0)$ is the exchange integral, and $\mathbf{S}_i$ and $\mathbf{S}_j$ represent the magnetic moments or spins at different lattice sites. Dirac actually gave an explicit proof for a simple case of electrons each confined to a different specified orthogonal orbital [3, 4]. This exchange interaction leads to a ferromagnetic alignment of spins, when the wave functions are orthogonal. There are two types of interactions between two electrons: (i) repulsive Coulomb interaction, and (ii) an exchange interaction, which arises as a result of the Pauli exclusion principle. The phenomenological Weiss molecular field generated in ferromagnets was correlated to this exchange interaction. This forms the basis of the Heisenberg model for the magnetism of localized spins. The relevance of the Heisenberg model expressed in the form $\mathbf{S}_i \cdot \mathbf{S}_j$ to the applications in magnetism had been first realized by Van Vleck [4, 5].

The importance of the exchange interaction had come to the forefront in the context of chemical bonding by Heitler and London [6] even before the introduction of Heisenberg theory. Heitler–London theory proposed that this interaction is the cause of covalent bonding of hydrogen molecules. There is an attractive Coulomb interaction between the electron of one hydrogen atom and the nucleus of the other atom along with a Coulomb repulsion between the two electrons. The non-zero overlap integral between two wave functions leads to a negative exchange integral J and a spin singlet ground state.

The applicability of the Heisenberg model in metals with their itinerant electron properties still remains a matter of debate. On the other hand in insulators (and in semiconductors) the spins and magnetic moments are localized, hence they are expected to be described by the Heisenberg model at least at the phenomenological level. Coming back to the metals, pioneering neutron scattering experiments carried out at Atomic Energy Establishment, Harwell, England in late 1950s and early 1960s showed that the spin-wave behaviour in some ferromagnetic metals like pure iron was very similar to that in various insulating magnets. Even in the case of insulators it was soon realized and accepted in the scientific community that this Heisenberg mechanism was very seldom the sole cause of ferromagnetism [4]. This is because in insulators the emerging results revealed that the interaction was almost always antiferromagnetic in nature. It was first suggested by Néel that the negative sign of exchange interaction could lead to a magnetic ground state where different sublattices of the spins in a crystalline solid could align themselves in an antiparallel arrangement [7]. This magnetic state was later named as antiferromagnetism [4, 8]. In this backdrop and being inspired by the works of Mott and Anderson, theorists started thinking that metals with their gas of electrons, and magnetic insulators with their magnetic moments at fixed lattice sites, were actually two extremes of the same thing. John Hubbard, who was based in Harwell in that period made this theoretical framework formal and mathematically concrete and introduced the now famous Hubbard model (see chapter 3). Hubbard later wrote that in formulating the Hubbard Hamiltonian he 'set up the simplest possible model containing the necessary ingredients' to describe the behaviour of correlated materials [9].

The field of magnetic insulators received strong impetus after 1940s due two experimental advances [4]:

1. The development of technically important insulating ferromagnetic materials: the ferrites and the iron-garnets. These materials are found to be *ferrimagnetic* in nature. The name *ferrimagnet* was coined by Néel to describe materials with antiparallel sublattices of spins with unequal magnetic moments, which results into a net resultant magnetic moment. Thus the predominantly antiferromagnetic exchange effect in insulators gives rise to a net ferromagnetic moment.

2. The neutron diffraction technique was recognized as an important experimental tool for observing microscopic magnetic structure i.e. ferro-, ferri- and antiferromagnetic structure in solid materials. In a pioneering experiment in 1949, Shull and Smart [10] studied the magnetic structure of transition metal monoxide MnO with neutron diffraction and demonstrated that MnO is an antiferromagnet. The nearest-neighbor Mn^{2+} ions in MnO interact via an intervening O^{2-}, and this interaction is antiferromagnetic in nature. This interaction is quite distinct from the direct exchange interaction in the Heisenberg model, and is known as superexchange.

In order to understand the magnetism of insulating solids, we may start with an interesting question: how could a magnetic moment be permanently fixed in a given region in such materials? The answer to this question can be found in the original idea of Mott that these insulating solids were magnetic because an additional energy U is required to create an ionized electronic excitation, which is the energy necessary to change the configurations of two distant atoms from $d^n + d^n$ to $d^{n-1} + d^{n+1}$ [11]. This extra energy U is the Coulomb interaction energy between two electrons at the same lattice site, which can be quite large. This was exemplified by Mott with the example of a simple dilute magnetic material $CuCl_2 \cdot 2H_2O$ [4, 12]. It is assumed that on each Cu ion there are nine d electrons in the Cu-d shell. In other words, there is one hole with spin $\frac{1}{2}$ in the d-shell, which has a magnetic moment of approximately $1\mu_B$. If the d holes in $CuCl_2 \cdot 2H_2O$ were holes in a d band rather than in an localized atomic d shell, it would have led to metallic conductivity with Pauli paramagnetic type magnetic response. This behaviour is not observed in such crystals even at very high temperatures. Thus, the electron band model is ruled out experimentally, but the question remained, why is this so? To this end it may be noted that the electronic structure of a partially filled band does involve ionized states like d^8 and d^{10} as well as d^9, which can be verified [4]. If one electron is taken away from a Cu^{++} ion (turning it into d^8 configuration), and put far away on another ion (while making it d^{10}), this electron will be repelled by nine other d electrons on the same ion (rather than by the eight electrons it saw in the previous ion). We have studied in chapter 3 that in such a situation the system as a whole will have to increase its Coulomb interaction energy by an amount U. The value of U may be estimated from the ionization potentials of the free ions, and is of the order 5–15 eV [4].

We have also studied in chapter 3 that this strictly localized approximation can be relaxed somewhat by allowing the electron to regain some energy by hopping from one lattice to other and thus wandering throughout the crystal. While in its localized state the electron has the energy of the average band state, the hopping electron may be placed in the most bottom state of the band and gaining an energy of half the bandwidth for the electron in that process. The extra hole may be allowed a similar energy gain by being placed in a freely wandering band state. This gain in the kinetic energy may outweigh the cost of electron Coulomb repulsion energy loss for s and p electrons and at high electronic densities. The resultant state in such cases is metallic, since now a very large number of electrons may get detached from their ions and move freely. In the opposite end of the dilute case, U outweighs the band energy. It is then energetically favorable for the electrons to remain localized at the corresponding lattice sites.

So it is clear that the two most important properties of these electrons are their kinetic energy or desire to delocalize themselves, and their Coulomb repulsion when they are too near each other [4]. The repulsion is represented by U, which is a Coulomb integral between two d-wave functions on the same atom (the exchange integral between the two orbitals would be an order of magnitude smaller) [4]. The next important thing is that electrons with parallel spin have to be away from each other due to the Pauli exclusion principle. So, whenever U predominates and prevents metallic conduction and the formation of bands, the opposite tendency to delocalize causes a necessarily antiferromagnetic interaction, and this was identified as superexchange [4]. The physical basis for this antiferromagnetic interaction is that the antiparallel electrons can gain energy by spreading into non-orthogonal overlapping orbitals, whereas parallel electrons cannot. We shall now elaborate more on this superexchange.

4.1 Anderson superexchange and magnetic insulators

The name superexchange arose, because such interaction involves the relative interaction of magnetic-ions over large distances in the medium occupied by diamagnetic ions and molecules. Kramers [13] in 1934 was the first to introduce the idea of superexchange, when he was studying the exchange interaction in paramagnetic salts. He suggested that the magnetic-ions could induce spin dependent perturbations in the wave functions of the ions inbetween, and in that process could transmit the exchange effect over large distances. In the early 1950s, P W Anderson built on the earlier ideas of Kramer to explain the antiferromagnetic order in MnO [14]. This idea can be illustrated by considering two Mn^{2+} and one O^{2-} ions arranged collinearly, and with a simple model involving four electrons [1]. The ground state will consist of a single unpaired electron in the d-orbital on each Mn^{2+} ion and two p electrons on the O^{2-} ion in its outermost occupied states. The dumbbell shaped oxygen p-orbitals lie in the same axis joining the two Mn^{2+} ions (see figure 4.1(a)). If the spins on Mn-ions are coupled antiferromagnetically then the ground state is represented by the spin (shown by arrows) configuration (i) in figure 4.1(b). Now, one of the two p electrons from the O^{2-} ion can go to one of the Mn^{2+} ions, because of the finite overlap of their wave functions. The two allowed excited states are shown in the configurations (ii) and (iii) of figure 4.1(b). The kinetic energy advantage of antiferromagnetism in this spin arrangement can be understood immediately, since the antiferromagnetic ground state configuration (i) in figure 4.1(b) can mix with the two allowed excited states shown in the configurations (ii) and (iii) of figure 4.1(b). This makes clear that the antiferromagnetic coupling between spins (or moments) in Mn-ion allows the delocalization of electrons over the whole Mn–O–Mn units, and thus there is an overall reduction in the kinetic energy of the system. For the sake of comparison, figure 4.1(c) presents the ground state and the excited states if the Mn ions are coupled ferromagnetically. It is obvious that the Pauli exclusion principle would have prevented mixing between these states, thus making ferromagnetic configuration energetically more costly.

This theory of Anderson to explain the antiferromagnetism in magnetic insulators, however, had certain problems. The exchange effect appears as third order in this perturbation theory, with the early terms (which do not describe magnetic effects) being quite large; as a result the theory was poorly convergent.

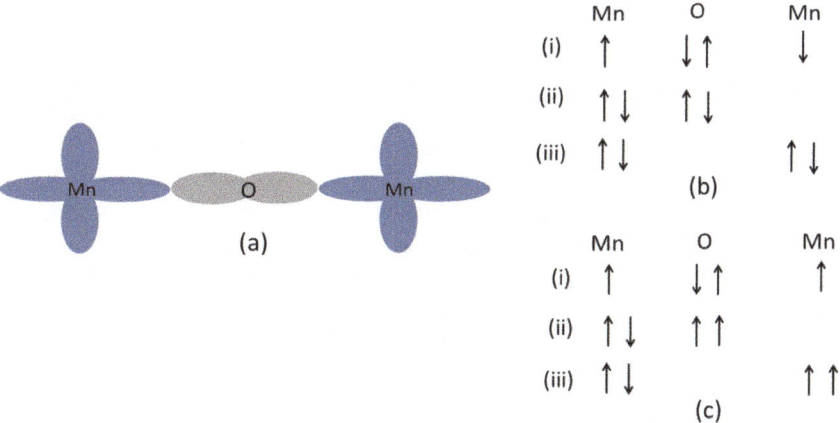

Figure 4.1. Superexchange process in magnetic insulator with the example of manganese monoxide (MnO). (a) Collinear arrangement of two Mn^{2+} and one O^{2-} ions. (b) Spin configurations leading to antiferromagnetic coupling between two manganese ions: (i) ground state; (ii) and (iii) excited states. (c) Spin configurations leading to ferromagnetic coupling between two manganese ions: (i) ground state; (ii) and (iii) excited states. (Reproduced from [1]. Copyright of OUP, Copyright 1992.)

This difficulty was circumvented in a new theory of superexchange interaction proposed by Anderson in 1959 [15]. In this theory the magnetic insulators were studied from a viewpoint in which the d (or f) shell electrons were placed in wave functions assumed to be exact solutions of the problem of a single d-electron in the presence of the full diamagnetic lattice. Inclusion of interactions amongst the electrons gave rise to three spin-dependent effects, which was termed by Anderson as: (i) superexchange; always antiferromagnetic in nature, (ii) direct exchange, always ferromagnetic, and (iii) indirect polarization effect. Anderson considered molecular orbitals formed by mixing of the localized $3d$ orbitals and the p orbitals of the surrounding negative ions. This resulted into two orbitals: (i) the bonding orbital, mainly occupied by p electron of a negative ion, and (ii) the antibonding orbital, which is partially occupied by $3d$ electrons and leading to the magnetism of the system. The wave function of the localized d orbitals thus extends over to the neighbouring negative ion, and there is a probability of transferring one $3d$ electron of the magnetic ion to the neighbouring $3d$ orbitals. The repulsive Coulomb interaction between two electrons, however, will tend to oppose such a transition. This situation can be described with the help of a model Hamiltonian in second quantized form in terms of creation (annihilation) and number operators [11]:

$$H = \sum_{(i,j)\sigma} t_{ij} c_{i,\sigma}^{+} c_{j,\sigma} + \sum_{i} U n_{i\uparrow} n_{i\downarrow} \tag{4.5}$$

where the summation is taken over the pair (i, j), $c_{j\sigma}^{+}$ and $c_{j\sigma}$ are the creation and annihilation operators (see appendix C) of electron with spin σ on the ion at j site, t_{ij} is the amplitude for the electron to hop from the lattice site i to the lattice site j, and U represents the Coulomb interaction energy between two electrons with different spin directions in the same atom. One may recognize this as the Hubbard model (discussed in chapter 3) if the hopping matrix elements are restricted to the nearest-neighbor, and all of them have the same value t. There is an energy increase by U, when one d electron of the magnetic ions hops into the unoccupied site of the neighboring magnetic ions. As we have already pointed out in chapter 3, this Hubbard Hamiltonian plays a very important role in the understanding of Mott physics in general. Here we will discuss it only in a qualitative manner in the context of superexchange and antiferromagnetism.

In this perturbation theory of Anderson, the first-order of the perturbation is an usual ferromagnetic exchange interaction. The superexchange is a second order process since it involves both O-ions and Mn-ions, hence is derived from second-order perturbation theory. We have discussed earlier in section 3.7 that in such a second-order perturbation theory, the energy involved is obtained from the square of the matrix element of the transition divided by the energy cost of getting the excited state. The transition matrix element here is controlled by the amplitude t_{ij} for the electron to hop from one lattice site to the other, and the energy cost of making an excited state is given by the Coulomb interaction energy U between two electrons with different spin directions. The resultant is an antiferromagnetic exchange interaction, which is expressed as [1, 11]:

$$J_{ij} = -\frac{2t_{ij}^2}{U}. \tag{4.6}$$

When $U \ll t$, electrons can propagate in the crystal as in the case of a metal. On the other hand, in the limit of $U \gg t$, electrons are localized at the lattice points in the solid, thus forming an insulator. Starting from the insulator as a limiting case, the superexchange interaction arises from a perturbation energy. The exchange interaction can be expressed in the more general form $J_{ij}\mathbf{S}_i \cdot \mathbf{S}_j$, for both direct and superexchange [4]. The exchange integral consists of two parts: potential exchange and kinetic exchange. The potential exchange represents electron repulsion and favours the ferromagnetic ground state. This term, however, is small when the ions are well separated. The kinetic exchange favours the antiferromagnetic ground state as is elaborated above with the case of Mn–O–Mn. This kinetic exchange term depends on the extent of orbital overlap. These terms, kinetic exchange and potential exchange, were chosen in order to emphasize that antiferromagnetism was the result of a gain in kinetic energy, and the ferromagnetism of a gain in potential energy. In the transition metal oxide systems like Mn–O–Mn, the hopping integral t is of order 0.1 eV and the Coulomb interaction U is in the range 3–5 eV [16]. The exchange interaction J depends sensitively on the ionic separation, and also on the Mn–O–Mn bond angle. In summary, the exchange interaction in various magnetic insulators is predominantly caused by the superexchange arising from the overlap of the localized orbitals of the magnetic electrons with those of intermediate diamagnetic atoms/molecules.

In the discussion so far on superexchange, it has been explicitly assumed that the oxygen ion lies between the two d-orbitals i.e. the 180° geometry. However, the situation changes entirely when the oxygen bridge between the two d-orbitals is 90° instead of 180°. By symmetry the hopping between the d- and the p-orbital is possibly only when they point towards each other (see problem 3 at the end of this chapter). The energy for the system, however, will depend on the relative orientation of the electron spins in the two d-orbitals, and the superexchange can also be ferromagnetic in nature under certain circumstance [17]. If the bond is between a filled orbital and a half-filled orbital or between half-filled orbital and an empty orbital, hopping of electron will be influenced by Hund's rule. There will be an energy advantage if the e_g electron arrives in the unoccupied orbital with its spin aligned with the spin of t_{2g} electrons. In this case superexchange will be ferromagnetic in nature (see problems 3 and 4 at the end of this chapter). But this interaction is relatively weak and less common than the antiferromagnetic superexchange.

Taking into account the occupation of the various d levels as dictated by the crystal field theory, some empirical rules were also developed [18, 19] for magnetic insulators, which are related to the formalism of Anderson [15] about the sign of superexchange. These empirical rules (widely known as Goodenough–Kanamori–Anderson rules) are quite useful in practice. These rules are narrated below [16]:

1. The exchange is strong and antiferromagnetic ($J < 0$) when two magnetic (M) ions possess lobes of singly occupied $3d$-orbitals pointing towards each other and have large overlap and hopping integrals. This is the case of 120°–180° M–O–M bonds as discussed above with the example of Mn–O–Mn.
2. The exchange is ferromagnetic ($J > 0$) but relatively weak when two M-ions have an overlap integral between singly occupied $3d$-orbitals, which is zero by symmetry. This is the case for ~90° M–O–M bonds.
3. The exchange is also ferromagnetic and relatively weak when two M-ions have an overlap between singly occupied $3d$ orbitals and empty or doubly occupied orbitals of the same type.

For a more detailed discussion on exchange interaction in general, as well as the evaluation of the magnitude and sign of exchange interactions, the readers are referred to the seminal review article by Anderson [4]. It may also be mentioned here that the situation becomes much more complex in orbitally degenerate systems. This is because: (i) both spin and orbital degrees of freedom are involved, (ii) the number of electrons or holes occupying the relevant orbitals will play an important role, and (iii) the presence of lattice instability in the form of Jahn–Teller distortions can change the physics considerably. In such situations a system is better described by Kugel and Khomskii-type Hamiltonians [20], where the M–O–M bond is replaced by an effective M–M bond.

4.2 Case studies of antiferromagnetic insulators: manganese monoxide and nickel monoxide

Manganese monoxide (MnO) is considered to be a classical antiferromagnet with Néel temperature T_N of 118 K [21]. It possesses a variety of interesting physical properties, and provides an interesting platform to understand magnetism and magnetic phenomena at the microscopic level. Shull and Smart [10] in 1949 investigated the magnetic properties of MnO using neutron-diffraction techniques, and that was followed by the determination the full magnetic structure of the antiferromagnetic state by Shull, Strausser and Wollan [22]. In the paramagnetic state MnO solid has the cubic NaCl-type structure. The transition to the anti-ferromagnetic state at T_N is accompanied by a cubic-to-rhombohedral lattice distortion, which is thought to be driven by nearest-neighbor and next-nearest-neighbor antiferromagnetic interactions. Neutron diffraction measurements of Shull *et al* [22] and the subsequent study of Roth [23] showed that the magnetic structure of MnO consists of Mn-spins aligned ferromagnetically within the (111) plane, and these planes are stacked antiferromagnetically in the direction normal to the given (111) plane (see figure 4.2). Thus MnO is classified as a type-II antiferromagnet.

In those early neutron diffraction studies, three different spin directions in the antiferromagnetic state were reported [24]: (i) along the cube axis [100], (ii) along the [111] direction i.e. perpendicular to the ferromagnetic planes, and (iii) perpendicular to the [111] direction i.e. parallel to the ferromagnetic planes. Later on, a high

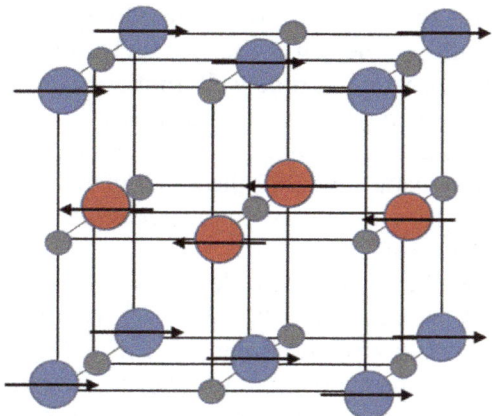

Figure 4.2. Schematic representation of magnetic structure in MnO. Blue and red spheres represent Mn-ions with opposite spin-direction; small grey atoms are oxygen-ions.

resolution neutron diffraction study suggested that the spin axis was perpendicular to the unique [111] direction [24]. While the lattice distortion associated with the onset of antiferromagnetic ordering in MnO was explained in terms of a lowering from cubic to rhombohedral symmetry, there was some question whether the antiferromagnetic structure could have true rhombohedral symmetry. This question was addressed more recently with the analysis of the total neutron scattering data with reverse Monte Carlo method [25]. The results of this study suggest that the magnetic moments in MnO are aligned ferromagnetically within (111) planes with the magnetization vectors of alternate planes along axes parallel and antiparallel to the $\langle 11\bar{2} \rangle$ directions, but with a small modulated out-of-plane component. The antiferromagnetic ordering is accompanied by small displacements of Mn and O modulated with the same periodicity, and both atomic and magnetic structures may be described in the monoclinic space group $C2$ [25].

The magnetic structure of MnO indicates that all exchange interactions between next-nearest neighbors are antiferromagnetic, which reflect the importance of the 180° Mn–O–Mn superexchange in MnO (see figure 4.1). Diffuse neutron-diffraction experiments revealed a significant amount of local order above T_N and also gave an estimation of 180° Mn–O–Mn interaction parameter $J/k_B = -4.6$ K [26, 27]. In the antiferromagnetic state the Mn–Mn interactions are repulsive within the ferromagnetic (111) layers. This introduces an exchange striction at temperatures $T < T_N$, which increases the Mn–Mn separation within these planes and simultaneously reduces the Mn–Mn separation between antiferromagnetically coupled adjacent planes. This results in a significant ($\sim$0.8%) rhombohedral distortion with the onset of antiferromagnetic ordering in MnO. Conventional one-electron band structure calculations using implicitly the Slater's theory of antiferromagnetism (see chapter 2, section 2.8 and chapter 3, section 3.8), predicted MnO to be an antiferromagnetic narrow gap semiconductor at $T = 0$, which would show metallic properties and Pauli paramagnetism above the antiferromagnetic transition temperature T_N [28]. However, contrary to such predictions, pure MnO is actually an insulator in the temperature regime well above T_N [29], with a resistivity of 10^9–10^{15} Ω cm at room-temperature [30]. In addition, the Mn atoms in MnO continue to carry the magnetic moment in the high temperature regime well above room temperature. Within the general framework of Mott physics, a Mott insulator must show the simultaneous existence of an insulating gap and a magnetic moment, and both the properties must exist even in the paramagnetic state. All the experimental results thus indicate that MnO fulfills the conditions necessary to be recognized as a classic exchange–correlation induced Mott insulator.

Until the mid 1980s, MnO was considered as a canonical example of Mott–Hubbard insulator [31]. However, in 1985, Zannen, Swatzky and Allen [32] argued that when the on-site electron–electron interaction U is greater than charge-transfer energy Δ (see section 3.9), the energy gap for charge excitations will be determined by the charge-transfer energy Δ, and not by the on-site Coulomb interaction Hubbard U. As a result MnO along with all the transition metal monoxides came under fresh scrutiny for the determination of exact nature of its insulating state, whether it is a (multiband) Mott–Hubbard insulator or a charge-transfer insulator. A careful photoemission spectroscopy study [33] on MnO revealed that the ligand-to-Mn d charge-transfer energy Δ is comparable to the on-site Coulomb interaction Hubbard $U \simeq 7.5$ eV, and MnO is indeed close to the boundary between the Mott–Hubbard ($U > \Delta$) and the charge-transfer ($U < \Delta$) regimes in the Zaanen–Sawatzky–Allen phase diagram [32]. Thus MnO is intermediate between the Mott–Hubbard and charge-transfer insulators, rather a marginally charge-transfer insulator [33]. There is still considerable interest in the exact calculation of ground state properties of MnO using various modern techniques of computational condensed matter physics (see [34] and references within). Such studies further underscore the importance of the proper understanding and determination of the electron exchange and correlation effects in order to explain various interesting physical properties of this benchmark transition metal compound.

Nickel monoxide (NiO) is an insulator [29, 35], and it undergoes a paramagnetic to antiferromagnetic transition at $T_N \approx 523$ K (see [36] and references therein). It forms in the NaCl-type crystal structure identical to MnO. The magnetic structure of NiO is also very similar to that of MnO, with ferromagnetic (111) layers stacked antiferromagnetically along the $\langle 111 \rangle$ direction (see [37] and references therein). The antiferromagnetic transition is again accompanied by a rhombohedral distortion. The magnetic moments of Ni in NiO lie in the (111) plane. The dominant exchange interaction in NiO is between the next-nearest neighbour Ni^{2+} ions, which arises due to the 180° Ni–O–Ni superexchange interaction caused by the overlap of the Ni-$3d$ orbitals with the O-$2p$ orbitals.

With the existence of magnetic moment and the insulating state in the temperature regime well above T_N, along with its sister compound MnO, NiO was also considered initially as a canonical example of Mott–Hubbard insulator [31]. However, the latter studies established NiO to be a charge transfer insulator, very close to the intermediate regime of the Zaanen–Sawatzky–Allen diagram [32] where $U \approx \Delta$. The physical properties of NiO (for that matter transition metal monoxides in general) continue to be a subject of considerable interest, which is evident in the recent studies on this system (see references [38] and [39] and the references therein). It is quite clear, however, that the conventional effective one-electron band theory with its modern developments like local (spin) density approximation are not quite adequate for studying such systems with strong electron correlations.

4.3 Double exchange mechanism in mixed-valence systems $La_{1-x}Sr_xMnO_3$ and Fe_3O_4

Double exchange is usually found in mixed-valence compounds. In the systems we have discussed so far, the lowest energy state had essentially the same number of electrons on every site, and in the presence of strong electron correlation the hopping of electron was strongly suppressed by the Coulomb repulsion energy U. In contrast, in a mixed valence system the number of electrons per site is non-integer, so even in the presence of large Coulomb repulsion, some sites will have more electrons than others. The electron hopping between such sites will be allowed without involving a cost of energy U. We shall now elaborate on the exchange interaction between magnetic ions in such mixed-valence systems, taking the example of (i) a mixed-valence compound containing Mn-ion that can exist in oxidation state 3+ or 4+ i.e. as Mn^{3+} or Mn^{4+}, and (ii) magnetite-Fe_3O_4. We will see that the resulting exchange interaction is a combination of potential and kinetic exchange.

4.3.1 $La_{1-x}Sr_xMnO_3$

The Mn-oxide compound we choose to discuss the double exchange mechanism is Sr-doped $LaMnO_3$, which forms in a perovskite structure. $LaMnO_3$ is a Mott-insulator, and it undergoes antiferromagnetic ordering via superexchange mechanism [40]. A Mn^{3+}-ion in $LaMnO_3$ has the electronic configuration $t_{2g}^3 e_g^1$. The e_g^1-electrons with spin 1/2 strongly hybridizes with the oxygen $2p$ state, and they are subject to the electron correlation effect. The t_{2g} electrons on the other hand are less hybridized with the oxygen $2p$ electrons and they are stabilized by crystal field splitting. They form a local spin 3/2, which is strongly coupled to spin 1/2 of e_g via strong on-site ferromagnetic Hund's coupling, thus giving total spin $S = 2$ for $LaMnO_3$. When doped with divalent Sr^{2+} a fraction x of manganese atoms in the $La_{1-x}Sr_xMnO_3$ compound goes to the Mn^{4+} state, while the remaining $(1 - x)$ fraction remain in the Mn^{3+} state. With Sr-concentration approximately above $x = 0.175$ the system becomes ferromagnetic around room temperature and shows metallic conductivity in the ferromagnetic state.

The ferromagnetism in the Sr-doped $LaMnO_3$ is understood in terms of double exchange mechanism giving rise to ferromagnetic coupling between Mn^{3+} and Mn^{4+} ions. This is exemplified with the help of figure 4.3. The e_g electron is allowed to hop to another Mn-ion site only if that site is not occupied by an electron with same spin direction. We recall here that no spin-flip process is allowed during hopping. A Mn^{4+} site with no e_g electron thus allow an electron to hop there from a Mn^{3+} site. However, the strong intra-site Hund coupling demands that the arriving e_g electron must be ferromagnetically aligned with the three existing t_{2g} electrons. Thus the hopping of an e_g electron to a neighbouring Mn-ion site where the t_{2g} electrons would have

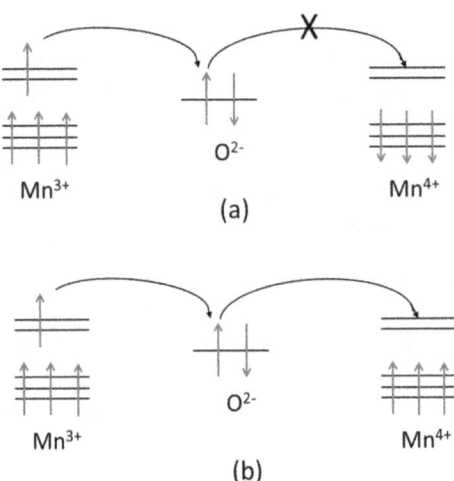

Figure 4.3. Double exchange mechanism leading to electron transfer and ferromagnetic coupling between Mn^{3+} and Mn^{4+} ions in Sr-doped $LaMnO_3$: (a) forbidden antiferromagnetic configuration of Mn-spins; (b) energetically favourable ferromagnetic configuration of Mn-spins.

antiparallel spin arrangement with respect to arriving e_g electron (see figure 4.3(a)) would be energetically unfavourable. The high-spin arrangement arising out of strong intra-site Hund's coupling therefore requires that both the donor and acceptor Mn-ions in this mixed valence $La_{1-x}Sr_xMnO_3$ compound must be ferromagnetically aligned. The ability of the electrons to hop results in kinetic energy saving. Thus, the ferromagnetic configuration shown in figure 4.3(b), which allows the hopping process to take place, reduces the overall energy of the system [1]. We can now see the double exchange mechanism between the Mn-ions via oxygen comprises of both the kinetic and potential components. Furthermore, this hopping of electrons in this ferromagnetic state of $La_{1-x}Sr_xMnO_3$ compound gives rise to metallic behaviour. We shall talk about this compound again in chapter 6 in the context of metal–insulator transition and collosal magneto-resistance.

4.3.2 Magnetite-Fe_3O_4

Magnetite Fe_3O_4, also known as loadstone, is the oldest known magnetic material. It forms in an inverted cubic spinel structure (see figure 4.4) with the chemical formula of $Fe^{3+}_{(Tetrahedral)}(Fe^{2+}Fe^{3+})_{(Octahedral)}$ O_4. In this structure the tetrahedral A sites contain one-third of the Fe ions surrounded by four oxygen ions, and the octahedral B sites contain the rest, two-third of the Fe ions surrounded by six oxygen ions. Fe(B)-O layers and Fe(A) layers are stacked alternately (see figure 4.4). The F ions at the A-site are trivalent Fe^{3+}, while B-sites contain equal number of trivalent Fe ions Fe^{3+} and divalent Fe ions Fe^{2+}, thus giving a mixed valent at the B-site with a formal average valence $Fe^{2.5}$.

Fe$_3$O$_4$ undergoes a ferrimagnetic transition below 860 K. In this ferrimagnetic state, the magnetic moments at the A-site have the opposite spin-directions to the moments at the B-site, with d-orbital occupations represented as $(t_{2g\uparrow})^3(e_{g\uparrow})^2$ and

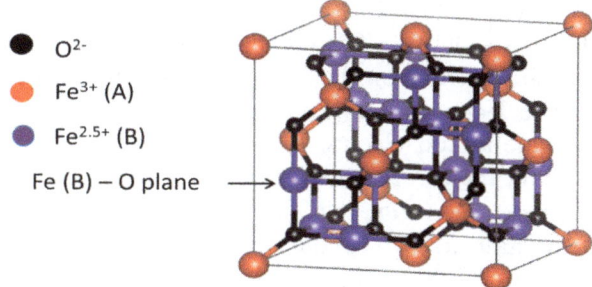

- O^{2-}
- Fe^{3+} (A)
- $Fe^{2.5+}$ (B)

Fe (B) – O plane ⟶

Figure 4.4. Inverse spinel crystal structure of Fe_3O_4 showing tetrahedral (A) and octahedral (B) Fe-sites.

$(t_{2g\downarrow})^3(e_{g\downarrow})^2(t_{2g\uparrow})^{2.5}$, respectively. The Fe^{3+} and Fe^{2+} ions at the octahedral-B sites align ferromagnetically via double-exchange mechanism. The Fe^{3+} ions at the tetrahedral-A sites do not participate in this double exchange interaction. They are coupled only to the Fe^{3+} ions at the octahedral-B sites through antiferromagnetic superexchange interaction. Thus, the spin contribution from these two sets of Fe^{3+} ions get canceled out, and the net moment in Fe_3O_4 arises only from the Fe^{2+} ions. The measured magnetic moment per formula unit in Fe_3O_4 is close to $4\mu_B$, which is expected from the contributions of the Fe^{2+} ions [1].

4.4 Nuclear fuel materials: uranium dioxide and plutonium

In the context of Mott physics we have so far discussed the cases involving only transition metal d-electron systems. There is another interesting class of system with open-electron shells namely f-electron systems, where the interplay of spin and unquenched orbital degrees of freedom of f-shells may lead to a rich variety of physical phenomena, including of course strong electron correlation effects and Mott physics [41]. We will discuss now some notable examples from the $5f$ actinide family, namely the famous nuclear fuel materials uranium dioxide (UO_2) and elemental plutonium (Pu). They are possibly the most investigated materials in the actinide family, both experimentally and theoretically (see [42, 43] and references therein), and they have still remained as subjects of great interest.

While the importance of UO_2 as nuclear fuel is an established fact, less well known is its possible applications in various other areas including electronics and thermoelectric power generations (see chapter 8 in part II of this book). There is also a fairly recent report on the observation of piezomagnetism and magnetoelastic memory in UO_2 [44]. To this end a comprehensive understanding of the fundamental electronic structure of UO_2, which determines such functional properties (including the energy efficiency as nuclear fuel), is therefore highly sought after. During the last decade both theoretical and experimental studies on UO_2 resulted in, amongst other things, the firm identification of UO_2 as a classic Mott–Hubbard insulator, where the electronic repulsion is responsible for the insulating state [45].

UO_2 is known to be a good insulator, with a conductance at room temperature $\sim 4 \times 10^{-3}\ \Omega\ cm^{-1}$ [46]. It forms in the CaF_2 crystal structure, in which the U^{4+} ions remain on a face-centred cubic (fcc) lattice with eight nearest-neighbor O^{2-} ions forming a cube (see figure 4.5(a)). UO_2 undergoes antiferromagnetic ordering below the Néel temperature $T_N \sim 30.2$ K. The magnetic dipole and the electric quadrupole moments of the two $5f$ electrons on the U^{4+}-ion form long-range antiferromagnetic order of the transverse 3-q type [41, 42]. In the antiferromagnetic state, (001)-type planes with ferromagnetic and electric ferro-quadrupolar order are stacked anti-ferromagnetically along the three equivalent $\langle 011 \rangle$ directions, and there is a Jahn–Teller distortion of the oxygen cage that surrounds each U^{4+}-ion (see figure 4.5(b)). In the paramagnetic state above T_N the long-range magnetic order, the electric quadrupole order, and the Jahn–Teller distortion of the oxygen cage all disappear. In recent times there is also some evidence of distinct magnetic transition at the surface of UO_2 in the Kosterlitz–Thouless universality class [47]. There are two major mechanisms to induce exchange coupling in UO_2 [48]: (i) superexchange and (ii) spin–lattice interaction. While super-exchange contributes to both dipole and quadrupole, spin–lattice interaction contributes to quadrupole only. A density functional based total energy calculation shows in the spin–orbit coupled system UO_2 the super-exchange tends to have ferromagnetic quadrupolar coupling rather than antiferromagnetic coupling [48].

The experimentally determined energy gap in UO_2 is of the order of 2 eV [49], which is comparable with other common semiconductors. The calculation using a

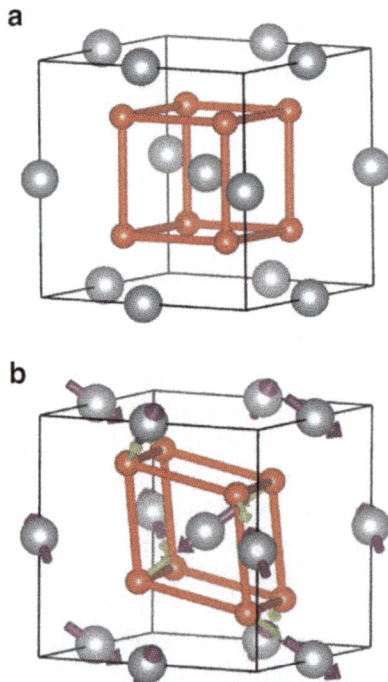

Figure 4.5. (a) Face-centred cubic unit cell of UO_2. (b) Antiferromagnetic state of UO_2 displaying the transverse 3-q type magnetic order of U-ions (violet arrows) and oxygen displacements (green arrows, not to scale) along the $\langle 111 \rangle$ directions. (Reprinted with permission from [44], from Nature Publishing (Nature Communications)).

combination of local density approximation (LDA) and dynamical mean-field theory (DMFT) revealed an energy gap of 2.2 eV between the lower (occupied) and upper (unoccupied) Hubbard bands in UO_2 [45]. However, in spite of a significant number of experimental and theoretical studies, there remained question on the exact nature of the insulating state, whether it is of f–f or f–d type. However, fairly recent x-ray absorption experiments in combination with first-principles calculations strongly suggest that UO_2 is possibly an f–f Mott–Hubbard insulator [50].

Another important nuclear material is metallic plutonium (Pu), which has six crystallographic allotropes [51]. At low temperatures it has α-structure with 16 atoms in elementary cell. It undergoes a series of phase transitions ending in relatively simple fcc-δ and bcc-ϵ phases at temperatures $T > 500$ K. There is a 25% increase in volume when Pu transforms from its α-phase (stable below 400 K) to the δ-phase (stable at $T \sim$ 600 K). This effect is important, and is related to the issues of long-term storage and disposal of Pu as a nuclear material. It has been suggested that the f-shell electrons in Pu are close to a Mott transition [52]. Pu does not show long-range magnetic order, and there are only local magnetic moments in Pu atoms [51]. More recent study [43] clearly indicates the presence of strong electron correlation in Pu, and it was suggested that α-phase and δ-phase are on the opposite sides of the interaction-driven localization–delocalization transition. In fact even the α-phase is not a weakly correlated phase; it is rather slightly on the delocalized side of the localization–delocalization transition [43]. The transport properties in the α-phase of Pu are reminiscent of heavy-fermion systems, with the anomalously large room temperature resistivity above the Mott limit (i.e. the maximum resistivity allowed in the conventional metal) [53].

4.5 Mott physics of molecular solid oxygen

The molecular oxygen O_2 is unique among the simple molecules because it carries a magnetic moment with a spin $S = 1$. An exchange interaction between O_2 molecules develops in the solid state, and it contributes to the cohesive energy in addition to the van der Waals force [54]. The solid oxygen has three phases in ambient pressure with different magnetic and crystal structures [54]. The high temperature ($T > 50$ K) γ-phase with A15-cubic crystal structure is paramagnetic in nature. This undergoes a cubic to rhombohedral transition with large volume contraction to β-phase around 44 K. The β-phase of solid O_2 is characterized by short-range antiferromagnetic correlation. With further reduction in temperature there is another transition around 24 K to α-phase with monoclinic crystal structure and long range antiferromagnetic order. With the application of external pressure, the antiferromagnetically ordered α-phase of solid O_2 transforms into another antiferromagnetically ordered δ-phase at 5.4 GPa, and that is followed by a non-magnetic ϵ-phase at 8 GPa [54]. Application of even higher pressure $P \sim 96$ GPa causes metallization of solid O_2, which is followed by an emergence of superconductivity below $T_C \approx 0.6$ K.

The possibility of Mott physics in purely p and s band systems is very appealing, because the standard framework of Mott physics suggests that in contrast with the narrow band d (or f) electrons, the kinetic energy of p and s electrons would be appreciable in comparison to the Coulomb interaction. To this end the pressure induced effects in the α-phase of solid molecular oxygen is quite reminiscent of strongly correlated, doped Mott insulating behaviour in some $3d$ transition metal oxide materials. There is a recent theoretical study on the insulator–metal transition in highly pressurized solid O_2 using first-principles local-density-approximation plus dynamical-mean-field calculations [55]. The result of this study suggests the existence of an orbital-selective Mott transition, and an incoherent metallic normal state arising from the Mott insulator via a weakly first-order transition as a function of pressure.

4.6 Interesting case of copper sulphate pentahydrate

In light of the discussion of Mott physics $CuCl_2 \cdot 2H_2O$ in the beginning of this section I am rather tempted to discuss the case of another interesting inorganic compound with Cu^{++} ions namely $CuSO_4 \cdot 5H_2O$ commonly known as blue vitriol. This compound has inspired beginner chemistry students over the generation all over the world by enabling them to grow beautiful blue crystals from evaporating solution of copper sulphate. In addition, there is a vast amount of practical applications of this material, which includes herbicide, wood impregnation and algae control [56]. In the artistic field, an artist Roger Hirons created a structure called Seizure, in which blue copper sulphate crystals covered a derelict housing apartment [57]. It is also interesting to note that Max von Laue and co-workers recorded the very first diffraction x-ray diffraction image from a single crystal sample of $CuSO_4 \cdot 5H_2O$ [56, 58].

The crystal structure of $CuSO_4 \cdot 5H_2O$ is triclinic with two Cu^{++} ions per unit cell. Each of these Cu^{++} is surrounded by a group of six oxygen atoms, four from H_2O molecules and two from SO_4 groups [59]. In the context of Mott physics, the enigmatic behaviour of $CuSO_4 \cdot 5H_2O$ has been pointed out by Anderson and Baskaran that in spite of having Cu^{++} ions it is not only insulating but transparent with a beautiful blue color at all reasonable temperatures [60]. In a series of papers in 1930s, Krishnan and Mookherjee [61–63] showed that the compound remains paramagnetic with interesting anisotropic magnetic response down to 90 K. Subsequent studies of heat capacity and magnetic properties by Geballe and Giaque [64] in the early 1950s confirmed the existence of this interesting paramagnetic state even at lower temperatures down to 0.25 K. However, the magnetic properties of $CuSO_4 \cdot 5H_2O$ did not attract further significant attention until early 2010, when it has been identified as a possible candidate for quantum many-body systems with spin-1/2 Heisenberg antiferromagnetic chain [56]. The exact ground state of this linear array of interacting magnetic moments is a macroscopic singlet state entangling all spins in the chain. This state of matter is termed as a spin liquid, where the spins are correlated, but at the same time fluctuates strongly even at low temperatures [65]. The idea of quantum spin-liquid involving a concept of 'resonating valence bond' is due to P W Anderson [66]. If two spins interact antiferromagnetically they can pair into a singlet state, and form a valence bond. When all the spins in a system form valence bonds, the ground state can be represented by the product of the valence bonds; this is a valence bond solid (VBS). A VBS state, however, is not a quantum spin-liquid, since it can break lattice symmetries and it lacks long-range entanglement [65]. In order to reach the quantum spin-liquid state, the valence bonds must be allowed to undergo quantum mechanical fluctuations. Here Anderson invoked the idea of a superposition of VBS states, which was earlier named by Linus Pauling as a resonating valence bond (RVB) [67]. Anderson [66, 68] proposed that in the triangular two dimensional spin-1/2 antiferromagnet the ground state is analogous to the precise singlet in the Bethe solution of the linear antiferromagnetic chain [69]. In such cases instead of forming a fixed array of spin singlets, strong quantum fluctuations lead to a superposition of singlet configurations. In other words the valence bond singlets resonate between different configurations. This resonating valence bond (RVB) state is clearly distinguishable from two other locally stable possibilities, the Neel antiferromagnetic state and the 'spin-Peierls' state consisting of a self-trapped localized array of singlet pairs [68]. The elementary excitations of this quantum many-body system—spin liquid—are fractional spin-1/2 quasiparticles called spinons, which have been created and detected in pairs in single crystals of $CuSO_4 \cdot 5D_2O$ through neutron scattering experiments [56].

Further reading

1. Blundell S J 2001 *Magnetism in Condensed Matter* (Oxford: Oxford University Press).
2. Anderson P W 1963 *Magnetism* ed G T Rado and H Suhl (New York: Academic) ch 2.
3. Coey J M D 2009 *Magnetism and Magnetic Materials* (Cambridge: Cambridge University Press).

Self-assessment questions and exercises

1. What are the origins of the kinetic exchange and potential exchange parts in the general exchange integral J_{ij}? Which one will favour an antiferromagnetic ground state and why?
2. Sketch a configuration with oxygen ion providing a 90° bridge between two magnetic ions with 3d-orbitals. Check whether a hopping between the d and the p-orbital is possible, as was in the case of 180° bridge (see figure 4.1). Now consider the possibility whether superexchange can be mediated via the Coulomb interaction on the connecting oxygen. Hint: a triplet state for two electrons in different orbitals on the same site is preferred to minimize the Coulomb interaction i.e. first Hund's rule.
3. The half filled Mott insulator $LaMnO_3$ forms in cubic perovskite structure. The five-fold degenerate 3d levels of Mn-ions in $LaMnO_3$ is split by the crystal field of oxygen octahedra in a three-fold degenerate t_{2g} levels and a higher energy two-fold degenerate e_g levels. A cooperative Jahn–Teller distortion of the oxygen octahedra within the ab plane lifts the degeneracy of the e_g levels [70]. Argue that the magnetic coupling between Mn-ions

via the oxygen bridge is ferromagnetic in the *ab*-plane, while the same is antiferromagnetic along the *c*-axis.

4. While tuning the properties of the oldest known magnetic system Fe_3O_4 with doping for various technological applications, it has been observed that ions such as Mn^{2+} and Zn^{2+} have a preference to occupy the tetrahedral A sites, while Ni^{2+} and Co^{2+} ions tend to sit at the octahedral B sites. In this situation what is likely to happen to the predominantly antiferromagnetic super-exchange between tetrahedral A-sites and octahedral B-sites of Fe_3O_4 in the event of say Mn^{2+} or Zn^{2+} substitutions?

5. In oxygen molecule O_2 there are eight $2p$ electrons, which are to be placed in the six available molecular orbitals originating from the 2 atomic states. The molecular states evolve from the atomic orbitals forming a system of bonding (σ, π) and antibonding (σ^*, π^*) molecular orbitals. Using the argument of exchange energy and Coulomb interaction energy, show that the placement of the oxygen $2p$ electrons in the molecular orbitals will make O_2 molecule strongly paramagnetic.

6. In light of the discussion of Mott insulating state in $CuCl_2\cdot2H_2O$ and $CuSO_4\cdot5H_2O$, it is quite instructive to check the possibility of the existence of Mott state in the compound $NiSO_4\cdot6H_2O$. Hint: the oxidation state of Ni in this compound is 2+.

References

[1] Blundell S J 2001 *Magnetism in Condensed Matter* (Oxford: Oxford University Press)
[2] Heisenberg W 1928 *Z. Phys.* **49** 619
[3] Dirac P A M 1929 *Proc. R. Soc.* A **123** 714
[4] Anderson P W 1963 *Magnetism* ed G T Rado and H Suhl (New York: Academic) ch 2
[5] Van Vleck J H 1932 *Theory of Electric and Magnetic Susceptibilities* (Oxford: Oxford University Press)
[6] Heitler W and London F 1927 *Z. Phys.* **44** 455
[7] Néel L 1932 *Ann. Phys.* **18** 5
[8] Hulthén L 1936 *Proc. Amsterdam Acad. Sci.* **39** 190
[9] Quintanilla J and Hooley C 2009 *Phys. World* **22** 32
[10] Shull C G and Smart J S 1949 *Phys. Rev.* **76** 1256
[11] Anderson P W 1978 *Science* **201** 307
[12] Mott N F 1949 *Proc. Phys. Soc. London* **862** 416
[13] Kramers H A 1934 *Physica* **1** 182
[14] Anderson P W 1950 *Phys. Rev.* **79** 350
[15] Anderson P W 1959 *Phys. Rev.* **115** 2
[16] Coey J M D 2009 *Magnetism and Magnetic Materials* (Cambridge: Cambridge University Press)
[17] Koch E 2012 *Correlated Electrons: From Models to Materials Modeling and Simulation* vol 2 ed E Pavarini, E Koch, F Anders and M Jarrell (Jülich: Forschungszentrum Jülich)
[18] Goodenough J B 1955 *Phys. Rev.* **100** 564
    Goodenough J B 1963 *Magnetism and the Chemical Bond* (New York: Interscience)
[19] Kanamori J 1959 *J. Phys. Chem. Solid* **10** 87

[20] Kugel K and Khomskii D 1975 *Sov. Phys. Solid State* **17** 285
[21] Boire R and Collins M F 1977 *Can. J. Phys.* **55** 688
[22] Shull C G, Strauber W A and Wollan E O 1951 *Phys. Rev.* **83** 333
[23] Roth W L 1958 *Phys. Rev.* **110** 1333
     Roth W L 1958 *Phys. Rev.* **111** 772
[24] Shaked H, Faber J Jr and Hitterman R L 1988 *Phys. Rev.* B **38** 11901
[25] Goodwin A L, Tucker M G, Dove M T and Keen D A 2006 *Phys. Rev. Lett.* **96** 047209
[26] Blecch I A and Averbach B L 1964 *Physics* **1** 31
[27] Goodenough J 1970 *Metallic Oxides*
[28] Terakura K, Oguchi T, Williams A R and Kubler J 1984 *Phys. Rev.* B **30** 4734
[29] De Boer J H and Veerwey E J W 1937 *Proc. Phys. Soc.* **49** 59
[30] Bhide V G and Dani R H 1961 *Physica* **27** 821
[31] Brandow B 1977 *Adv. Phys.* **26** 651
[32] Zaanen J, Sawatzky G A and Allen J W 1985 *Phys. Rev. Lett.* **55** 418
[33] Fujimori A, Kimizuka N, Akahane T, Chiba T, Kimura S, Minami F, Siratori K, Taniguchi M, Ogawa S and Suga S 1990 *Phys. Rev.* B **42** 7580
[34] Schrön A, Rödl C and Bechstedt F 2010 *Phys. Rev.* B **82** 165109
[35] Morin F J 1958 *Bell Syst. Tech. J.* **37** 1047
[36] Slack G A 1960 *J. Appl. Phys.* **31** 1571
[37] Chatterji T, McIntyre G J and Lindgard P A 2009 *Phys. Rev.* B **79** 172403
[38] Panda S *et al* 2016 *Phys. Rev.* B **93** 235138
[39] Panda S, Jiang H and Biermann S 2017 *Phys. Rev.* B **96** 045137
[40] Tokura Y and Nagaosa N 2000 *Science* **288** 400
[41] Santini P, Carretta S, Amoretti G, Caciuffo R, Magnani N and Lander G H 1989 *Rev. Mod. Phys.* **81** 807
[42] Wilkins S B, Caciuffo R, Detlefs C, Rebizant J, Colineau E, Wastin F and Lander G H 2006 *Phys. Rev.* B **73** 060406(R)
[43] Savrasov S Y, Kotliar G and Abrahams E 2001 *Nature* **410** 793
[44] Jaime M *et al* 2017 *Nat. Commun.* **8** 99
[45] Yin Q and Savrasov S Y 2008 *Phys. Rev. Lett.* **100** 225504
[46] Castell M R, Muggelberg C, Briggs G A D and Goddard D T 1996 *J. Vac. Sci. Technol.* B **14** 966
[47] Langridge S, Watson G M, Gibbs D, Betouras J J, Gidopoulos N I, Pollmann F, Long M W, Vettier C and Lander G H 2014 *Phys. Rev. Lett.* **112** 167201
[48] Pi S-T, Nanguneri R and Savrasov S 2014 *Phys. Rev.* B **90** 045148
[49] Schoenes J 1978 *J. Appl. Phys.* **49** 1463
[50] Yu S-W, Tobin J G, Crowhurst J C, Sharma S, Dewhurst J K, Olalde-Velasco P, Yang W L and Siekhaus W J 2011 *Phys. Rev.* B **83** 165102
[51] Dai X, Savrasov S Y, Kotliar G, Migliori A, Ledbetter H and Abrahams E 2003 *Science* **300** 353
[52] Johansson B 1974 *Philos. Mag.* **30** 469
[53] Boring A M and Smith J L 2000 *Los Alamos Sci.* **26** 91
[54] Freiman Y A and Jodl H J 2004 *Phys. Rep.* **401** 1
[55] Craco L, Laad M S and Leoni S 2017 *Sci. Rep.* **7** 2632
[56] Mourigal M, Enderle M, Klöpperpieper A, Caux J-S, Stunault A and Rønnow H M 2013 *Nat. Phys.* **9** 435

[57] Morton T, Charlesworth J J and Lingwood J 2008 *Roger Hiorns: Seizure* (London: Artangel)
[58] Friedrich W, Knipping P and Laue M 1912 *Proc. Bavarian Acad. Sci.* 303
[59] Beevers C A and Lipson H 1934 *Proc. R. Soc. Lond.* **146** 570
[60] Anderson P W and Baskaran G 1977 *A Critique of "A Critique of Two Metals"* (arXiv: cond-mat/9711197)
[61] Kishnan K S and Mookherji A 1934 *Phys. Rev.* **50** 860
[62] Kishnan K S and Mookherji A 1938 *Phys. Rev.* **54** 533
[63] Kishnan K S and Mookherji A 1938 *Phys. Rev.* **54** 841
[64] Geballe T H and Giaque W F 1952 *J. Am. Chem. Soc.* **74** 3513
[65] Balents L 2010 *Nature* **464** 199
[66] Anderson P W 1973 *Mater. Res. Bull.* **8** 153
[67] Pauling L 1949 *Proc. R. Soc. Lond.* A **196** 343
[68] Anderson P W 1987 *Science* **235** 1196
[69] Bethe H A 1931 *Z. Phys.* **71** 205
[70] Wall S, Prabhakaran D, Boothroyd A T and Cavalleri A 2009 *Phys. Rev. Lett.* **103** 097402

Chapter 5

Mott metal–insulator transition

Mott metal–insulator transition is observed in various inorganic and organic materials, including transition metal and actinide based oxide systems, transition metal sulphides and layered organic compounds [1–3]. In spite of possessing very different crystal structures, orbital degeneracies, and electron band structures, these materials have similar high temperature phase diagrams. A first-order phase transition from a high-conductivity phase to a high-resistivity phase is a common feature in all these systems. In contrast, at low temperatures these materials may exhibit very different ordered phases: some remain as paramagnetic insulators down to the lowest temperature, some order magnetically, and some are superconductors.

Mott metal–insulator transition arises in a strongly correlated electron system, due to the competition between the optimization of the kinetic energy and the potential (correlation) energy—that is the wave-like and particle-like character of electron in a solid. The systems display anomalous properties around the Mott metal–insulator transition, such as metallic conductivity smaller than the minimum predicted within the framework of electron band theory, unusual optical conductivity and spectral functions beyond what electron band theory can describe. In this chapter we shall discuss theoretical approaches to understand this metal–insulator transition within the framework of Mott physics. The Coulomb interaction energy (U) and the matrix elements or kinetic energy (t) that describe the electron hopping from site to site are the basic inputs necessary to understand and calculate the phase diagrams obtained through various experimental studies on materials across Mott metal–insulator transitions. An important question is, how the local electron density of states varies as the ratio of the correlation energy U and the kinetic energy t (or bandwidth W) increases? When electron is delocalized, its spectral function closely resembles the local density of states of electron band theory. On the other hand, the density of state peaks at the ionization energy and the electron affinity of the atom when the electron is localized. Such atomic like states lead to the Hubbard bands. In the intermediate correlation energy region between the two extreme limits of total localization and delocalization, the electrons will have the features of both quasi-particle and the Hubbard band.

The discussions in the subsequent part of this chapter will be kept mostly at the phenomenological level so that it is understandable to an audience of advanced (fresh) undergraduate (graduate) students of physics, chemistry, materials science and electrical/electronics engineering. The main aim will be to highlight the competition between the kinetic energy and potential energy of the electrons leading to the metal–insulator transition in various real materials systems with strong electron–electron correlation. A particular theoretical development of the last two decades namely dynamical mean-field theory (DMFT) will be discussed in more detail. Adequate references of specialized books and review articles will be made available for the interested advanced graduate students and professional researchers for further studies.

5.1 Bandwidth-control and filling-control Mott transition

Mott transition, which is defined as a transition between a (correlated) metal and a Mott insulator, can be of two types: band-width control transition (BC-MIT) and filling-control transition (FC-MIT) [2]. In this regard it is important to note down the two important parameters in the Hubbard model: (1) the ratio U/t (or U/W) of electron correlation energy and kinetic energy (or bandwidth), and (2) electron concentration (or band filling) n. Figure 5.1 shows the schematic of a metal–insulator phase diagram in terms of these important characteristic control parameters. In the case of a solid with nondegenerate band, the $n = 0$ and $n = 2$ fillings correspond to the band insulator [2]. When $n = 1$ i.e. the half-filled case, if the Coulomb repulsion U between electrons is stronger than the kinetic energy t, then it can inhibit electrons from delocalizing to form bands or hopping through the lattice (see figure 5.2(a)). The competition between the tendency of electron localization and delocalization can drive a transition between insulating and metallic states. Such a transition can be induced by tuning the magnitude of delocalization energy i.e. kinetic energy t, hence the bandwidth W. Reducing the lattice constant by external

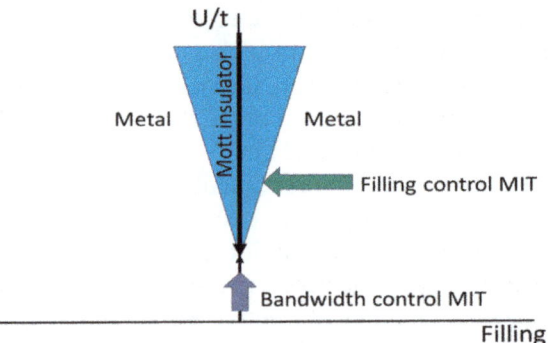

Figure 5.1. Schematic of a metal–insulator phase diagram in the plane of U/t and band filling n. (Reprinted figure with permission from [2]. Copyright (1998) by the American Physical Society.)

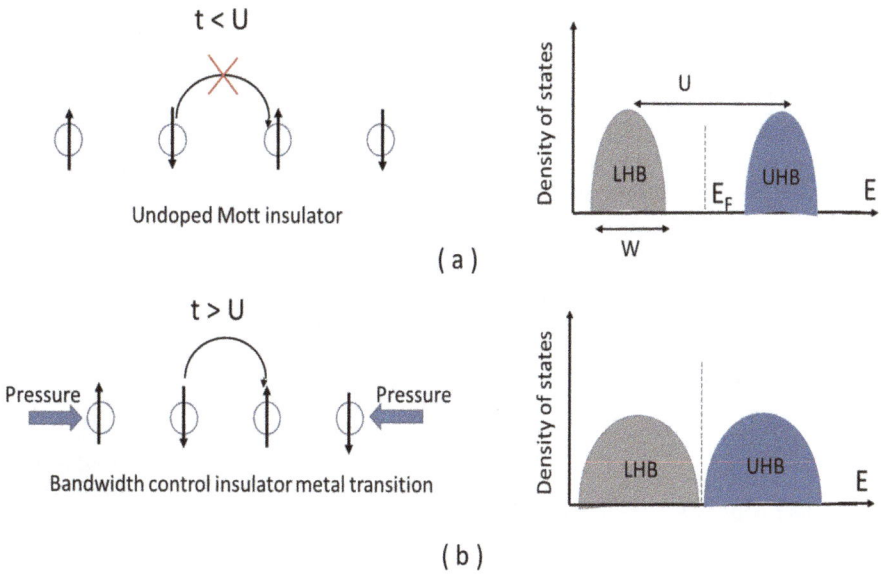

Figure 5.2. Schematic representation of Mott metal-to-insulator transitions: (a) intrinsic Mott insulator in the half-filled case, (b) bandwidth-control Mott metal–insulator transition (BC-MIT).

pressure can increase W leading to a MIT (see figure 5.2(b)) at a critical value of U_C. This transition at a finite U_C is known as a bandwidth-control Mott metal–insulator transition (BC-MIT). It may, however, be noted that in the case of perfect nesting the critical value U_C becomes zero. According to Mott the BC-MIT is a first-order transition when the long-range nature of Coulomb force (not included in the Hubbard model) is taken into account [4, 5]. The reasoning of Mott was based on the argument that with the decrease in carrier density with increasing U/t, the screening of long-range Coulomb forces by other carriers becomes ineffective. This in turn gives rise to the formation of an electron–hole bound pair at a finite U and leads to a first-order transition to the insulating state. Independent to this argument of Mott, a coupling to lattice degrees of freedom may also cause a first-order transition by increasing the transfer amplitude t discontinuously in the metallic phase.

The effective correlation energy U between electrons can be changed in a system with the addition of electrons or removal of electrons (i.e. introducing holes). In such cases the energy cost for hopping is less for certain electrons, thus reducing the effective Coulomb repulsion energy U (see figures 5.3(a) and (b)). This point may be also be understood from the fact that the added carriers can more effectively screen the electron–electron interaction, and hence reduce U. This kind of additional carrier-induced metal–insulator transition is known as filling-control Mott metal–insulator transition (FC-MIT) (see figure 5.3). The discovery of high-temperature superconductors, apart from other things, highlighted this concept of 'carrier doping' or filling-control in the parent Mott insulator compounds. FC-MIT is now generally recognized as one of the important aspects of the MIT in the $3d$

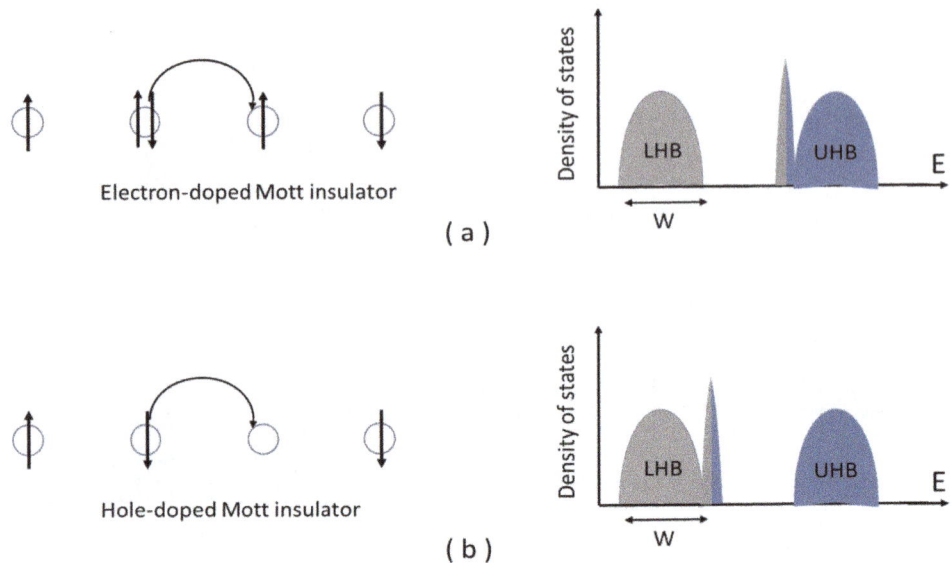

Figure 5.3. Schematic representation of filling control Mott metal-to-insulator transitions: (a) electron-doping case, and (b) hole-doping case.

electron systems [2]. The filling at non-integer filling n usually leads to the metallic phase. In contrast to the BC-MIT, the mechanisms for a first-order transition are not effective for the FC-MIT [2]. The absence of carrier compensation does not allow the electron–hole bound states to make an insulator by themselves. The coupling to the lattice is also ineffective because usually it does not couple to the electron filling [2]. The MIT may therefore be allowed to be continuous in the case of an FC-MIT.

It may be noted here that both BC-MIT and FC-MIT are controlled by quantum fluctuations rather than by temperatures. The FC-MIT has as control parameter the electron concentration, and the BC-MIT has as a control parameter the ratio of the electron interaction to the bandwidth U/t. The quantum fluctuation is enhanced for small U/t, which stabilizes the metallic phase i.e. the quantum liquid state of electrons. Large U/t, on the other hand, makes the localized state more stable. Since these control parameters are intrinsically quantum mechanical in nature, Mott metal–insulator transition is also a subject of research in the more general field of quantum phase transitions [2].

5.2 Theoretical approaches in Mott metal–insulator transition

5.2.1 Fermi-liquid based descriptions

We have seen in section 2.7 that the Fermi-liquid description is a conventional way to study the electron correlation effects by assuming the adiabatic continuity of the paramagnetic metallic phase with the non-interacting electrons. Within this framework, the Mott insulating phase may arise as a consequence of the opening of a charge gap due to symmetry breaking in spin or orbit degrees of freedom. This was the pioneering work of Slater (see chapter 2, section 2.8 and chapter 3, section 3.7).

Here the Mott insulating state arises due to the simultaneous formation of the magnetic moment and the onset of long-range antiferromagnetic order. However, it is now well known that in many real systems, the Mott insulating state continues to exist in the temperature regime well above the antiferromagnetic transition temperature, and the local magnetic moment exists in the paramagnetic state as well. It is also known that there are practical difficulties of applying the density functional theory (DFT) to strongly correlated systems, which arise from the difficulty in calculating $V_{ex}(\mathbf{r})$ into which all the many-body effects are included [2]. In strongly correlated systems, the assumption that $V_{ex}(\mathbf{r})$ is determined from the local electron density $n(\mathbf{r})$ may not be valid in general.

We have seen in section 2.6 that the local density approximation (LDA) can be a useful tool for calculating the band structure of metals in the ground state when the correlation effect is not important. But it is well known that while dealing with Mott insulating state as well as anomalous metallic states near the Mott MIT, the simple LDA is not very good in reproducing the experimental results in MnO, NiO, NiS and many other strongly correlated oxide systems, including high T_C superconducting oxides [2]. To improve upon the LDA the local spin-density approximation (LSDA) was introduced, where the spin-dependent electron densities $n \uparrow (\mathbf{r})$ and $n \downarrow (\mathbf{r})$ were taken into account separately to allow for possible spin-density waves or antiferromagnetic states. LSDA approach was successful in explaining the presence of the band gap at the Fermi level for MnO by allowing antiferromagnetic order [6] and reproducing the antiferromagnetic insulating state of $LaMO_3$ (M = Cr, Mn, Fe, Ni) [7].

The LSDA approach, however, failed to reproduce the antiferromagnetic ground state of many strongly correlated compounds such as NiS, La_2CuO_4, $YBa_2Cu_3O_6$, $LaTiO_3$, and $LaVO_3$ [2]. Even in the case of MnO, the observed strongly insulating behavior well above the antiferromagnetic transition temperature indicates that the charge-gap structure does not arise as a direct consequence of antiferromagnetic order, while antiferromagnetic order is an essential ingredient to open a charge-gap in the LSDA calculation. In an attempt to improve further on LDA/LSDA, the generalized gradient approximation (GGA) method was introduced [8, 9], which resulted in partial improvements of the LDA results in the transition-metal compounds FeO, CoO, FeF_2, and CoF_2 [2]. GGA was also applied to transition-metal oxides $LaVO_3$, $LaTiO_3$, and YVO_3, where it showed improvements in reproducing the band gap in $LaVO_3$. However, in the case of YVO_3 and $LaTiO_3$ the estimated ground state was not correct. The limitations of LDA or GGA type of approach are also known in explaining the ground state of strongly correlated nuclear fuel materials like UO_2 and elemental Pu [10].

In LDA and LSDA calculations, because of the drastic approximation in the exchange-interaction term, the self-interaction is not canceled between the contributions from the Coulomb term and the exchange interaction term [2]. This is a serious problem, and produces a physically unrealistic self-interaction leading to a poor approximation in strongly correlated systems. Perdew and Zunger [11] proposed a method of self interaction correction, which was applied to the transition-metal oxides with quantitative improvement of the band gap and the magnetic moment in MnO, FeO, CoO, NiO, and CuO [12]. While the Kohn–Sham equation provides the procedure to calculate exactly the ground state at the Fermi level (see section 2.6), it does not really guarantee that excitations will be well reproduced.

In order to improve the estimates for the excitation spectrum, one possible way is to combine the Kohn–Sham equation with a standard technique for many-body systems by introducing the self-energy correction to the electron density of states obtained from the LDA-method of calculations. This attempt is called the 'GW' approximation, and was applied to NiO, although not many systematic studies on transition-metal compounds and other strongly correlated systems are available (see [2] and references therein).

In an attempt to solve the problem of underestimating the band gap of the Mott insulator in LDA-type calculations, a combination of the LDA and a Hartree–Fock type approximation called the LDA + U method was introduced by Anisimov *et al* [13]. The Coulomb interaction between electrons for double occupancy of the same site requires a large energy U, when one is dealing with the localized as compared to extended states. This aspect is not adequately considered in the LDA method, where the exchange interaction is estimated using information from the uniform electron gas. This is appropriate only for extended wave functions. In the LDA + U approach the contribution of the interaction proportional to U is added to the LDA energy in an ad hoc way when the orbital is supposed to be localized [2]. The LDA + U method, however, does not necessarily provide a reliable way to treat this interaction term U, while all the important effects of strong correlation are attributed to this term. In addition the estimate of U may depend on the choices of local atomic orbitals such as the linear muffin-tin orbital (LMTO) and linear combination of atomic orbitals (LCAO). This may give different U because in the presence of strong hybridization the status of the orbital becomes ambiguous. However, the DFT + U (and its refinements over the period) is still considered to be quite useful in the study of strongly correlated electron systems, and interested readers are referred to a fairly recent perspective article by Kulik [14].

5.2.2 Mott physics and metal–insulator transition

We have studied in chapter 3 that Mott physics starts with the basic premise that a periodic system with odd number of electrons per unit cell could be a Mott-insulator when the electron–electron interaction energy U is much larger than the band-width W (or kinetic energy t). In this system adding and removing electrons from an atom in the periodic lattice site leaves it in an excited configuration. The internal degrees of freedom like orbital angular momentum and spin in the remaining atoms in the system will scatter these excited configurations. These states propagate through the system incoherently and broaden to form bands, which are called the lower and the upper Hubbard bands. These Hubbard bands describe propagating empty and doubly occupied sites in a half-filled lattice. On the other hand, electron band theory rightly predicts that the system must be metallic in the limit of weak electron–

electron interactions. So at some critical ratio of the electron–electron interaction to the band-width, there must be a metal–insulator transition—the Mott transition.

The problem of modeling Mott metal–insulator transition, however, is really quite difficult, as one is away from the two well-understood extreme limits and deals with materials made up of electrons that are neither propagating as Bloch waves in the simple metallic solids nor fully localized on their atomic sites. This dual particle-wave character of the electron in such strongly correlated materials needs the involvement of both the real-space and momentum space pictures for a proper description. To this end one of the simplest models of correlated electrons to start with is of course the Hubbard Hamiltonian. However, an accurate numerical solution of Hubbard Hamiltonian on a three dimensional lattice is not within the reach of available computers [15]. In the region of metal–insulator transition where U and t are comparable, a perturbative treatment to study Hubbard Hamiltonian fails. Hence, to make further progress one needs to look for approximate solutions of non-perturbative many-body physics problem or study solvable models, which still contain the physics of original Hamiltonian. We will summarize here some basic approaches to understand the Mott metal–insulator transition. The discussion will be more on the qualitative level and follow closely the presentation in two related articles [15, 16] and the book by Martin–Reining–Ceperley [17]. (Some mathematical details concerning the Hubbard model involving Green's function formalism will be provided in appendix D.) This hopefully will enable the readers to understand/appreciate the experimental results as well as applications of Mott insulators presented in the subsequent chapters of the book. For details of such theoretical methods, the readers are referred to the excellent review articles [2, 18–20] and the books by Martin–Reining–Ceperley [17], Gebhard [21] and Freerick [22].

There are two general approaches to understand a Mott metal–insulator transition without a broken symmetry, which are associated with the works of Hubbard [23, 24], and Gutzwiller [25], and that was further developed by Brinkman and Rice [26].

Gutzwiller [25] performed an approximate variational calculation of the ground state wave function for the Hubbard Hamiltonian with a single tight-binding band by taking into consideration only intra-atomic Coulomb interactions between the electrons. The Gutzwiller approach has two separate approximations. The first one is the Gutzwiller variational wavefunction. The second Gutzwiller approximation is for the estimation of the kinetic energy for this variational wavefunction while neglecting the correlations between the lattice sites. The Gutzwiller variational function discourage multiple electron occupancy of a lattice site, and for a one-band model it is expressed as [17]:

$$|\Psi_G\rangle = \prod_i^N [1 - (1 - g)n_{i\uparrow}n_{i\downarrow}]|\Psi_{HF}\rangle \qquad (5.1)$$

where N is the number of lattice sites, $(1 - g)n_{i\uparrow}n_{i\downarrow}$ is a projection operator which for $0 < g \leqslant 1$ reduces the number of doubly occupied sites, and Ψ_{HF} is a single determinant like the Hartree–Fock wavefunction. The strength of the correlation is determined by the parameter g. There is no correlation when $g = 1$. On the other hand, the quantity within the square bracket is 1 if either of $n_{i\uparrow}$ and $n_{i\downarrow}$ is zero, or it is equal to g if the lattice site is doubly occupied with $n_{i\uparrow}n_{i\downarrow} = 1$. Reduced probability of unlike-spin electrons occupying the same lattice site lowers the interaction energy. This, however, will increase the kinetic energy because the probability of hopping is also reduced. The best wavefunction is found in the variational method by minimizing the expectation value of the Hamiltonian with respect to g [17]:

$$E = \frac{\langle \Psi_G | H | \Psi_G \rangle}{\langle \Psi_G | \Psi_G \rangle}. \tag{5.2}$$

This will give the tightest upper bound to the ground-state energy for the Gutzwiller trial function [17]. The next step in the Gutzwiller approach is the approximation for the energy and other properties of the Gutzwiller wavefunction. The total energy is expressed as [17]:

$$\frac{E_G}{N} = q_\uparrow \bar{E}_\uparrow + q_\downarrow \bar{E}_\downarrow + Ud \tag{5.3}$$

where,

$$\bar{E}_\sigma = \frac{1}{N_k} \sum_k (E_{k,\sigma} - \mu). \tag{5.4}$$

Here d is the fraction of doubly occupied sites and $\bar{E}_\sigma$ is the kinetic energy per cell for uncorrelated particles. It may be noted that $E_{k,\sigma} - \mu < 0$ for the occupied states, hence $\bar{E}_\sigma < 0$. The reduced hopping due to repulsive interactions is taken into account by the term $q \leqslant 1$. Now the Gutzwiller approximation assumes that there is no correlation of spin and charge on adjacent lattice sites, and that leads to [17]

$$q_\sigma = \frac{([(n_\sigma - d)(1 - n_\sigma - n_{-\sigma} + d)]^{1/2} + [(n_{-\sigma} - d)d]^{1/2})^2}{n_\sigma(1 - n_\sigma)}. \tag{5.5}$$

Minimization of the energy in equation (5.3) with respect to d gives the following interesting results. For any filling of the band other than $\frac{1}{2}$, Gutzwiller approximation results into a renormalized metal with reduced kinetic energy. This is interpreted as an increased effective mass by the factor $\frac{1}{q}$, and a Fermi surface same as the original unrenormalized Fermi surface.

In the absence of any spin polarization, for a half-filled band equation (5.5) reduces to $q = 8d(1 - d)$. In this situation if U is less than a critical value U_C^G, minimization of the energy gives:

$$d = \frac{1}{4}\left(1 - \frac{U}{U_C^G}\right) \tag{5.6}$$

and

$$q = 1 - \left(\frac{U}{U_C^G}\right)^2. \tag{5.7}$$

This is turn gives the ground state energy:

$$\frac{E_G}{N} = -|\bar{E}|\left(1 - \frac{U}{U_C^G}\right)^2. \tag{5.8}$$

Here the kinetic energy for the system without any correlation is represented by $\bar{E}$. If U is less than U_C^G, the solution is a metal with a well-defined Fermi surface. With the increase in the interaction U, there is a decrease in kinetic energy by a factor q given in equation (5.7). This indicates a narrowing of the band and an increase in the effective mass given by $m^* \approx 1/q$, until the mass diverges at $U = U_G^C$. For the semicircular non-interacting electron density of states represented by the equation [17]:

$$\rho(w) = \frac{2}{\pi D^2}\sqrt{D^2 - w^2} \tag{5.9}$$

the kinetic energy for a half-filled band can be estimated as $|\bar{E}| = \frac{4}{3}\pi$ and $U_G^C/D = 32/3\pi \sim 3.4$. Here D is half-width of the band. This density of state is customarily called 'semicircular' because $\sqrt{D^2 - w^2}$ is the shape of a semicircle.

Now, if $U > U_G^C$ the solution is $d = q = E_G = 0$; that means there is no double occupancy, no hopping, and thus no gain in kinetic energy. While this is not a physically meaningful solution at half-filling there is a solution for any other filling; that is a metal. Thus, for U greater than the critical value the approach to the metal–insulator transition can be visualized as the divergence of the electron mass as the filling approaches limit of half-filled band. Building on the work of Gutzwiller, Brinkman and Rice [26] studied the metal–insulator transition starting from the metallic phase. Brinkman and Rice described the metallic phase as a strongly renormalized Fermi-liquid with a characteristic energy scale of renormalized Fermi energy ϵ_F^*. This energy scale decreases with the increase in the strength of electron correlation energy U and eventually goes to zero at a critical value of interaction U_C. In the Brinkman–Rice framework, the Mott metal–insulator transition is driven by the disappearance of the Fermi-liquid quasi-particle. The quasi-particle residue Z vanishes as $Z \approx U - U_C$ and the quasi-particle effective mass diverges as $m^* \approx 1/(U - U_C)$. In this framework there is a well defined peak at the Fermi level with a finite width, which gets narrowed down gradually as the metal–insulator transition is approached. This feature is shown schematically in figure 5.4. This evolution of electron density of state can be studied through one-particle spectral function, which is experimentally accessible through the photo-emission and inverse photo-emission spectroscopy measurements.

In the Hubbard approach the interaction between electrons is formulated in terms of the spectrum of excitations within the framework of Green's function method (see appendix D). This is in line with the proposal by Mott that the energies for addition

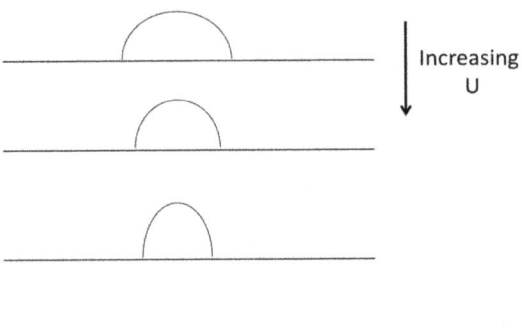

Figure 5.4. Schematic presentation of the one-particle spectral function with the variation in the electron correlation energy U in the Brinkman–Rice picture.

and removal of electrons from the atoms compared with the hopping matrix elements between atoms, determines metal–insulator transition. Hubbard first treated the interacting-electron problem for the system of decoupled atoms by including interactions and then added the coupling between the atoms as a perturbation. In the decoupled limit (i.e. no hopping and $t = 0$), the spectrum consists of peaks at E_0 and $E_0 + U$. These are the lower Hubbard band and upper Hubbard bands. We recall here that the upper Hubbard band corresponds to adding an electron at a site occupied by an opposite-spin electron, whereas the lower Hubbard band corresponds to the situation of sites with absence of electron. In the next step to include the hopping of electron t, Hubbard proposed that the interacting system was analogous to an alloy where each electron propagated in a disordered array of sites. The energy of electron increased by U if a site was occupied by an electron with opposite spin. The resultant picture is the same as the coherent potential approximation (CPA) widely used for studying actual alloy systems. However, the electrons are not fixed like atoms in an alloy. To take into account of this fact, Hubbard added a resonance-broadening correction. In the Hubbard picture the band is not narrowed, rather the spectrum of energies broadens with the increase in U until a gap opens indicating the onset of a metal–insulator transition. In the case of the semicircular density of states Hubbard found the critical value of correlation energy $U_C = \sqrt{3}\,D$. Summarizing we can say that in the Hubbard picture, for large correlation energy U the lower and upper Hubbard bands of width W are separated by an energy gap of the order $U - W$. With the variation of U eventually a critical value of correlation energy U_C is reached, where the two Hubbard bands merge to form a metallic state. This state, however, is not a Fermi liquid state. The evaluation of the one-particle spectral function within the Hubbard picture is shown schematically in figure 5.5. The spectral function for an electron in a crystal is given by $A(k, w) = -2\,\mathrm{Im}\,G(k, w)$, where $G(k, w)$ is the retarded Green function describing the motion of the electron (see appendix D).

In spite of the fact that the use of model Hamiltonian somewhat simplifies the problem, even the simple experimental properties such as the phase diagram of a

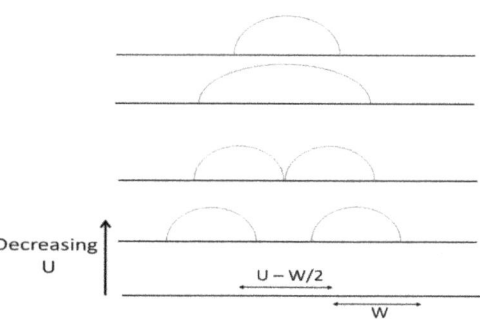

Decreasing
U

$U - W/2$

W

Figure 5.5. Schematic representation of the one-particle spectral function with the variation in the electron correlation energy U in the Hubbard picture at half-filling.

strongly correlated electron system were still difficult to calculate exactly. In the late 1980s Metzner and Vollhardt [27] introduced a new limit to the correlated electron problem namely the infinite lattice coordination, where each lattice site is imagined to have infinitely many neighbors. This infinite dimensional theory (also known as *dynamical mean-field theory*) treats the dynamic fluctuations in the system correctly when spatial fluctuations can be ignored [2]. In this approach the competition between kinetic energy and Coulomb interaction of electrons is treated adequately, and at the same time the computation is simplified quite significantly. A second advance in this field took place when Georges and Kotliar mapped the Hubbard model onto a self-consistent quantum impurity model [28]. This later model is represented by a set of local quantum mechanical degrees of freedom that interact with a bath or continuum of non-interacting excitations. These developments provided the basis of the dynamical mean-field theory (DMFT) of the correlated electron systems [18, 19], which allowed many-body theorists to formulate and solve a variety of model Hamiltonian on the lattice using analytic techniques as well as numerical techniques like quantum Monte Carlo. The DMFT solutions become exact as the number of neighbours increases, and the theory has been relatively successful in explaining the experimental results in various strongly correlated d and f-electron materials. We will now provide a brief (mostly qualitative) introduction to DMFT, and for the details of DMFT the readers are referred to the review articles [18, 19] and the books by Martin, Reining and Ceperley [17], Freerick [22] and Wills *et al* [29].

In the framework of the DMFT, a mean-field theory reduces a many-body lattice problem to a single-site problem with effective parameters. Here it may be quite instructive to recall the classical theory of magnetism as an analogy, where electron spin is the relevant degree of freedom at a single lattice site and the complete system is represented by an effective magnetic field i.e. the classical mean field [16]. The degrees of freedom at a single site in the correlated electron system, are the quantum states of the atom inside a selected central unit cell of the crystal. A reservoir or bath of non-interacting electrons represents the rest of the crystal lattice, and the electrons can be emitted or absorbed in the atom. This local description of a correlated electron solid in terms of an atom embedded in a medium of non-interacting

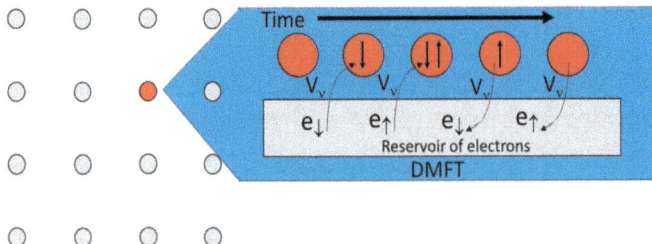

Figure 5.6. Schematic representation of a correlated-electron solid, where the full lattice of atoms and electrons is replaced with a single impurity atom imagined to be embedded in a reservoir of electrons. The effect of the environment on the impurity atom site enables the electrons to hop in and out of that site via the hybridization V_ν.

electrons corresponds to the celebrated Anderson impurity model, but with an additional self-consistency condition [16]. Figure 5.6 pictorially represents the essence of Anderson impurity model, in which the emission or absorption of electron is mediated via quantum mechanical amplitude V_ν or the hybridization between the atomic and the bath electrons. The effect of the environment on the impurity atom site is to allow the atom to make transitions between different configurations. In other words the electrons may hop in and out of that site via the hybridization V_ν.

We rewrite here the model lattice Hubbard Hamiltonian in second quantized form for a single band expressed in terms of creation (annihilation) and number operators as:

$$H = -\sum_{(i,j)\sigma} t_{ij} c_{i,\sigma}^{+} c_{j,\sigma} + \sum_{i} U n_{i\uparrow} n_{i\downarrow} + \epsilon_o \sum_{i\sigma} n_{i\sigma} \qquad (5.10)$$

where the summation is taken over the pair (i, j), $c_{j\sigma}^{+}$ and $c_{j\sigma}$ are the creation and annihilation operators of electron with spin σ on the ion at j site, t_{ij} is the amplitude for the electron to hop from the lattice site i to the lattice site j, U represents the Coulomb interaction energy between two electrons with different spin directions at the same lattice site, n is the number operator and ϵ_0 is the on-site energy. When the infinite lattice coordination limit is taken, the parameters need to scale in a definite way. This is necessary to avoid the domination of a single term in the Hamiltonian. To this end in the Hubbard model on a d-dimensional hypercubic lattice with nearest-neighbor hopping, the hopping parameter t_{ij} needs to be scaled as $1/\sqrt{d}$, while the on-site interaction U remains unaltered. In this way both the kinetic and potential energy per site remains finite.

Following the work of Metzner and Vollhardt [27], Brandt and Mielsch [30] derived the exact solution of the infinite-dimensional Falicov–Kimball model, which is a simplified Hubbard model where intersite hopping is confined to only one of the two spin species. This work was advanced further to study the Hubbard model in the pioneering work of Geroges and Kotliar [28] with the aim to find a set of equations allowing the calculation of the self-energy Σ defined from the interacting single-electron Green function.

We introduce here a single-electron Green function defined at finite temperature for a representative site $G_{i\sigma}(\tau - \tau') \equiv -\langle c_{i\sigma}(\tau)c_{i\sigma}^\dagger(\tau')\rangle$ and its Fourier transform $G_i(iw)$. This specifies the probability amplitude required to create an electron with spin $\sigma(\uparrow$ or $\downarrow)$ at a site i at time τ and destroy it at the same site at a later time τ'. Here the imaginary time τ runs between 0 and $\beta = 1/T$. This Green's function is a local quantity, which is coupled to an effective bath i.e. rest of the lattice. The local representative site can be represented by the Anderson impurity model Hamiltonian, which is written as [16]:

$$H_{AIM} = H_{atom} + \sum_{\nu,\sigma} \epsilon_\nu^{bath} n_{\nu,\sigma}^{bath} + \sum_{\nu,\sigma}(V_\nu c_{0,\sigma}^\dagger a_{\nu,\sigma}^{bath} + h.c) \qquad (5.11)$$

where $h.c$ stands for the Hermitian conjugate, and $c_{0,\sigma}$ and $a_{\nu,\sigma}^{bath}$ represents atomic and bath electrons, respectively. The Anderson impurity model yields the exact local Green function (which contains information about the local one-electron photo-emission spectrum) in DMFT when the V_ν fulfills a self-consistency condition [16]. Starting from a general Hamiltonian, one can separate atomic degrees of freedom at a lattice site described by H_{atom}, from the remaining degrees of freedom treated as a bath of electrons with energy levels ϵ_ν^{bath}. However in contrast with the classical case, in which the effect of the medium on the central site is represented by a number (the effective magnetic field), in the present quantum case a hybridization function $\Delta(w)$ is required to capture the ability of an electron to enter or leave an atom on a time scale w. This hybridization function is expressed in terms of the parameters ϵ_ν^{bath} and V_ν as [16]:

$$\Delta(w) = \sum_\nu \frac{|V_\nu|^2}{w - \epsilon_\nu^{bath}}. \qquad (5.12)$$

$\Delta(w)$ plays the role of a mean field, and its frequency dependence makes it a dynamic mean field [28].

The electron hopping to and from the impurity is described by the bath Green function given by:

$$G_0(iw) = \frac{1}{iw + \mu - \epsilon_0 - \Delta(w)}. \qquad (5.13)$$

One now needs to get an effective field $G_0(\tau, \tau')$ in terms of a local quantity. We now define the local self energy $\Sigma_{imp}(iw) \equiv G_0^{-1}(iw) - G^{-1}(iw)$ and the lattice Green function $G(k, iw)$ and self energy as $\Sigma(k, iw)$ as:

$$G(k, iw) = \frac{1}{iw + \mu - \epsilon_0 - \epsilon_k - \Sigma(k, iw)}. \qquad (5.14)$$

In the limit of infinite coordination the self-energy does not depend on momentum and $\Sigma(k, iw) \equiv \Sigma_{imp}(iw)$. Thus, all the information on single-electron properties is enclosed in a function of frequency only. This is because in the limit $z \to \infty$ the spatial fluctuations are frozen, but the nontrivial dynamics of temporal on-site fluctuations between the four possible states still remains. These states are: (i) state with zero electron occupancy, (ii) state with a single up-spin electron, (iii) state with a single down-spin electron and (iv) state with a double occupancy with electrons up-spin and down-spin. Thus, it is possible to get all one-electron properties from the study of a single-site problem, which describes the effective dynamics of these fluctuations. To this end, the Anderson impurity model serves as a reference system for the Hubbard model. Since the bath describes the same electrons as those on the local site, $\Delta(w)$ needs to be determined from the self-consistency condition [16]:

$$G[\Delta(w)] = \sum_k \left\{ w - \sum[\Delta(w)] - t_k \right\}^{-1} \tag{5.15}$$

where the self-energy term $\sum[\Delta(w)]$ now represents a frequency-dependent potential, and t_k is the Fourier transform of the hopping matrix elements t_{ij} of the solid.

An exact functional of both the charge density and the local Green function of the correlated orbital can be written as [16]:

$$\begin{aligned}
\Gamma[\rho(r),\, G] &= T[\rho(r),\, G] + \int V_{ext}(r)\rho(r)d^3r \\
&+ \frac{1}{2} \int \frac{\rho(r)\rho(r')}{|r - r'|} d^3r d^3r' + E_{XC}[\rho(r),\, G].
\end{aligned} \tag{5.16}$$

This functional [16, 31] has a similar form as in the density functional theory (DFT) (see chapter 2, section 2.6). Here $T[\rho(r),\, G]$ is the kinetic energy of a system with given density $\rho(r)$ and the local Green function G, $V_{ext}(r)$ the potential energy of the crystal system and E_{XC} denotes the exchange and correlation energy. It may, however, be noted that in contrast with DFT, the kinetic energy is no longer that of a non-interacting electron system.

In the limit of small hybridization V_ν, the electron is almost entirely localized at a lattice site, and moves only virtually at short durations compatible with the Heisenberg uncertainty principle. In the other extreme limit of large hybridization, the electron is able to move throughout the crystal system. The mapping of the lattice model onto an impurity model simplifies the spatial dependence of the electron correlations, and also takes into account the local quantum fluctuations missed in static mean-field treatments like the Hartree–Fock approximation. Overall, this leads to a simpler local picture for the competition between itinerant and localized tendencies in d and f electrons underlying the various interesting phenomena exhibited by correlated electron materials. This in turn enabled system-specific modeling of materials combining ideas from both one-electron band theory and many-body theory.

We will now address to the question, how do the local electron density of states or the one-particle spectral function behave as the ratio of the electron correlation strength U to bandwidth W increases? In the one extreme when $U = 0$ i.e. the electron is delocalized, its spectral function closely resembles the local density of states as obtained within the electron band theory. This is shown schematically in figure 5.7(a). The electron density of states has the form of a half ellipse with the

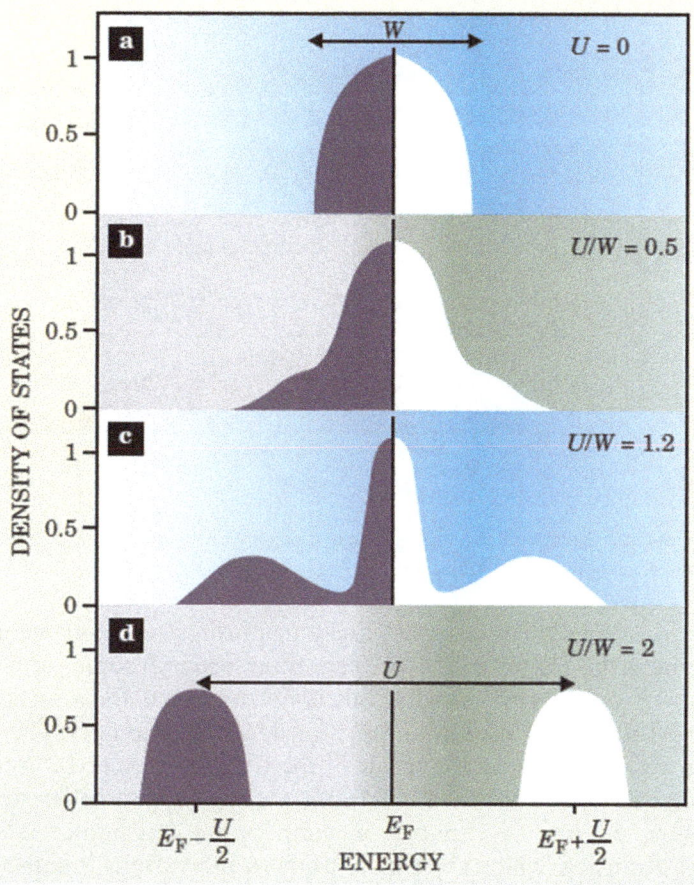

Figure 5.7. Schematic representation of the one-particle spectral function or density of states (DOS) of electrons as a function of electron interaction energy U: (a) the case of non-interacting electrons, (b) weak correlation regime, small U, (c) spectrum for strongly correlated metals exhibiting a characteristic three-peak structure and (d) the Mott metal–insulator transition, where U is much larger than the bandwidth W. (Reprinted from [16] with the permission of AIP Publishing.)

Fermi level E_F, located in the middle of the band, which is the characteristic of a metal. On the other extreme when the electron is localized i.e. U is much larger than W, the density of states peaks at the ionization energy and the electron affinity of the atom. Those atomic-like states lead to the Hubbard bands (see figure 5.7(d)). In the intermediate correlation region between two extreme limits, the one particle-spectral function has features of both quasiparticle and Hubbard bands (see figures 5.7(b) and (c)). In the weakly correlated regime with small U (see figure 5.7(b)), the electrons can be described as quasi-particles whose density of states still resemble free electrons, and the Fermi liquid model (see section 2.7) accounting for the narrow nature of the peak. The one-particle spectral function displays a characteristic three-peak structure in the regime of strongly correlated metals (see figure 5.7(c)). The Hubbard bands, which originate from local 'atomic' excitations are

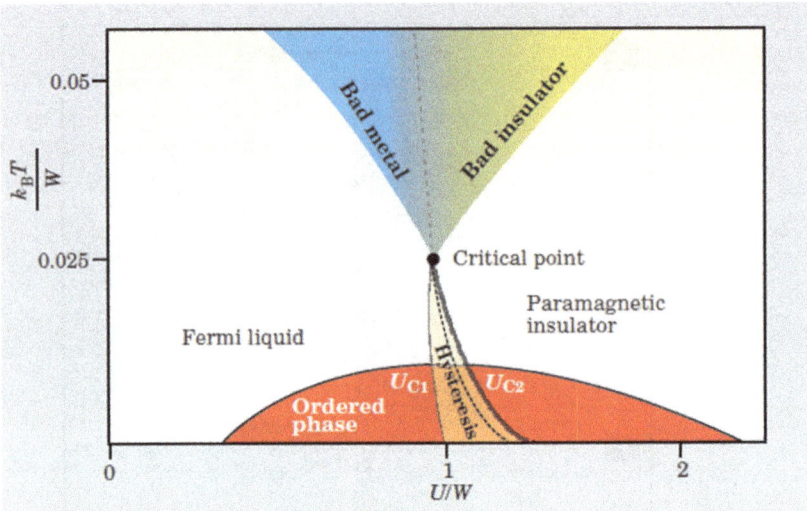

Figure 5.8. Schematic phase diagram of a material undergoing a Mott metal–insulator transition. (Reprinted from [16] with the permission of AIP Publishing.)

broadened by the hopping of electrons away from the atom, and there is the quasi-particle peak near the Fermi level [16]. This three-peak structure appears naturally within the Anderson impurity model, but is not expected for a lattice model that describes the Mott metal–insulator transition [16]. In summary, the Mott metal–insulator transition appears as the result of the transfer of spectral weight from the quasi-particle peak to the Hubbard bands of the correlated metallic state. Externally applied pressure, doping, or variation in temperature can induce this one-particle spectral weight transfer, which eventually leads to the various interesting properties observed around the Mott metal–insulator transition.

It is now quite clear that the Coulomb interactions and the matrix elements that describe electron hopping from site to site are the basic ingredients necessary to calculate the experimental phase diagram of materials near a Mott transition [16]. Figure 5.8 presents qualitatively the features of a partially frustrated Hubbard model in different temperature regimes in the form of a phase diagram, where $\frac{k_B T}{W}$ is plotted as a function of $\frac{U}{W}$. The disappearance of metallic coherence and the closing of the high-energy Mott–Hubbard gap are distinct phenomena that take place in different regions of this phase diagram. These distinct regions are separated at low temperatures by lines of energies U_{C1} and U_{C2} that encompass a hysteretic region around the value U/W. Such hysteresis is a typical characteristic feature of a first order phase transition. One of the significant contribution of the DMFT is the identification of a specific signature of electron correlation in the region near to the Mott MIT line. While a clear gap exists between the lower and upper Hubbard bands on the insulating side, on the metallic side a distinct quasiparticle peak appears in the gap at the Fermi energy. This is exemplified in figure 5.7(c) in the characteristic three-peak structure of one-electron spectral function or density of states obtained for a representative strongly correlated metal with a value of $U/W \sim 1.2$. At high temperature above a second-order critical point, there are two distinct crossover lines that gradually separate the metallic and the insulating phases. Experimentally the Mott transition line and the high temperature part of this phase diagram seem to be universal, but the low temperature part depends on materials and can have various types of long-range magnetic order.

Within the DMFT framework the phase diagram presented in figure 5.8 can be obtained from an electronic model alone without any coupling of electrons to the lattice. Lattice changes, if any, across a first-order phase transition between a paramagnetic insulator and a paramagnetic metal are the consequence, rather than the cause of this discontinuous metal–insulator transition.

We have now seen that the Hubbard model is very suitable for explaining the basic features of the phase diagram of correlated electron systems. However, the Hubbard model does not take into account of the detailed physics of real materials like electronic and lattice structure of the systems. On the other hand we have seen in chapter 3 that the DFT with its local density approximation (LDA) do not need empirical input parameters, and they are quite popular techniques for the electronic structure calculation of the real materials. But DFT/LDA alone is not suitable for describing strongly correlated electron materials like Mott insulators. To this end a combination of the power of DFT/LDA and the model Hamiltonian approach can be very useful for the investigations of real materials [20, 32]. In this LDA + DMFT computational scheme electronic band structure calculations in the local density approximation (LDA) are combined with many-body physics via the local Hubbard interaction and Hund's rule coupling terms, and then the corresponding correlation problem is tackled by DMFT. The many-electron model within the LDA + DMFT scheme comprises of two parts: (i) a kinetic energy which describes the specific band structure of the uncorrelated electrons, and (ii) the local interactions between the electrons in the same orbital as well as in different orbitals [20, 32]. This many-particle problem with various energy bands and local interactions is then solved using DMFT, usually by the application of quantum Monte-Carlo (QMC) techniques [32]. The LDA + DMFT scheme correctly describes the correlation induced dynamics near a Mott–Hubbard MIT and beyond, and it also reproduces the LDA results in the limit of weak Coulomb interaction U.

In fact all the models/frameworks of Mott metal–insulator transition discussed so far are of purely electronic origin and do not take into consideration of any other type of interactions. However, in many real materials showing characteristic properties of Mott insulator, the energy of many other kinds of interactions can be of the same order of magnitude with electron correlation energy U and kinetic energy t. These interactions include electron–lattice, spin–spin, and other types of interactions. Such interactions may also become important in determining the overall electronic properties of such materials. The interplay between electronic, lattice, charge, and other degrees of freedom is in line with the observation that most of the electronic phase transitions in transition metal oxides are accompanied by structural and other types of transition, while intrinsic Mott metal–insulator transitions are purely electronic in origin and not assisted by any other degree of freedom [33]. For example, the *d* orbitals in transition metal oxides are five-fold degenerate in spherical potential, and in a tetrahedral or octahedral crystal field split into two levels because of the broken symmetry: (i) e_g level with two-fold degeneracy, and (ii) t_{2g} level with three-fold degeneracy (see figure 5.9). The localized *d*-electrons in metal-oxide systems can occupy multiple possible orbitals, adding an orbital degree of freedom to the system. Further, a change in crystal structure in a material can influence the electronic band structure, which in turn can be coupled with Mott transitions. For example, the Jahn–Teller effect resulting in the elongation of metal–oxygen bonds along the *z*-axis of the MO_6 octahedron (see figure 5.10(a)) can lift the degeneracy in e_g and t_{2g} levels leading to metal-to-insulator transitions for certain electron configurations [33]. We have earlier discussed the Jahn–Teller effect in chapter 4 (section 4.4) in the context of the 5*f*-Mott insulator system UO_2. The interaction between metal ions in the edge-sharing MO_6 octahedra also modifies the band structures in oxide systems with rutile structure. This is shown in figure 5.10(b). The $d_\parallel$ orbital has electron lobes pointing along the rutile *c*-axis, whereas the π^* orbitals have their electron lobes in the plane perpendicular to the *c*-axis. This aspect of the electron band structure plays an important role in the Mott transition in the metal oxide system VO_2 (to be discussed in chapter 6). In addition, long-range charge or spin ordering can also take place in various materials, and this is shown in figures 5.11(a) and (b) with the examples of charge ordering and spin ordering in oxides with d^7 electron configurations. Charge ordering usually takes place when the electron density is commensurate with the number of lattice sites. A well-known example of charge order state in magnetite Fe_3O_4 was discovered originally by Verwey [34], where charge ordering leads to spatially varying metal–oxygen bond length and insulating properties (see chapter 4, section 4.3.2 and chapter 6, section 6.3.1). The other possible charge order systems are Ti_4O_7 and several perovskite

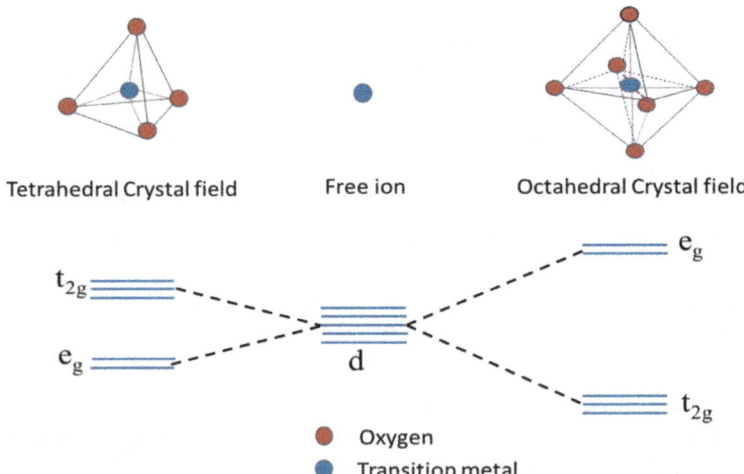

Figure 5.9. Schematic representation of the interplay between various degrees of freedom in Mott insulators: crystal field splitting of the originally five-fold degenerate *d*-orbitals into a two-fold degenerate e_g and a three-fold degenerate t_{2g} orbitals.

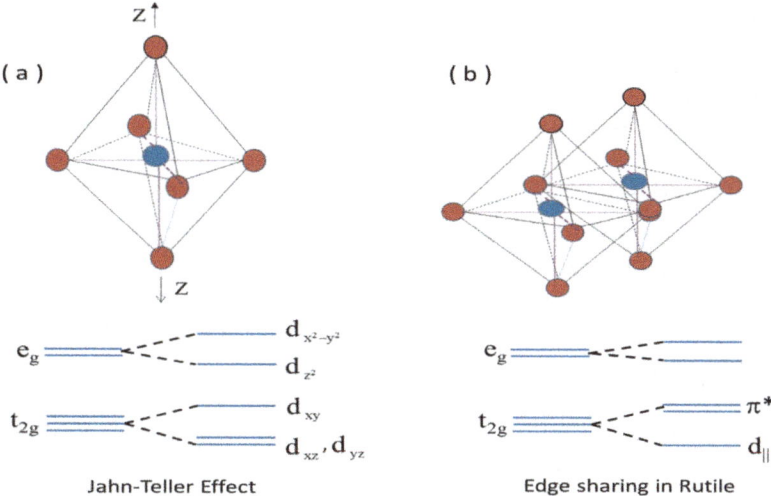

Figure 5.10. Schematic representation of the interplay between various degrees of freedom in Mott insulators and the dependence of the relative energy of e_g and t_{2g} levels: (a) the Jahn–Teller effect and (b) metal-to-metal interaction in oxide materials with rutile structure.

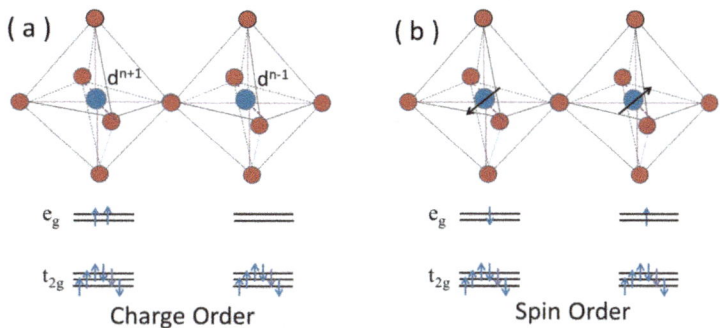

Figure 5.11. Schematic representation of the interplay between various degrees of freedom in Mott insulators and the dependence of the relative energy of e_g and t_{2g} levels: (a) charge ordering and (b) spin ordering.

oxides [2]. In a similar way spin ordering may also affect the electron transport properties and can be a natural result of charge localization. A metal–insulator transition involving double exchange mechanism has been a subject of detail investigations in systems like $R_{1-x}A_xMnO_3$ (where R = La, Nd, Pr etc, and A = Ca, Sr, Ba, Pb etc) showing colossal magneto-resistance (see chapter 4, section 4.3.1 and chapter 6, section 6.3.2 and also [2]). It is now well accepted that in many real systems, in addition to the strong electron correlation the charge, lattice, spin, and orbital degrees of freedom are coupled in Mott transition. This gives rise to complex phase diagrams, which are better explained by the cooperation between various interactions instead of strong electron correlation as the lone driving force.

5.2.3 What is the nature of Mott metal–insulator transition?

The exact nature of Mott metal–insulator transition still remains a subject of considerable debate [35]. In a pure Mott-MIT the involved metal and insulator phases share the same symmetries, and the clear cut distinction between them is apparent only at zero temperature. If a direct and continuous transition between a paramagnetic metal and a paramagnetic Mott insulator is established at $T = 0$, then it would be an obvious example of a quantum critical point, and that is beyond the realm of standard Landau theory of phase transition involving a mechanism of spontaneous symmetry breaking. However, as discussed in the section above, in most of the experimental situations, the Mott MIT is often accompanied by magnetic, charge, structural, or orbital ordering. The characteristic temperature scale T_C, below which many of such ordered phases form, is relatively small in comparison to Fermi energy E_F representing the quantum fluctuations and the Coulomb repulsion. This actually results in the very sharp crossover between metallic and insulating behavior in the experimental measurements of various physical quantities in the temperature range even well above T_C. The important question to be answered here is: what is the main physical mechanism behind this finite-temperature Mott-MIT observed in various systems? Should it be viewed as a quantum critical phenomenon influenced by appropriate order-parameter fluctuations, or is it a dynamical phenomenon not directly related to any ordering tendency as originally suggested by Mott and Anderson [35]. In order to address this question, all ordering tendencies need to be suppressed at least in the relevant temperature range, and then achieve an understanding of the remaining physical processes controlling the resultant finite-temperature crossover regime pertinent to Mott-MIT and the associated quantum critical region if there is any. This is rather a formidable problem from the theoretical point of view, especially for realistic model systems. Some theoretical approaches beyond mean-field theories will be discussed in the next section. There has also been some progress in the experimental front in recent times to understand the nature of metal–insulator transition in pure Mott systems without any intervening orderings, and that will be discussed in chapter 6.

5.2.4 Beyond mean-field theories

The various theoretical approaches to the Mott metal–insulator transition described above, including Fermi-liquid based approaches, the Hubbard approximation, the Gutzwiller approximation and the infinite-dimensional approach of DMFT, have a common feature: they all rely on the mean-field approximation. The range of mean-field approximations encompasses dynamically correct treatments in DMFT to simple static approximations on the Hartree–Fock level. The magnetic transition in metals is described by mean-field fixed points for antiferromagnetic transitions at dimensions $D \geqslant 3$ and for ferromagnetic transitions at $D \geqslant 2$ or $D \geqslant 3$ under certain assumptions [2]. In general, the mean-field approximations are justified if the dimension is high enough. However, in comparison to simple magnetic transitions of the Ising model, the estimation of upper critical dimension for the validity of the mean-field is not so straightforward in the case of the Mott metal–insulator transition of fermionic models. Here the infinite-dimensional approach of DMFT provides an example where the mean-field approximation becomes exact at $D = \infty$ by taking account of dynamic fluctuations correctly. Other mean-field approximations do not correctly take account of the on-site dynamic fluctuations and incoherent excitations. This leads to failure of such methods even in the limit of large dimensions, because incoherent excitations become very significant near the Mott metal–insulator transition point.

Temporal, as well as spatial, fluctuations originating from the thermal and quantum effects become important in low-dimensional systems. The mean-field theories for phase transitions become invalid in low dimensional systems below the upper critical dimension due to large spatial fluctuations. In this situation phase transitions are believed to be described in general by scaling theory [2]. In fact, scale invariance is one of the most interesting concepts in physics, a prominent example of which is the critical end point of the liquid–gas transition of water. At this point the correlation length diverges and fluctuations take place at all length scales. Moreover, the bulk response of the many particle systems showing phase transition belongs to certain 'universality classes' irrespective of the variety involved in such systems, which may include atoms, molecules, electrons and their spins [36, 37]. Continuous metal–insulator transition has been analyzed in terms of the scaling concept in the Anderson localization problem in the 1970s [38, 39]. On the other hand, theoretical studies on Mott-insulator indicated that Mott-transition belong to the Ising universality class [18, 40]. This theoretical prediction received experimental support from the study on the canonical Mott-insulator system $(V_{1-x}Cr_x)_2O_3$ [41]. This system shows a distinct critical end point in the pressure (P)–temperature (T) phase diagram (see figure 6.1), and the conductivity measurements near the critical end point showed a mean-field behavior with a crossover to scaling properties with 3D-Ising critical exponents. However, a question was then raised on this matter with the experimental findings of an unconventional critical response in a quasi-2D organic charge-transfer compound [42, 43]. It was subsequently suggested that the apparent disagreement between theory and experimental results on a quasi-2D organic charge-transfer compound, can possibly be sorted out if the experimental results were analyzed with a modified scaling law in the vicinity of critical end point [44]. (The experimental results will be discussed in more detail in chapter 6.)

In this theoretical approach, starting with the singular part of Gibbs free energy for 2D-Ising universality class, it was shown that the Gruneisen parameter diverges at a critical point. And then a new scaling function for expansivity was derived [44]. Here the expansivity is expressed as:

$$\alpha_p = V^{-1}\frac{\partial V}{\partial P} \tag{5.17}$$

and the Gruneisen parameter is defined as:

$$\Gamma_p = \frac{\alpha_p}{C_p} \tag{5.18}$$

where C_p is the specific heat at constant pressure.

The identification of the pressure p and the temperature T as independent parameters to tune the phase transition, already implies that the change of volume ΔV at the Mott transition is proportional to the order parameter m. This is a consequence of the fact that the volume V and pressure p are conjugate variables, and in turn it implies that the change in volume across the Mott transition should scale as $\Delta V \propto (T_c - T_0)^{1/\beta}$, where $\beta = 1/8$ for the 2D Ising universality class [44]. This is quite similar to the well known liquid–gas transition, but with a different value of the critical exponent β.

The singular part of Gibbs free energy $f(t, h)$ is expressed as:

$$f_s(t, h) = \frac{t^2}{8\pi} \ln t^2 + |h|^{\frac{d}{y_h}} \Phi\left(t/|h|^{\frac{y_t}{y_h}}\right).$$ (5.19)

Here $t = (T - T_c)/T_c$, $h = (p - p_c)/p_c$ are temperature and pressure like scaling variables, Φ is the scaling function, T_c and p_c are critical temperature and pressure of Mott transition, and d is the dimension of the system. The scaling law for the Gruneisen parameter with $h = \pm 0$ and $t < 0$ was derived as [44]:

$$\Gamma \propto \text{sgn}(h)(-t)^{-1+\alpha+\beta}.$$ (5.20)

It is clear from equation (5.20) that for a 2D Ising model ($\alpha = 0$, $\beta = 0.25$) the Gruneisen parameter Γ diverges at the critical point. In the next step the scaling form of the expansivity was derived with a new scaling function in the vicinity of critical point [44]:

$$\alpha_p(t, h) \propto \text{sgn}(h)|h|^{-1+(d-y_t)/y_h} \Psi(t/|h|^{y_t/y_h}).$$ (5.21)

Here Ψ is the new scaling function and it is defined as:

$$\Psi(x) = \frac{d - y_t}{y_h} \Phi'(x) - \frac{y_t}{y_h} x \Phi''(x).$$ (5.22)

$\Phi(x)$ is obtained numerically for the 2D Ising universality class, for which $y_t = 1$ and $y_h = 15/8$. It was assumed that in the vicinity of critical point, the scaling law could be expressed with a linear combination of temperature and pressure, and accordingly t and h was written as $t = (T - T_c - \zeta(p - p_c))/T_0$ and $h = (p - p_c - \lambda(T - T_c))/p_0$. So, in addition to the critical temperature T_c and pressure p_c, the scaling variables now depend on the four non-universal constants T_0, p_0, ζ and λ. With the assumption that the non-singular background contribution to the thermal expansivity is approximately linear, the scaling function for thermal expansivity in the 2D Ising universality class can be written as [44]:

$$\alpha(T, P) = A \, \text{sgn}[p - p_c - \lambda(T - T_c)] \times [p - p_c - \lambda(T - T_c)]^{-7/15}$$
$$\times \Psi\left(\frac{B[T - T_c - \zeta(p - p_c)]}{[p - p_c - \lambda(T - T_c)]^{8/15}}\right) + C + D[T - T_c - \zeta(p - p_c)].$$ (5.23)

This scaling function for thermal expansivity was successfully used to analyze the experimental results on thermal expansivity [43], and thus supports the conjecture that the Mott transition in the quasi-2D organic charge-transfer compounds belongs to the 2D Ising universality class. It may, however, be noted that in contrast with thermal expansivity, electrical conductivity has also been used as the order parameter to determine critical exponents in the experimental results reported for Cr-doped V_2O_3 [41] and quasi-2D organic charge-transfer compounds [42]. So, the question on the order parameter associated with Mott transition is yet to be settled completely.

In addition, various numerical methods have also been developed for the purpose of obtaining insight to the Mott metal–insulator transition without involving many approximations. Such methods include the quantum Monte Carlo method, exact diagonalization and second order perturbative calculations. Interested readers may consult the article by Imada *et al* [2] (and references within) to know more on such numerical techniques.

Further reading

1. Mott N F 1990 *Metal-Insulator Transitions* (London: Taylor and Francis).
2. Gebhard F 1998 *Mott Metal–insulator Transition* (Berlin: Springer).
3. Edwards P P and Rao C N R (ed) 1995 *Metal-insulator Transitions Revisited* (London: Taylor and Francis).
4. Martin R M, Reining L and Ceperley D M 2016 *Interacting Electrons: Theory and Computational Approaches* (Cambridge: Cambridge University Press).
5. Freerick J K 2006 *Transport in Multi Layered Nanostructures: The Dynamical Mean-field Theory Approach* (London: Imperial College Press).
6. Wills J M, Alouani M, Andersson P, Delin A, Eriksson O and Grechneyv O 2010 *Full Potential Electronic Structure Method* (Berlin: Springer).

Self-assessment questions and exercises

1. What is the control parameter in bandwidth-control Mott transition and how can it be tuned to induce the transition?
2. What is the control parameter in filling-control Mott transition and how can it be tuned to induce the transition?
3. What are the practical difficulties of applying density functional theory to Mott insulators?
4. What function represents the mean-field in dynamical mean field theory (DMFT), and why it is called dynamic? What does this mean-field in DMFT take into account, which is not captured in static mean-field treatments like the Hartree–Fock theory?
5. Is symmetry breaking across the transition a necessary condition for Mott transition?
6. How does the one-electron spectral function obtained within DMFT in the correlated metals just beyond the Mott transition differs from the predictions of simple Fermi liquid model?

7. What are the possible order parameters to distinguish a Mott metal–insulator transition?

References

[1] Edwards P P and Rao C N R (ed) 1995 *Metal-insulator Transitions Revisited* (London: Taylor and Francis)

[2] Imada M, Fujimori A and Tokura Y 1998 *Rev. Mod. Phys.* **70** 1040

[3] Edwards P P, Johnston R L, Henstell F, Rao C N R and Tunstall D P 1999 *Solid State Physics* vol 52 ed F Seitz and D Turnbull p 229

[4] Mott N F 1956 *Can. J. Phys.* **34** 1356

[5] Mott N F 1990 *Metal-Insulator Transitions* (London: Taylor and Francis)

[6] Terakura K, Oguchi T, Williams A R and Kübler J 1984 *Phys. Rev.* B **30** 4734

[7] Sarma D D, Shanthi N, Barman S R, Hamada N, Sawada H and Terakura K 1995 *Phys. Rev. Lett.* **75** 1126

[8] Perdew J P 1986 *Phys. Rev.* B **33** 8822

[9] Becke A D 1988 *Phys. Rev.* A **38** 3098

[10] Savrasov S Y and Kotliar G 2000 *Phys. Rev. Lett.* **84** 3670

[11] Perdew J P and Zunger A 1981 *Phys. Rev.* B **23** 5048

[12] Svane A and Gunnarsson O 1988 *Phys. Rev.* B **37** 9919
Svane A and Gunnarsson O 1988 *Phys. Rev. Lett.* **65** 1148

[13] Anisimov V I, Zaanen J and Andersen O 1991 *Phys. Rev.* B **44** 943

[14] Kulik H J 2015 *J. Chem. Phys.* **142** 240901

[15] Kotliar G 1995 The Mott transition: recent results, more surprises *Metal-insulator Transitions Revisited* ed P P Edwards and C N R Rao (London: Taylor and Francis) p 317

[16] Kotliar G and Vollhardt D 2004 *Phys. Today* **57** 53

[17] Martin R M, Reining L and Ceperley D M 2016 *Interacting Electrons: Theory and Computational Approaches* (Cambridge: Cambridge University Press)

[18] Georges A, Kotliar G, Krauth W and Rozenberg M 1996 *Rev. Mod. Phys.* **68** 13

[19] Pruschke T, Jarrell M and Freericks J K 1995 *Adv. Phys.* **44** 187

[20] Kotliar G, Savrasov S Y, Haule K, Oudovenko V S, Parcollet O and Marianetti C A 2006 *Rev. Mod. Phys.* **78** 865

[21] Gebhard F 1998 *Mott Metal-insulator Transition* (Berlin: Springer)

[22] Freerick J K 2006 *Transport in Multi Layered Nanostructures: The Dynamical Mean-field Theory Approach* (London: Imperial College Press)

[23] Hubbard J 1963 *Proc. R. Soc.* A **276** 238

[24] Hubbard J 1964 *Proc. R. Soc.* A **281** 401

[25] Gutzwiller M C 1965 *Phys. Rev.* **137** A1726

[26] Brinkman W F and Rice T M 1970 *Phys. Rev.* B **2** 4301

[27] Metzner W and Vollhardt D 1989 *Phys. Rev. Lett.* **62** 324

[28] Georges A and Kotliar G 1992 *Phys. Rev.* B **45** 6479

[29] Wills J M, Alouani M, Andersson P, Delin A, Eriksson O and Grechneyv O 2010 *Full Potential Electronic Structure Method* (Berlin: Springer)

[30] Brandt U and Mielsch C 1989 *Z. Phys.* B **75** 365
Brandt U and Mielsch C 1990 *Z. Phys.* B **79** 295
Brandt U and Mielsch C 1991 *Z. Phys.* B **82** 37

[31] Kotliar G and Savarasov S Y 2002 *Dynamical Mean Field Theory, Model Hamiltonians and First Principles Electronic Structure Calculations* (arXiv: cond-mat/0208241)

[32] Vollhardt D 2012 *Ann. Phys.* **524** 1

[33] Zhou Y and Ramanathan S 2015 *Proc. IEEE* **103** 1289

[34] Verway E 1939 *Nature* **144** 327

[35] Vucicevic J, Terletska H, Tanaskovic D and Dobrosavljevic V 2013 *Phys. Rev.* B **88** 075143

[36] Goldenfeld N 1992 *Lectures on Phase Transitions and the Renormalization Group* (Boston, MA: Addison-Wesley)

[37] Cardy J L 1996 *Scaling and Renormalization in Statistical Physics* (Cambridge: Cambridge University Press)

[38] Wegner F 1976 *Z. Phys.* B **25** 327

[39] Abrahams E, Anderson P W, Licciardello D C and Ramakrishnan T V 1979 *Phys. Rev. Lett.* **42** 673

[40] Park H, Haule K and Kotliar G 2008 *Phys. Rev. Lett.* **101** 186403

[41] Limelette P, Georges A, Jerome D, Wzietek P, Metcalf P and Honig J M 2003 *Science* **302** 89

[42] Kagawa F, Miyagawa K and Kanoda K 2005 *Nature* **436** 534

[43] de Souza M, Brühl A, Strack C, Wolf B, Schweitzer D and Lang M 2007 *Phys. Rev. Lett.* **99** 037003

[44] Bartosch L, de Souza M and Lang M 2010 *Phys. Rev. Lett.* **104** 245701

IOP Publishing

Mott Insulators
Physics and applications
Sindhunil Barman Roy

Chapter 6

Experimental studies on Mott metal–insulator transition

We have studied in chapters 3 and 4 that Mott insulators represent a large class of materials with half-filled d or f orbitals, which are supposed to be metallic according to the one-electron band theory, but are found to be insulating in nature. In these systems, the on-site electron–electron or Coulomb repulsion energy U, which is not taken into account properly in conventional band theory, gives rise to an insulating ground state when U is larger than the one-electron bandwidth (W) or kinetic energy. Then we learnt in chapter 5 that an interesting characteristic of Mott insulators is that the external perturbations may induce a metal–insulator transition (MIT). For example, MIT can take place with the application of physical or chemical pressure or by the variation of temperature. Application of an external pressure on a Mott insulator causes a volume contraction, which increases the kinetic energy or bandwidth W and thus reduces the effect of Coulomb repulsion energy. The other possible way is to change the effective U between electrons by introducing electrons/holes into a Mott insulator. In this situation the energy cost for hopping is reduced for certain electrons, hence reducing U. Thus, a deviation from half band-filling caused by electron/hole doping may also induce an MIT. Formally, Mott-MIT can be classified into two categories: (i) bandwidth control (BC)-MIT, which changes the effective electron correlation strength represented by the ratio U/W, and (ii) band-filling control (FC)-MIT, which changes the integer electron number n per atom site by chemical or electrostatic doping. Figure 5.8 in chapter 5 indicates that the temperature controlled Mott-MIT occurs in a narrow window around $U/W \approx 1$. It may be noted that this Mott-MIT is between a low temperature metal and a high temperature insulator state, which is in contrast with the more usual transition between low temperature insulator and a high-temperature metal. In the sections below we will discuss these classes of Mott-MIT taking examples of a few well studied materials. Several classes of Mott-insulators showing MIT have actually been identified during the last three decades, and hence the examples of Mott insulators presented in this chapter, are by no means exhaustive. There is a huge amount of literature available on Mott insulators. In fact there are more than 500 papers on the Mott-insulator vanadium sesquioxide V_2O_3 alone, even in the pre-2000 period [1]. The materials chosen in this chapter for discussion are therefore just to give a flavour of the experimental results, which will be helpful in the subsequent discussions on the real applications of Mott-insulator in chapter 8.

The material NbO_2 with a MIT taking place at a relatively high temperature of 1083 K is an interesting material from a device applications points of view (see section 8.4.2). While this MIT was earlier thought to be a Peierls type MIT, we will discuss below the recent studies, which would reveal that Mott physics plays an important role in this compound.

We have also discussed in chapter 5 that in strongly correlated electron systems the interplay between the spin, charge, and orbital degrees of freedom can give rise to versatile electronic phases as well as possible electronic phase separation, all of which play important roles in the MIT. In addition there can be coupling with the lattice degree of freedom. In this situation the fractional cases beyond the half-band filling can also have an insulating ground state accompanying the regular ordering of localized charge known as charge ordering. This state is often accompanied by spin and orbital ordering. In this regard we will first discuss the interesting properties of magnetite (Fe_3O_4; the oldest known magnet) including an ambient pressure metal–insulator Verwey transition around 122 K. We will then discuss MITs in Mn-oxide compounds including the ones showing colossal magneto-resistance (CMR) and in a sulphide compound namely TaS_2.

The metal insulator transition in Sr_2IrO_4 will be discussed to highlight the debate whether the low temperature antiferromagnetic insulating state is a Mott insulator or a Slater insulator, where the insulating gap is associated with the onset of long range antiferromagnetic order. Lastly, we will discuss the possibility of a Mott-insulating state involving the surface atoms of covalently bonded band semiconductors.

A detail discussion of the experimental results of the pre-2000 period on various classes of Mott insulators and related systems can be found in the excellent review articles by Imada *et al* [1] and Edwards *et al* [2] and the book edited by Edwards and Rao [3]. However, the field has expanded rapidly since then. Apart from the study of the newer materials, there are also even re-investigations of many of the already well studied Mott-insulators highlighting newer interesting features of those systems.

6.1 Temperature induced and bandwidth control Mott metal–insulator transition

6.1.1 Canonical Mott insulator Cr-doped V_2O_3

The canonical Mott insulators are those systems where the Mott-insulating state arises solely due to strong electron correlation and without involving any other degree of freedom. The most well studied canonical Mott insulator is possibly the transition metal oxide compound $(V_{1-x}Cr_x)_2O_3$, which has been investigated over the last 50 years through various kinds of macroscopic [4–7] and microscopic [8] experimental techniques. Figure 6.1 shows the phase diagram of $(V_{1-x}M_x)_2O_3$, where M = Cr and Ti. In this system, a change in the V/M ratio by 1% is equivalent to the application of an external pressure of ≈4 kbar [4]. The experimental phase diagram of $(V_{1-x}Cr_x)_2O_3$ matches well with the theoretically predicted phase diagram (figure 5.8) for MIT in Mott-insulators. Figure 6.2(a) shows that the application of a moderate pressure induces a bandwidth control (BC)-MIT in $(V_{1-x}Cr_x)_2O_3$ with $x = 0.0375$. It may be noted that in spite of the strong reduction of volume associated with this BC-MIT indicating the first order nature of this phase transition, there is no structural change involved in this transition. Both the lower pressure paramagnetic Mott-insulator phase and the higher pressure paramagnetic metallic phase remains in the trigonal R-3c phase. Thus, this BC-MIT takes place without any crystal lattice symmetry breaking. The Mott-MIT line separating a Mott insulating phase from a metallic phase terminates to critical end point around 450 K. The presence of quasi-particle peak in the Mott–Hubbard gap as observed in the photo-emission spectra in the paramagnetic metallic side, just beyond the BC-MIT line in pure V_2O_3, established the strongly correlated nature of the para-magnetic metal [9]. The BC-MIT line is not quite vertical, and for a narrow region of very small Cr-concentration around 1% there is a transition from the low temper-ature paramagnetic metal to high temperature paramagnetic Mott-insulator state, again without any crystal lattice symmetry breaking. This transition is termed as temperature control metal–insulator transition or TC-MIT. This is exemplified in figure 6.2(b) with the resistance versus temperature plots for $(V_{1-x}Cr_x)_2O_3$ with $x = 0.006$ and 0.012. As shown in the inset of figure 6.2(b) this transition is observed only in a narrow range of Cr-doping in the pressure-temperature phase diagram. This is again in line with the theoretical prediction of such temperature induced MIT in a narrow region around $U/W \sim 1$ (see figure 5.8). Finally, further down the temperature there is a transition from the paramagnetic metallic state with trigonal crystal structure (R-3c) to a low temperature antiferromagnetic state with a monoclinic crystal structure (I2/a) both in pure and Cr-doped V_2O_3. In pure V_2O_3

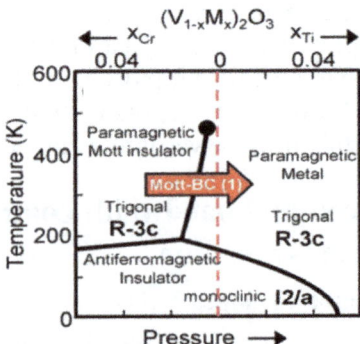

Figure 6.1. Pressure-temperature phase diagram of $(V_{1-x}M_x)_2O_3$ with M = Cr and Ti, highlighting the bandwidth-control metal–insulator transition (BC-MIT) line between paramagnetic Mott-insulator and paramagnetic-metal. (Reproduced from [10] with the permission of John Wiley & Sons.)

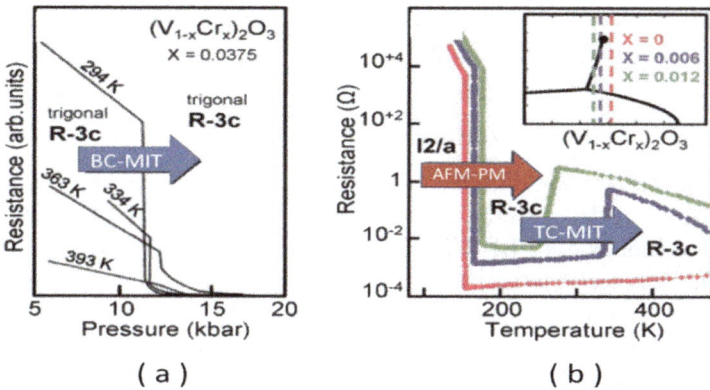

Figure 6.2. (a) Resistance versus pressure plots across BC-MIT in $(V_{0.9625}Cr_{0.0375})_2O_3$ at various constant temperatures. (b) Resistance versus temperature plots for V_2O_3 and $(V_{1-x}Cr_x)_2O_3$ with $x = 0.006$ and 0.012, highlighting the transition from low temperature insulating antiferromagnetic state to paramagnetic metallic state to paramagnetic Mott-insulator state. The last transition is observed only in a narrow range of Cr-doping in the pressure-temperature phase diagram as shown in the inset of the figure. (Reproduced from [10] with the permission of John Wiley & Sons.)

this transition takes place around 165 K. In many quarters this transition, which is associated with a crystal lattice symmetry breaking and apparently driven by an additional mechanism that is magnetic ordering in this case, is not considered to be pure Mott transition [10]. However, it may also be recalled that we have discussed earlier in section 3.7 that Mott insulators can be antiferromagnet when neighboring spins are oppositely aligned, because there can be a gain an energy $4t^2/U$ by virtual hopping. In figure 6.2(b) this transition is marked as insulating antiferromagnet (AFM) to metallic paramagnet (PM) transition. Moreover, relatively recent experimental results reveal a much richer phenomenology including path dependent electronic phase separation on a microscopic scale in V_2O_3, which needs to be taken into account in the complete picture of this model Mott-insulator system [8].

The critical behavior of the conductivity has been studied in detail near the Mott insulator to metal critical endpoint of Cr-doped V_2O_3 as a function of pressure and temperature [7]. The pressure-dependent data-sets for many different temperatures could be collapsed onto a single curve, which indicated the existence of a scaling law. The results obtained near the critical end point revealed mean-field behavior, with a crossover to scaling properties with 3D Ising critical exponents (see chapter 5, section 5.2.3). The observed behaviour is quite similar to the familiar liquid–gas transition.

Overall with all these characteristic experimental features, the Cr-doped V_2O_3 system is presently considered to be a canonical Mott-insulator. Apart from Cr-doped V_2O_3, by the early 2000 few other systems like $NiS_{2-x}Se_2$, $RNiO_3$ (R = La, Nd, Pr etc) $Ca_{1-x}Sr_xVO_3$ were identified to be canonical Mott-insulators showing BC-MIT. These systems have been covered in considerable detail in the review articles by Imada *et al* [1]. Instead of discussing these systems, we will rather discuss here two newer canonical Mott-insulators namely the 2D molecular system

κ-(BEDT-TTF)$_2$X and the series of chalcogenides, the AM$_4$Q$_8$ compounds (A = Ga, Ge; M = V, Nb, Ta, Mo; Q = S, Se, Te).

6.1.2 Case study of organic Mott-insulators κ-(BEDT-TTF)$_2$X

The family of quasi-two-dimensional layered charge-transfer salts κ-(BEDT-TTF)$_2$X provide model systems for the investigation of the Mott metal–insulator transition in 2-dimensions. BEDT-TTF stands for bis-(ethylenedithio)tetrathiaful-valene and X represents various kinds of anions. These organic compounds have layered structures comprising of the conducting BEDT-TTF layers and the insulating X layers. The conducting layer consists of an anisotropic triangular lattice formed by the BEDT-TTF dimers. Two degenerate highest occupied molecular orbitals (HOMOs) belonging to the neighboring BEDT-TTF's are split into bonding and antibonding HOMOs in a dimer [11]. They form respective bands due to the inter-dimer transfer integrals, and according to the band-structure calculations the two bands are separated in energy [11]. The anion X introduces one hole to one dimer, and hence the antibonding HOMO band, which is the upper band and is half filled [12].

The compound κ-(BEDT-TTF)$_2$Cu[N(CN)$_2$]Cl, commonly known in its abbreviated form κ-Cl, has a very rich phase diagram with paramagnetic insulating, antiferromagnetic insulating, semiconducting, and various metallic phases with the variation of pressure and temperature [13–16]. Figure 6.3 shows the

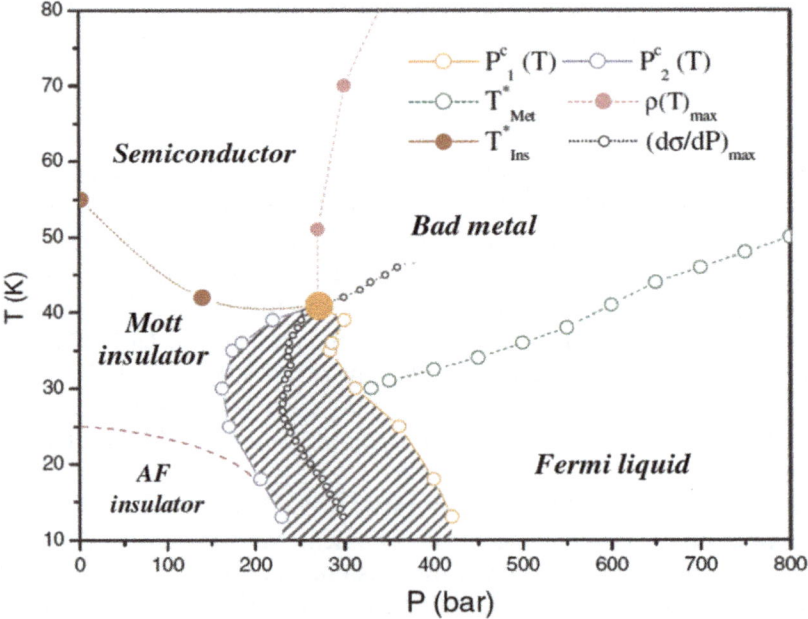

Figure 6.3. Pressure–temperature phase diagram of κ-(BEDT-TTF)$_2$Cu[N(CN)$_2$]Cl compound showing the first order Mott-MIT line and its critical end point. See text for details. (Reprinted figure with permission from [16]. Copyright (2003) by the American Physical Society.)

pressure–temperature phase diagram obtained with the help of in-plane electrical resistivity measurements [16]. Within the paramagnetic region the transition lines distinguish four different phases, namely a Mott-insulator phase, a semiconducting phase, a bad metal phase and a Fermi-liquid metal phase. The Mott-MIT is observed when the pressure is varied over a range of a few hundred bars. A first order transition line is identified in terms of the points where the variation of the electrical conductivity with respect to the applied pressure is maximum. As predicted by the theory (see figure 5.8) and as was observed experimentally in $(V_{1-x}Cr_x)_2O_3$, this first order phase transition line terminates into a critical end point. The hatched region in figure 6.3 marks the region of coexistence of Mott-insulator and correlated metallic phases. Such phase-coexistence is a characteristic of first order phase transition (see chapter 8). In the pressure region below 250 bar, κ-Cl shows the important characteristic of canonical Mott-insulator namely it is a paramagnetic insulator which does not undergo a temperature induced insulator to metal transition with further increase in temperature. However, in a narrow region of pressure beyond 250 bar this phase diagram indicates the possibility of a temperature induced first order phase transition from the low-temperature paramagnetic correlated metallic state to a high temperature paramagnetic Mott-insulator state. There is also a transition line in the phase diagram from the paramagnetic Mott-insulator to a low temperature antiferromagnetic insulating phase. This phase transition line was added from the results of an earlier study [15]. A similar pressure-temperature phase diagram with a first order Mott-MIT line terminating into critical end point was also obtained by Kagawa et al [11]. This later phase diagram also indicated the existence of a superconducting state below about 15 K in certain pressure regime.

The critical phenomena has been studied in detail in the κ-(BEDT-TTF)$_2$Cu[N (CN)$_2$]Cl compound near the critical end point of P–T phase diagram through conductivity measurements [17]. The analysis of the results indicated the presence of novel critical phenomena in the quasi-2D compound that perhaps belonged to a new universality class. However a subsequent study of thermal expansivity measurements on quasi-2D compound κ-(BEDT-TTF)$_2$Cu[N(CN)$_2$]Br and a detailed scaling theory [18] showed that the critical behaviour in this family of quasi-2D Mott-insulator could be rationalized within the framework 2D-Ising universality class.

The generic pressure–temperature phase diagram for both Cr-doped V_2O_3 and κ-(BEDT-TTF)$_2$X compound, however, reveals some striking differences with the corresponding liquid–solid pressure–temperature phase diagram. Liquid usually transforms to a solid on cooling or compression, and the well-ordered solid crystal phase has a reduced entropy. The continuous rotational and translational symmetry that characterizes the liquid is spontaneously broken. In contrast, the behaviour is opposite in the canonical Mott-insulators, which is quite similar to what has been observed in the well known Fermi liquid system ^{3}He below $T = 0.3$ K. In ^{3}He, heating the fluid under a pressure of about 3 MPa transforms it into a solid. This peculiarity of the melting curve is known as the Pomeranchuk effect, and the Clausius–Clapeyron relation suggests that solid-^{3}He has a larger entropy than the liquid phase [19]. In a similar way the existence of the low temperature metallic thermodynamic state of canonical Mott-insulators indicates that the mobile electrons in the Fermi liquid state are supposed to have less entropy than those localized in the higher temperature Mott-insulating state. This peculiar feature of electronic systems in Mott-insulators is really difficult to observe experimentally because of the intervention of magnetic order in most of the Mott-insulator systems. In this respect the recent suggestion of the existence of a quantum spin liquid state in several organic Mott insulators (see [20] and references within) provides an opportunity to explore the generic phase diagram of canonical Mott-insulators in further detail.

Three well-characterized organic Mott insulators, β'-EtMe$_3$Sb[Pd(dmit)$_2$]$_2$ (EtMe) (where EtMe$_3$Sb stands for ethyltrimethylstibonium, and dmit for 1,3-dithiole-2-thione-4,5-dithiolate), κ(BEDT-TTF)$_2$Ag$_2$(CN)$_3$(AgCN) (where BEDT-TTF stands for bis(ethylenedithio)tetrathiafulvalene) and κ-(BEDT-TTF)$_2$Cu$_2$(CN)$_3$(CuCN) have been identified as potential quantum spin liquid systems and investigated for prototype realizations of the single-band Hubbard model in the absence of magnetic order. The Hubbard bands are mapped by optical spectroscopy study, which provides an absolute measure of the electron interaction strength U and electron bandwidth W [20]. A generic phase diagram for canonical Mott-insulators is obtained (see figure 6.4) in terms of the interaction strength (U/W) and temperature

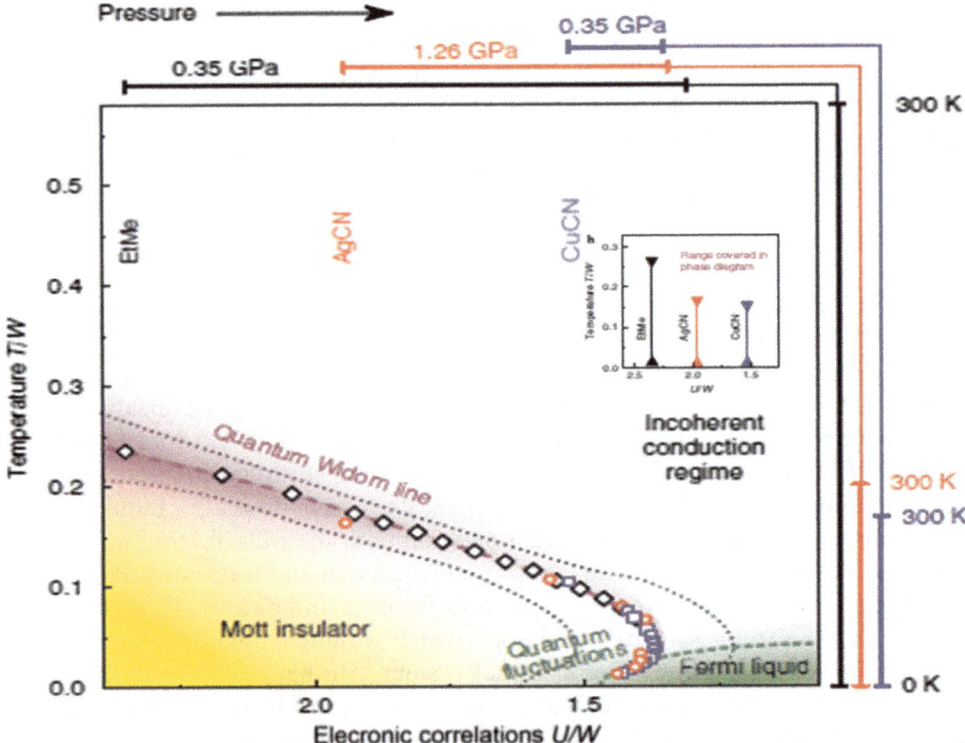

Figure 6.4. Normalized temperature versus electronic correlations generic phase diagram of canonical Mott-insulators, generated from the experimental results on organic Mott-insulators EtMe, AgCN and CuCN. (Reprinted from [20] with the permission of Springer Nature.)

(T/W) normalized to the bandwidth W. In this generic phase diagram, the three organic compounds are quantitatively arranged in the order EtMe ($U/W = 2.35$), AgCN (1.96) and CuCN (1.52) on the descending horizontal U/W scale [20] (see inset of figure 6.4). The interaction energy $U \approx 220$ meV is rather small in EtMe, but the very narrow bandwidth of $W \approx 90$ meV makes it the most-correlated electron system in this group of organic Mott-insulators. The experimentally accessed temperatures (5–300 K) also cover a much broader vertical T/W range in comparison to the κ-phase compounds [20]. In this generic phase diagram, the Mott-insulator state is separated from the Fermi-liquid state by a first-order transition line at low temperatures. There is a crossover above the critical endpoint where the phase transition line gets converted into a quantum Widom line (see figure 6.4); this arises from the interplay of U, W and T and separates the Mott insulating state from an incoherent conduction state of the material. The Widom line is generally associated with the thermodynamic transition of a pure fluid but has been observed in various other areas including magnetic phase transitions (see [21] and references therein). It is defined as the set of states with a maximum correlation length of the fluid [21, 22]. The optical conductivity study in the above mentioned organic Mott-insulators revealed the signature of metallic quantum fluctuations inside the Mott-insulator state as a precursor of the Mott insulator-metal transition [20]. In the effort for extending such a generic phase diagram to the transition metal compounds like Cr-doped V_2O_3, it has been argued that the much larger energy scales that characterize these compounds push the first-order Mott-transition all the way up to room temperature and the Widom line beyond the scales accessible in the laboratory [20].

6.1.3 Case study of the AM_4Q_8 compounds (A = Ga, Ge; M = V, Nb, Ta, Mo; Q = S, Se, Te)

These ternary chalcogenide AM_4Q_8 compounds are relatively new materials to be identified as canonical Mott-insulators (see [10] and references therein), which have considerable potential for future device applications (see chapter 8). They belong to an interesting class of transition metal compounds, which have a cubic structure (see the inset of figure 6.5) of the $GaMo_4S_8$-type often termed as a cation-deficient spinel, because the tetrahedral sites are only half-occupied by the A atoms (see [23] and references therein). The most interesting structural feature compared to the regular spinel structure, however, is the shift of the metal atom M from the center of the octahedral site. This leads to the formation of tetrahedral M_4 clusters containing 7–11 electrons. For example, each M_4 cluster in GaM_4Q_8 compounds have one unpaired electron among 7 (M = V, Nb, Ta) or 11 (M = Mo) d electrons [24]. The large distance between the M_4 clusters inhibit metallic bonding, but the M–M distance within these clusters is favourable for metallic bond. These interesting structural aspects give rise to molecular like electronic states within the clusters. These AM_4Q_8 compounds are insulators at ambient pressure. All these AM_4Q_8 compounds show the important characteristics of canonical Mott insulators (see [10] and references therein) namely they are paramagnetic insulators above 55 K and there is no temperature induced insulator to metal transition up to 800 K (see the

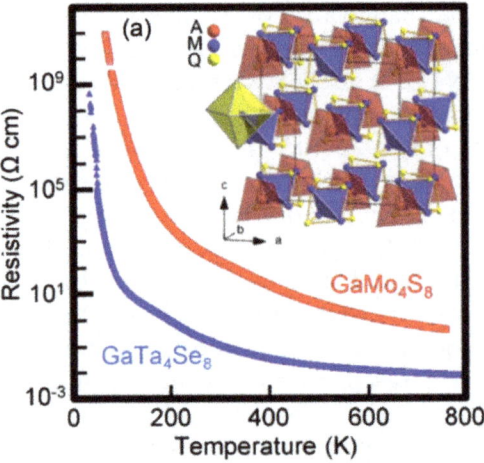

Figure 6.5. Resistivity versus temperature plots for the Mott-insulators GaMo$_4$S$_8$ and GaTa$_4$Se$_8$ showing the existence of Mott-insulating state even up to 800 K. Inset shows the crystal structure of AM$_4$Q$_8$ compounds with M$_4$ tetrahedral clusters. (Reproduced from [10] with permission from John Wiley & Sons.)

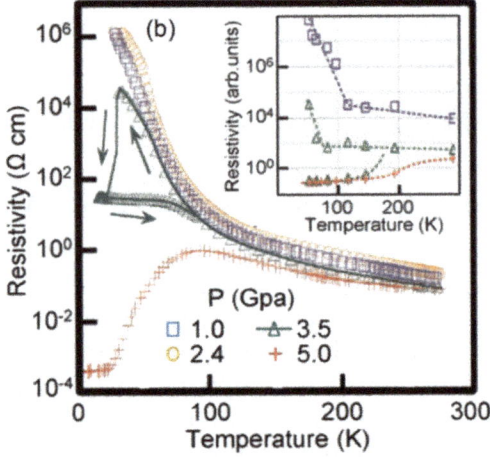

Figure 6.6. Resistivity versus temperature plots in GaTa$_4$Se$_8$ in the presence of different applied pressure. The observed thermal hysteresis in the resistivity obtained in the presence of 3.5 GPa pressure indicates the first order nature of the Mott-MIT. Inset shows the results of theoretical studies within the framework of LDA + DMFT. (Reproduced from [10] with permission from John Wiley & Sons.)

mainframe of figure 6.5). The compounds GaTa$_4$Se$_8$ and GaNb$_4$Q$_8$ undergo pressure induced BC-MIT.

The main frame of figure 6.6 shows resistivity versus temperature plots at different pressures in GaTa$_4$Se$_8$. The Mott-MIT is clearly visible in the resistivity plots obtained with the applied pressures 3.5 and 5 GPa. The distinct hysteresis in the temperature dependence of the resistivity between the decreasing and increasing temperature cycles obtained with applied pressure of 3.5 GPa clearly highlights the first order nature of Mott-MIT. Furthermore, the optical conductivity experiments revealed the signature of a quasi-particle peak emphasizing the strongly correlated nature of the metallic state just beyond the Mott-MIT line [25]. All these results again are in qualitative argument within the theoretical picture obtained within the framework of DMFT. To provide further support, LDA + DMFT numerical calculations were performed using the realistic electronic structure of $GaTa_4Se_8$ [10, 26]. The results of such theoretical studies are in clear agreement with the experimental results (see the inset of figure 6.6), thus highlighting again the impact of Mott physics in this compound.

6.2 Filling control Mott metal–insulator transition

When an electron is added to an electron-band insulator it goes to the next available higher energy band. On the other hand, removal of an electron leads to an empty state in an otherwise topmost filled band, which is termed as a 'hole' with the same spin as an electron but opposite charge. This information indicates that the resultant metallic behaviour from these two different channel is expected to have rather different properties, because the electrons and holes occupy different atomic orbitals. In this respect the meaning of electrons and holes in Mott insulators is quite different.

We have studied in chapters 3–5 that insulating behaviour arises in a Mott insulator with half-filled band, because hopping of electrons from one lattice site to another creates double occupancy, and this is strongly discouraged by electron–electron Coulomb repulsion. A small number of unoccupied sites are generated if the band filling is slightly reduced. On the other hand, addition of electrons generates sites with double electron occupation. It is to be noted here that in either of these cases hopping of electron does not change the number of doubly occupied lattice sites. Hence, there is less inhibition by the Coulomb interaction towards electron hopping, and eventually an insulator-to-metal transition may occur. Unlike electron-band insulators, conduction takes place in a doped Mott insulator in the same band regardless of whether electrons have been removed or added.

There are several systems showing filling control Mott-MIT, and the literature again is really vast. We will present here five representative systems namely $R_{1-x}A_xVO_3$, $R_{1-x}A_xTiO_3$, and high T_C superconductors (HTSC) $La_{2-x}Sr_xCuO_4$, $Nd_{2-x}Ce_xCuO_4$ and $YBa_2Cu_3O_6$.

6.2.1 Case study of $R_{1-x}A_xTiO_3$

The rare-earth titanates $RTiO_3$, where R is a trivalent rare-earth ion, are prototype Mott insulators. They form with Ti^{3+} $3d^1$ configuration, where inter site Coulomb interaction U splits the half-filled $3d$ conduction band into upper and lower Hubbard bands. The Mott–Hubbard gap collapses upon hole doping, and the family of $R_{1-x}A_xTiO_3$ compounds (where A is a divalent alkaline-earth ion) is one of the most appropriate systems for experimental investigations of the Mott filling control (FC)-MIT [1]. They form in an orthorhombical distorted perovskite crystal structure of the $GdFeO_3$ type. Filling control in the $3d$ band is achieved by partial replacement of the trivalent R ions with divalent Sr or Ca ions. The Sr or Ca content x represents a nominal 'hole' concentration per Ti site, which is related to the $3d$ band filing n as $n = 1 - x$. It is possible to form a solid solution for an arbitrary R/A ratio, and hence the band filling n can be varied from 1 to 0. The Ti–O–Ti bond angle distortion affects the kinetic energy t of the $3d$ electron or the one-electron bandwidth W, and it depends critically on the ionic radii of the (R, A) ions. As an example, the Ti–O–Ti bond angle is $157°$ for $LaTiO_3$ but decreases to $144°$ (ab plane) and $140°$ (c-axis) for $YTiO_3$ [1]. This causes a reduction in W of the concerned $3d$ state by as much as 20%. The necessary doping concentration for this FC-Mott MIT scales with the magnitude of the $GdFeO_3$-type distortions [27]. The compound $LaTiO_3$ ($YTiO_3$) undergoes a transition to an insulating antiferromagnetic (ferromagnetic) state around 140 K (30 K).

Figure 6.7 shows the resistivity as a function of temperature for $La_{1-x}Sr_xTiO_3$ and for $Y_{1-x}Ca_xTiO_3$ near the FC-MIT. Electron correlation is relatively weak in $LaTiO_3$, and a few % of hole doping is sufficient to destroy the insulating antiferromagnetic state and cause the MIT. Non-stoichiometry ($\delta/2$) of oxygen was also utilized to control the filling on a finer scale [28]. The relatively large value of U/W in YTO_3 allows the insulating phase to persist up to $x = 0.4$, although ferromagnetic spin ordering disappeared around $x = 0.2$ [1]. Near the metal–insulator phase boundary metallic compounds $La_{1-x}Sr_xTiO_3$ for $x > 0.05$ and $Y_{1-x}Ca_xTiO_3$ for $x > 0.4$, showed signatures of filling dependent renormalization of the effective mass of charge carrier in the heat capacity and magnetic susceptibility measurements. Such a critical enhancement of the carrier effective mass near the MIT phase boundary was also probed by Raman spectroscopy [1]. Furthermore, a spectral weight transfer from the correlation gap or Mott–Hubbard excitations to the inner gap excitations with hole doping (which is a typical signature of FC-MIT) was observed in optical conductivity and photo-electron spectroscopy in both $La_{1-x}Sr_xTiO_3$ and $Y_{1-x}Ca_xTiO_3$ systems.

There is evidence in some experimental studies of thermal hysteresis and phase-coexistence around the FC-MIT in these titanate compounds [1, 29]. This is

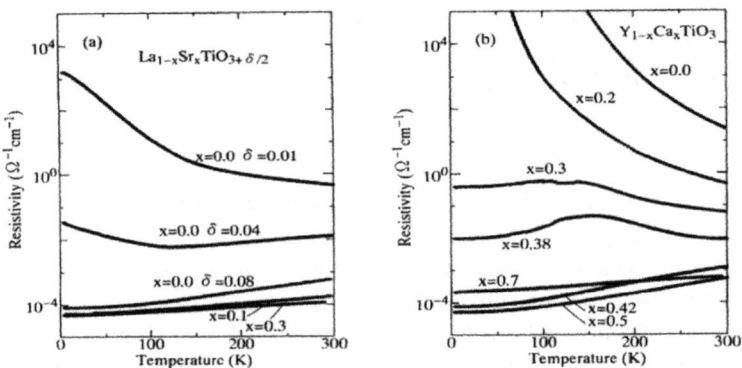

Figure 6.7. Resistivity versus temperature plots showing the FC-MIT: (a) $La_{1-x}Sr_xTiO_3$; (b) for $Y_{1-x}Ca_xTiO_3$. (Reprinted figure with permission from [28]. Copyright (1994) by the American Physical Society.)

indicative of an underlying first order phase transition, which is often associated with some kind of symmetry breaking. This raises the question on the exact role of lattice in the observed FC-MIT in $R_{1-x}A_xTiO_3$. This is especially so in the presence of $GdFeO_3$ type orthorhombical distortion. This question has been addressed in a relatively recent study correlating the atomic scale structure in $Sm_{1-x}Sr_xTiO_3$ with the electrical properties across the FC-MIT [30]. A high resolution scanning transmission electron microscopy technique was used to study the $GdFeO_3$-type distortions and the results showed that with the increase in the Sr concentration starting from the parent compound $SmTiO_3$, such distortions were reduced gradually and uniformly without any phase coexistence. A significant amount of distortion remained even in the metallic state. While the change in the structure is gradual across the FC-MIT, the electrical properties change by a large amount. This result clearly indicates that the FC-MIT is not coupled to a particular symmetry of lattice and is driven by strong electron correlation, which continues to remain important even in the metallic phase just beyond the FC-MIT point.

6.2.2 Case study of $R_{1-x}A_xVO_3$

The $R_{1-x}A_xVO_3$ system, where R stands for a trivalent rare-earth ion and A for divalent alkaline-earth ion, is one of the early systems identified to be undergoing FC-Mott-MIT. The end compound RVO_3 has V^{3+} ions with $3d^2$ configuration. The crystal structure of RVO_3 is tetragonal for R = La and Ce, and for the rest of the compounds it is of $GdFeO_3$ type with orthorhombic distortion [1]. The one-electron bandwidth W decreases with decreasing radius of the R ion [1]. RVO_3 are Mott insulators undergoing transition to a canted antiferromagnetic state in the temperature range 120–150 K.

Substitution of divalent A ions (Sr or Ca) for the rare-earth R ions causes hole doping in the $R_{1-x}A_xVO_3$ system. $La_{1-x}Sr_xVO_3$ is one of the very well studied member of this family of compounds. Figure 6.8(a) presents the variation of the MIT temperature in $La_{1-x}Sr_xVO_3$ as a function of hole doping. The insulating phase disappears around $x \approx 0.2$ [31, 32]. Figure 6.8(b) presents resistivity versus temperature plots for the polycrystalline samples of $La_{1-x}Sr_xVO_3$. This reveals that the conductivity is of thermal activation type in the high temperature regime above the antiferromagnetic transition temperature for the compounds with $x < 0.15$. The thermal activation energy was estimated to be 0.15 eV for $LaVO_3$, which decreased rapidly with the increase in x with the critical variation of the form of $(X_c - x)^\zeta$ where $\zeta = 1.8$ [33]. This led Mott [34] to argue that the critical variation with $\zeta = 1.8$ negated the simple scenario of the disorder induced Anderson localization with $\zeta = 1$. In the concentration region of $0.05 < x < 0.15$, the $La_{1-x}Sr_xVO_3$ compounds show a variable-range hopping type of conduction at low temperatures below the antiferromagnetic transition temperature. Here also the critical exponent for the localization length is found to be $\zeta = 0.6$ instead of the expected value of Anderson localization [35]. The critical behavior of MIT in $La_{1-x}Sr_xVO_3$ was subsequently investigated thoroughly in single crystal samples using electrical transport magnetization and heat capacity measurements [36]. It was observed that the precursor to

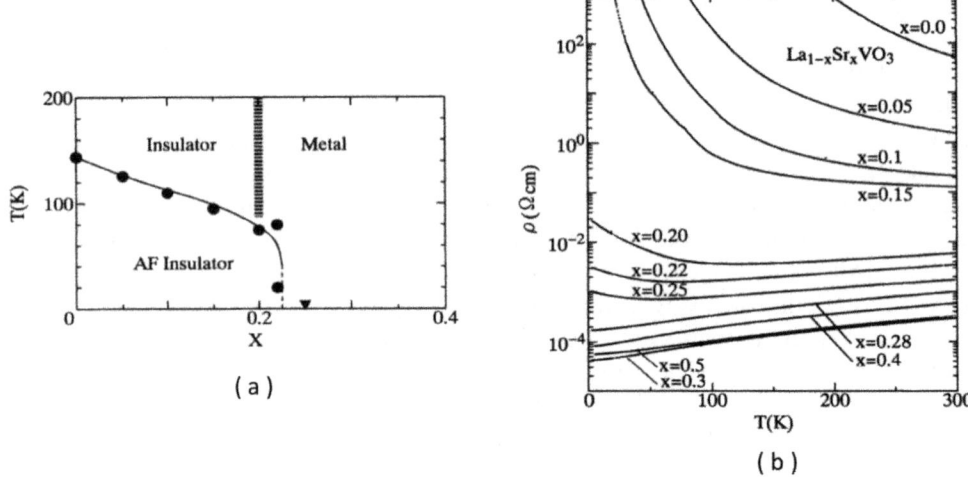

Figure 6.8. (a) Phase diagram showing the variation metal–insulator transition temperature as a function of hole-doping in $La_{1-x}Sr_xVO_3$. (b) Resistivity versus temperature plots showing the FC-MIT in $La_{1-x}Sr_xVO_3$. (Reprinted figure with permission from [32]. Copyright (1995) by the American Physical Society.)

the MIT manifested itself as an enhancement of the effective mass in the paramagnetic metal phase.

The FC-Mott-MIT and the associated change in the electronic structure showed up clearly in the optical spectra of $La_{1-x}Sr_xVO_3$ (see [1] and references therein). The Mott–Hubbard gap closes with the increase in hole doping. In the low-energy region, with hole doping the spectral weight is first increased in the gap region and eventually creates the inner-gap absorption band. In the barely metallic samples the conductivity spectra showed the signature of strong electron-correlation across the FC-MIT, whereas the spectra for $x > 0.3$ are typical that of a metal. In the photo-electron spectroscopy experiments starting from the large x limit of $La_{1-x}Sr_xVO_3$ as the band filling was increased the intensity of the remanent of the lower Hubbard band was found to be increased, while the quasiparticle band in the photo-emission and inverse photo-emission spectra lost its intensity [1]. In $Y_{1-x}Ca_xVO_3$ the FC-MIT takes place around $x \approx 0.5$ with smaller value of kinetic energy or bandwidth W. In optical conductivity measurements the evolution of the inner-gap excitation with doping was also observed for $Y_{1-x}Ca_xVO_3$ at a higher hole doping level than $La_{1-x}Sr_xVO_3$. This possibly reflects the presence of stronger electron correlation in $Y_{1-x}Ca_xVO_3$ [1]. It may be noted here that in $R_{1-x}A_xVO_3$ both the $3d$ band filling $n - x$ and (U/W) decrease with hole doping.

6.2.3 Case study of high T_C oxide superconductors

The discovery of high-temperature superconductivity (HTSC) in various copper oxide compounds (popularly known as cuprates) in the late 1980s provided a great impetus in the study of Mott physics. Until then only a relatively small community of condensed matter physicists were interested in the physics of strong electron correlation. There was of course a big community studying magnetism and magnetic materials. However, the emphasis was more on the phenomenology and applied aspects, although one of the initial aims of the Hubbard model was actually to understand the microscopic magnetic properties of transition metals and alloys.

It was soon realized that superconductivity was only one aspect of a rich phase diagram of the cuprate high-temperature superconductors (HTSCs), which needed to be understood in its totality [37]. All the HTSCs consist of a layered structure made up of one or more copper–oxygen plane. It is now universally accepted that the parent compound of all the HTSC-cuprates is a Mott insulator. Taking the example of La_2CuO_4, the parent compound of one of the first reported HTSCs, the copper ion in the copper–oxygen plane is doubly ionized and is in a d^9 configuration, so that there is a single hole in the d-electron shell per unit cell. From our study in the previous chapters we now know that there is a strong repulsive energy U when two electrons or holes are put on the same ion, and if U dominates over the kinetic energy or hopping energy t, the ground state will be an insulator. We also learnt in section 3.7 that when the spins in the neighboring lattice sites are anti-parallel there is a gain in energy $\approx 4t^2/U$ by means of virtual hopping and the ground state is antiferromagnetic. This gain in energy is known as exchange energy J. The parent compound La_2CuO_4 is indeed an antiferromagnetic insulator (see [37] and references therein) with an ordering temperature $T_N \approx 300$ K. This value of ordering temperature is much lower than that expected from the value of exchange energy. This is because the ordering temperature in La_2CuO_4 is governed by a small inter-layer coupling, and moreover that is frustrated in La_2CuO_4. In fact the exchange energy J is quite high, of the order of 1500 K, and that is reflected in the presence of antiferromagnetic correlation in La_2CuO_4 in the temperature regime much higher than T_N. It may be mentioned here that strictly speaking cuprates are actually considered to be a charge-transfer insulator rather than a pure Mott–Hubbard insulator [38, 39].

When some of the trivalent La in the parent compound La_2CuO_4 is substituted by divalent Sr, holes are added to the copper–oxygen (Cu–O) plane in $La_{2-x}Sr_xCuO_4$. The situation is reverse in the compound $Nd_{2-x}Ce_xCuO_4$ [40], where Ce substitution causes introduction of electrons to the Cu–O plane. Thus, $La_{2-x}Sr_xCuO_4$ is known as a hole-doped system and $Nd_{2-x}Ce_xCuO_4$ is known as a electron-doped system. Most of the HTSCs are hole-doped systems. Figure 6.9 shows the generic phase diagram of HTSC-cuprate compounds showing the results of hole doping and electron doping. In the beginning of this section we mentioned that in a doped Mott insulator conduction takes place in the same band regardless of whether electrons have been removed or added. Thus, a high degree of electron–hole symmetry was expected in the phase diagram, which is certainly not the case in the HTSCs as shown in figure 6.9. The antiferromagnetic order is rapidly suppressed on the hole-doped side of the phase diagram and is completely destroyed by hole concentration $x = 0.3$–0.05. The superconductivity appears almost immediately after the destruction of the antiferromagnetic state with x ranging from 0.06 to 0.25. The dome-shape of the superconducting transition temperature (T_C) line is a universal feature for all hole-doped HTSCs, with the maximum T_C varying from about 40 K in

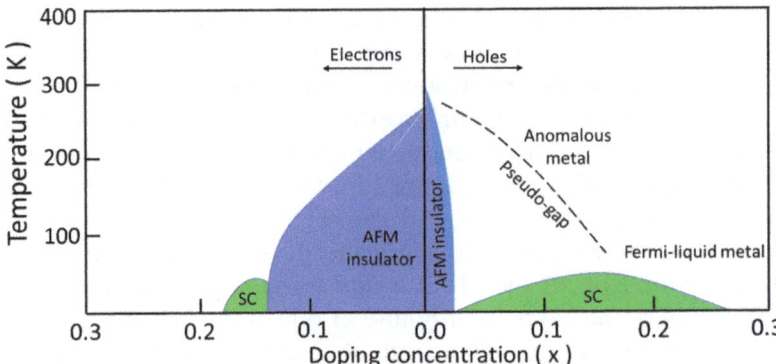

Figure 6.9. Schematic phase diagram of high T_C cuprate superconductors showing the effect of hole doping and electron doping.

the La$_{2-x}$Sr$_x$CuO$_4$ family to 93 K and higher in other HTSC families like YBa$_2$Cu$_3$O$_{66+\delta}$ and Ba$_2$Sr$_2$CaCu$_2$O$_{8+\delta}$ [37]. Needless to say that the HTSC superconductivity is a very interesting subject by itself, and its correlation with Mott physics is well documented [37, 41–43]. On the hole-doped side the region in the phase diagram below the concentration smaller than that of the maximum T_C is commonly known as the under doped region. The metallic state above T_C in the under doped region shows many unusual properties including evidence of a pseudo-gap probably not seen before in any other metal. The pseudo-gap phase is not a well defined phase, and its origin is still a matter of intense debate. It does not have a definite finite-temperature phase boundary, and the line in figure 6.9 just indicates a crossover region. We will not talk any more about superconductivity and pseudo-gap here, and will rather focus on Mott-MIT. The normal state above the optimal T_c also shows unusual properties, namely the resistivity is linear in T and the Hall coefficient is temperature dependent [37]. These properties are taken as examples of non-Fermi-liquid behavior. The concentration region beyond the peak of the superconducting dome in the hole-doped part of the phase diagram is termed the over doped region. The normal state or the metallic state here shows the usual Fermi-liquid behaviour.

The antiferromagnetic state is more robust on the electron-doped side of the phase diagram and remains up to $x \approx 0.15$ (see figure 6.9); a relatively small region of superconductivity appears beyond this concentration. The enigmatic pseudo-gap state here is either weaker or not present at all [37]. A systematic and careful comparison between electron and hole doping was not possible for a long time due to the absence of a single material in which both can be realized. It was ultimately possible to cross the zero-doping state in an ambipolar cuprate compound Y$_{1-z}$La$_z$(Ba$_{1-x}$La$_x$)$_2$Cu$_3$O$_y$ (YLBLCO) (see [39]). In this compound the sign of the charge carriers can be changed without altering the crystal structure. Substitution of La for Ba in YLBLCO provides electrons. The charge carriers can be controlled for $x = 0.13$ and $z = 0.62$ from 7% of holes to 2% of electrons per planar Cu through the zero-doping state. This is achieved by changing the oxygen content with the removal of oxygen from the Cu–O chain site [39]. Thus, YLBLCO provides a unique opportunity to investigate the very low-doping region around the undoped Mott insulating state covering both hole and electron-doped regions. The electronic structure, crystal structure, electrical and magnetic properties of this compound have been studied in detail using various experimental techniques, including x-ray photo-emission spectroscopy and neutron scattering [39]. The antiferromagnetic ordering temperature is found to be increasing initially both with the hole and electron doping reaching a peak before dropping, which is of course more rapid in the case of hole-doping. The neutron scattering study revealed the antiferromagnetic structure to be different in electron and hole-doping cases. However, no abrupt change in the crystal structure including the lattice constants was found at the crossover from electron to hole-doping region. The doping asymmetry was rationalized in terms of spectral-weight-transfer asymmetry expected for cuprates as charge-transfer-type Mott insulators [39]. The valence and conduction bands are primarily oxygen (O) and copper (Cu) bands. If the system has N cells, the spectral weights of oxygen and copper bands are $2N$ and N, respectively. Due to the on-site Coulomb repulsion the original spectral weight $2N$ of copper is split into the upper and lower Hubbard bands. If an electron is added to the undoped insulating state, it will create one doubly occupied copper site on which the two electrons have the same energy. The new filled state will have the spectral weight of two, whereas the spectral weight of both the upper and lower Hubbard bands becomes $N - 1$. The doping of a hole on the other hand, just reduces the spectral weight of the oxygen band to $2N - 1$. This implies that the rate of the spectral-weight transfer is twice as rapid for electron doping, which causes an inherent asymmetry between the electron and hole-doping in the localized limit. Thus, the observed doping-rate asymmetry in the HTSC cuprate compounds may have its origin in the fundamental spectral-weight-transfer asymmetry associated with the charge transfer insulator nature of these compounds.

So, HTSCs provide another example of FC-MIT, although the presence of numerous interesting properties including superconductivity, anomalous metallic behaviour and pseudo-gap, does not make it very clear whether this FC-MIT is driven by strong electron correlation alone.

6.3 Systems with additional degrees of freedom beyond strong electron correlation

We have earlier seen in chapter 5 (section 5.2.3) that in many materials showing characteristic properties of Mott insulator, the energy of many other kinds of interactions can be on the same order of magnitude with electron correlation energy U and kinetic energy t. These interactions include electron–lattice, spin–spin, and other types of interactions. In this section we discuss some interesting representative systems belonging to this class of materials.

6.3.1 Interesting properties of magnetite Fe_3O_4 and Verwey transition

Magnetite Fe_3O_4, also known as loadstone, is the oldest known magnetic material. It forms in an inverted cubic spinel structure (see figure 4.4) with the chemical formula of $Fe^{3+}_{(Tetrahedral)}(Fe^{2+}Fe^{3+})_{(Octahedral)}O_4$. In this structure the tetrahedral A sites contain one-third of the Fe ions surrounded by four oxygen ions, and the octahedral B sites contain the rest, two-third of the Fe ions surrounded by six oxygen ions. Fe(B)–O layers and Fe(A) layers are stacked alternately (see figure 4.4). The F ions at the A-site are trivalent Fe^{3+}, while B-sites contain a host number of trivalent Fe ions Fe^{3+} and divalent Fe ions Fe^{2+}, thus giving a mixed valent at the B-site with a formal average valence $Fe^{2.5}$.

Fe_3O_4 undergoes a ferrimagnetic transition below 860 K. In this ferrimagnetic state, the magnetic moments at the A-site have the opposite spin-directions to the moments at the B-site, with d-orbital occupations represented as $(t_{2g\uparrow})^3(e_{g\uparrow})^2$ and $(t_{2g\downarrow})^3(e_{g\downarrow})^2(t_{2g\uparrow})^{2.5}$, respectively. Over the years Fe_3O_4 has been subjected to numerous experimental and theoretical studies to understand its electronic, magnetic, and structural properties in extreme pressures and temperatures, because of its usefulness in technological applications as magnetic and electronic materials and also its natural geological existence in the Earth's crust and mantle. For an outline of this vast experimental and theoretical works the readers are referred to the review articles [1, 44, 45]. In the present context the most interesting observation is a first order phase transition known as Verwey transition at 120 K where the dc electrical conductivity abruptly changes by two orders of magnitude [46, 47]. Figure 6.10 shows the variation of electrical conductivity of Fe_3O_4 as a function of temperature. Since its discovery in the late 1930s by Verwey [46, 48], this transition, now commonly known as the Verwey transition, has been a subject of continued interest. Even today it has remained as a model system for studying the classic condensed matter physics problem of the electron motion in a solid crystalline material when the kinetic energy of electron, electron–electron interactions, and electron–lattice interaction are of comparable strength.

It was originally suggested by Verwey that the high-temperature state above the Verwey transition temperature T_V was a bad metal state, and the electrical

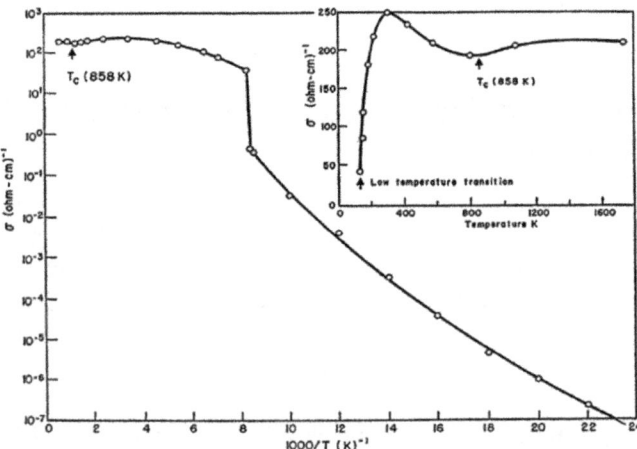

Figure 6.10. Conductivity variation as a function of temperature in Fe_3O_4. (Reprinted figure with permission from [47]. Copyright (1957) by the American Physical Society.)

conductivity in the bad metal was due to the thermally activated fast electron hopping between Fe^{3+} and Fe^{2+} cations randomly distributed over the same octahedral B-sites of inverted spinel cubic structure. The drastic lowering in the conduction at the Verwey transition temperature originates in the onset of a long-range spatial order of the equal number of Fe^{3+} and Fe^{2+} ions at the B sites, i.e. charge ordering among the extra $t_{2g\uparrow}$ electrons. This is accompanied by a decrease in the lattice symmetry from cubic to tetragonal. This charge ordering arises due to the Coulomb repulsion between the electrons, which effectively localizes them and thus prevents the motion of the charge carriers [49].

In spite of the experimental and theoretical investigations of the past 80 years, several questions about the Verwey transition, such as the exact nature and origin of the charge ordering, the nature of the electronic state and the mechanism for the formation of the insulating gap below the Verwey transition, have remain unresolved. The proposal of charge ordering as the origin of Verwey transition and the physical properties at low temperatures below the transition in various refined forms continues to be considered seriously [44]. However, in between it was demonstrated by neutron diffraction that the lowering of lattice symmetry predicted in Verwey's original charge ordering model may not be entirely correct [50]. Subsequently, alternative charge ordering possibilities have been put forward through high resolution neutron and x-ray powder diffraction experiments, which are consistent with the low-temperature monoclinic crystal symmetry [51–53]. These experimental results were supported by local density approximation (LDA) + Hubbard U band structure calculations, which suggested that below the Verwey transition temperature Fe_3O_4 formed into an insulating charge-orbital-ordered state [54]. The idea of this insulating charge-orbital-ordered state got further support from the results of resonant x-ray diffraction measurements [55].

It is well known that the physical properties of Fe_3O_4 in extreme conditions can be quite sensitive to the stoichiometry, twinning, and presence of impurities and defects in the starting sample, and the metal–insulator transition near the T_{Verwey} may occur continuously or discontinuously depending on the stoichiometry of the Fe_3O_4 samples [45]. In addition the complexity associated with the twinning of crystal domains in the low-temperature structure makes it very difficult for diffraction studies to obtain robust conclusive evidence of charge ordering. This has even led to doubts whether there is a charge ordered state at all in Fe_3O_4 (see [56] and references therein). In fact an alternative mechanism of Verwey transition has also been proposed, which was primarily driven by a change in the lattice symmetry due to strong electron–phonon interaction [56]. In that framework a weak charge ordering appears only as a consequence of that change in lattice symmetry.

A fairly recent study of full low-temperature superstructure of Fe_3O_4 with the help of high-energy x-ray diffraction from an almost single domain of Fe_3O_4 sample suggests that Verwey's hypothesis of charge-ordering is correct at least to a first approximation [57]. Without going into further debate on this charge-orbital ordered state in Fe_3O_4, we will now focus on the influence of strong electron correlation and Mott physics on the Verwey transition and the low temperature insulating state of Fe_3O_4. Photo-emission spectroscopy studies indicated that the electrical conduction in Fe_3O_4 was dominated by the correlation energy of the $3d$ electrons and the compound was a Mott–Hubbard type insulator [58]. A mean-field study by Seo *et al* [59] for a three-band model appropriate for d electrons of Fe ions at B sites in Fe_3O_4 suggested that the Verwey transition was a bond dimerization due to the interplay of strong electron correlation and electron–lattice interaction. Strong on-site Coulomb interaction between different t_{2g} orbitals stabilized the ferro-orbital ordered state in a wide temperature range, and resulted in an effectively one-dimensional electronic state. The Mott insulating state can be visualized as if each charge localizes in every two sites i.e. dimer. The bond dimerization is induced by the Peierls instability (see chapter 1) in a one-dimensional state due to the electron–lattice interaction [59]. Subsequent high-energy x-ray diffraction study in an almost singled-domain Fe_3O_4 sample, however, suggested, that the localized electrons are distributed over linear three-iron site units, which are called 'trimerons' [57]. At ambient pressure, the Verwey transition is accompanied by the trimeron ordering, the monoclinic structural transition, and the metal–insulator transition. The formation of the trimerons induce substantial atomic displacements, which affect the coupling between electrical polarization and magnetization. It was also suggested that in Fe_3O_4 the trimerons may become important quasiparticles even above the Verwey temperature [57]. High pressure studies indicate that with increasing pressure the trimeron ordered phase gets separated from the metal–insulator transition, and a drastic change in the electrical transport property is no longer observed [45]. Figure 6.11 presents a pressure–temperature phase diagram of Fe_3O_4. At 300 K Fe_3O_4 undergoes a pressure-induced transition from a charge disordered metallic state to a trimeron ordered state at approximately 8 GPa [45]. This trimeron ordered state would result in a less symmetrical inverse spinel structure with a small distortion along the [110] direction [45].

A very recent study [60] reported that the Verwey transition in Fe_3O_4 nanoparticles accompanied a size-dependent thermal hysteresis in magnetization, [57]Fe

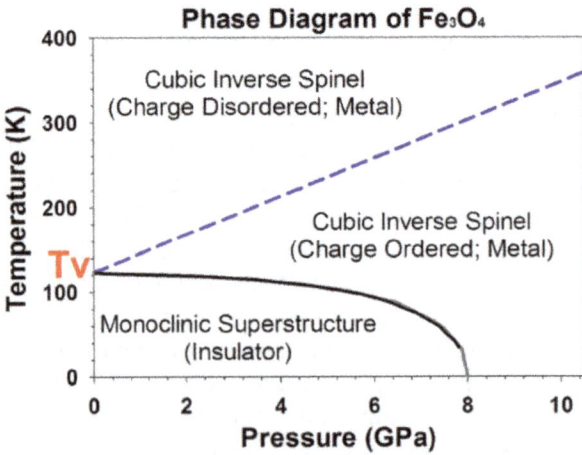

Figure 6.11. Pressure–temperature phase diagram of Fe_3O_4. (Reproduced from [45] with permission of Springer Nature.)

NMR, and x-ray diffraction measurements. The observed behaviour was interpreted as the result of a correlation between charge ordering and spin ordering in a single domain of Fe_3O_4 [60].

6.3.2 Charge and orbital orderings in Mn-oxide compounds with colossal magneto-resistance

$LaMnO_3$ is a Mott insulator with spin $S = 2$ and orbital degrees of freedom [61]. The spin $S = 2$ is represented by the t_{2g} spin 3/2 strongly coupled to the e_g spin 1/2 via ferromagnetic Hund's coupling J_H. A pseudo-spin $\mathbf{T}$ represent two possible choices of the orbitals. The z-component of pseudospin $T_z = 1/2$ when the e_g orbital $d_{x^2-y^2}$ is occupied and $T_z = -1/2$ when the e_g orbital $d_{3z^2-r^2}$ is occupied. The transfer integral between the two adjacent Mn atoms in $LaMnO_3$ is influenced by the overlaps of the d orbitals with the p orbital of the oxygen atom between them. The overlap between the $d_{x^2-y^2}$ and p_z orbitals is zero because of the different symmetry. This introduces anisotropy in the system, and the electron in the $d_{x^2-y^2}$ orbital cannot hop along the z-axis [61]. In the hole-doped system $La_{1-x}Sr_xMnO_3$, because of the intra-atomic strong ferromagnetic Hund's coupling between e_g and t_{2g} spins, transfer integral is maximum for parallel spins and is zero for antiparallel spins. Hence, the kinetic energy gained by the doped holes will be maximum for parallel spins, and this gives rise to the ferromagnetic interaction between the spins. As discussed in chapter 4 section 4.3, this mechanism is known as double-exchange interaction.

The orbital interaction in Mn-oxide compounds supports (or does not support) ferromagnetic and antiferromagnetic interaction in an orbital direction dependent manner, and that leads to a complex spin–orbital coupled state. The A-type antiferromagnetic state in pure $LaMnO_3$ is the manifestation of the anisotropic super-exchange interactions, which due to the orbital ordering is ferromagnetic within the plane and antiferromagnetic between the plane (see also exercise 3 in chapter 4) [61].

In the hole-doped Mn-oxide compounds the strength of double-exchange interaction depends on the level of doping, and various orbital-ordered/disordered states emerge accompanying the spin-ordering states. Taking the example of $Nd_{1-x}Sr_xMnO_3$, hole-doping in the parent compound melts the orbital ordered state into a quantum-disordered state, and the system shows the ferromagnetic-metallic (F) state for $0.3 < x < 0.5$ [61]. The configuration of such a ferromagnetic (F) state is shown schematically in the top panel of figure 6.12(a). With further increase in doping concentration the kinetic energy of the carriers decreases, and a 2D metallic state with layered-type antiferromagnetic ordering (A) emerge in the concentration regime $0.5 < x < 0.7$ (see figure 6.12(a) middle panel). The magnetic state changes to chain type antiferromagnet (C) in the concentration regime $x > 0.7$ (see figure 6.12(a) bottom panel). Figure 6.12(b) shows the temperature dependence of the resistivity in the three hole-dopings concentration regime of $Nd_{1-x}Sr_xMnO_3$. This interesting phase diagram can be reproduced and understood within the framework of a mean field approximation applied to the generalized Hubbard model [61, 62]. The antiferromagnetic A state emerges as a compromise

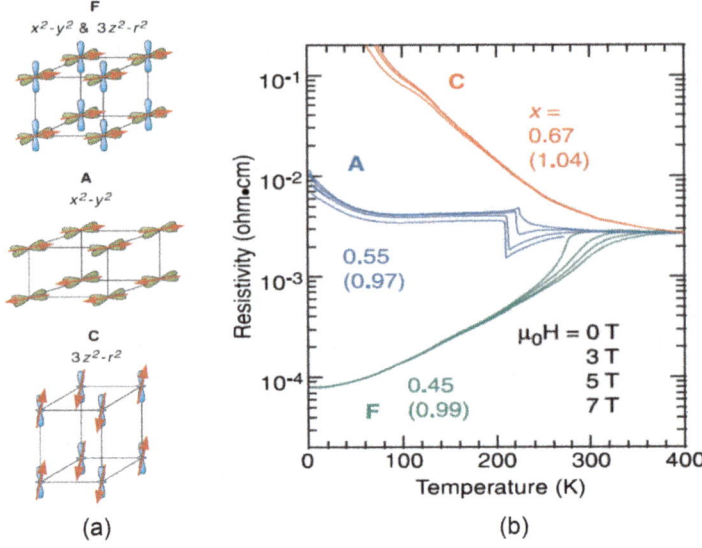

Figure 6.12. (a) Schematic diagrams showing ferromagnetic-metallic (F; top panel), layered type antiferromagnetic (A; middle panel) and chain type antiferromagnetic (C; bottom panel) type states. (b) Resistivity versus temperature plots in the different doping-concentration regions of $Nd_{1-x}Sr_xMnO_3$. (From [61]. Reprinted with the permission from AAAS.)

between the antiferromagnetic super-exchange interaction between the t_{2g} spins and the double-exchange interaction through the ferromagnetic order of (x^2-y^2) orbital [62]. The electron transfer is almost prohibited in cubic perovskite along the c-axis due to the (x^2-y^2) orbital order. This is the origin for the interplane antiferromagnetic coupling. The C-type antiferromagnetic state in the concentration regime $x > 0.7$ (see figure 6.12(b)) is accompanied with the $(3z^2-r^2)$ orbital, and this state shows insulating behaviour, possibly due to the charge ordering [61].

The orbital ordering in the Mn-oxide compounds indeed occasionally accompanies the concomitant charge ordering [61]. The most archetypal CE type order (see figure 6.13) is observed at a doping level of $x = 0.5$. The experimentally observed spin/charge/orbital ordered state for $Pr_{0.5}Ca_{0.5}MnO_3$ has been successfully explained using band calculation with the local density approximation (LDA) combined with the on-site Coulomb interaction U [63]. In this compound the transition temperature of charge ordering is higher than that of antiferromagnetic transition. This suggests that possibly the charge-ordering is the driving force behind the spin/charge/orbital ordering [61].

Application of magnetic field can influence the CE-type orbital/charge-ordered state in the Mn-oxide compounds. Figure 6.14 presents a $H-T$ phase diagram showing the variation of the orbital/charge-ordered state in the perovskite Mn-oxide compounds $R_{0.5}^{3+}A_{0.5}^{2+}MnO_3$ (R and A are trivalent rare-earth and divalent alkaline-earth ions, respectively) with the application of magnetic field. The change in the average size of the R and A ions at that site changes the amount of deviation of the Mn–O–Mn bond angle from 180°, and thus controls the e_g electron-hopping energy t

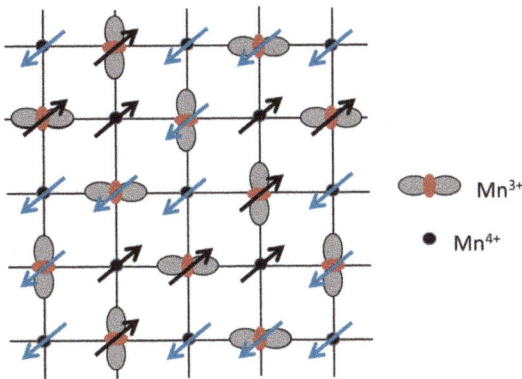

Figure 6.13. Schematic diagrams showing orbital, charge, and spin arrangement in the CE-type state with top view of two different 3d orbitals.

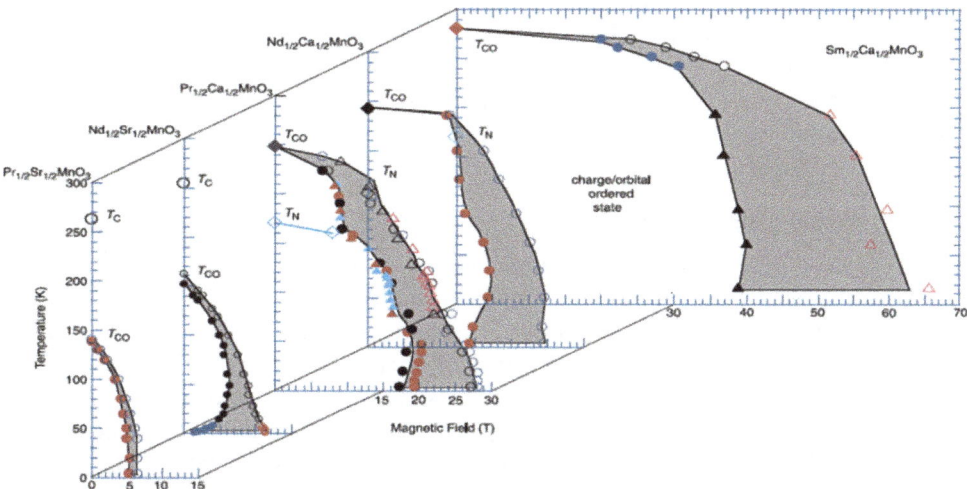

Figure 6.14. $H–T$ phase diagram of $R^{3+}_{0.5}A^{2+}_{0.5}MnO_3$ (R and A are trivalent rare-earth and divalent alkaline-earth ions, respectively) compounds showing transition to CE-type charge/orbital ordered phase. (From [61]. Reprinted with the permission from AAAS.)

via a change in manganese $3d$ and oxygen $2p$ hybridization [61]. It can be seen from the $H–T$ phase diagram in figure 6.14 that the extent of stable orbital/charge-ordered state is increased with the decrease of the ionic radius from say (Nd, Sr) to (Sm, Ca).

In Mn-oxide compounds there are two types of orbital/charge/spin phase diagram [64]. In $Nd_{0.5}Sr_{0.5}MnO_3$, which is a relatively wide-bandwidth system, the ferromagnetic ordering first takes place with the reduction in temperature at the critical temperature T_C in the cooling process, and that is followed by the concomitant CE-type orbital, charge, and antiferromagnetic spin ordering at a lower temperature $T_{CO} = T_N$ (see figure 6.14). This is the type-I behaviour, and the type I systems undergo the CE-type orbital/charge/spin-ordering transition only at the doping concentration very close to $x = 0.5$ or 1/2 [61]. In the smaller bandwidth system, say $Pr_{0.5}Ca_{0.5}MnO_3$, the concomitant orbital/charge-ordered state appears first at $T_{CO} > 250$ K, and the antiferromagnetic spin ordering takes place subsequently at a lower temperature T_N (see figure 6.14). This is the type-II behaviour. In type-II systems the ferromagnetism and metallic state can be realized only on the application of an external magnetic field. The orbital–charge correlation is a source of the high-resistance in the phases above the ferromagnetic transition temperature T_C. The insulator to metal transition takes place in such Mn-oxide compounds under the application of magnetic field, and that is the origin of the colossal magneto-resistance (CMR) observed in such systems.

The orbital fluctuation has been observed in the doped Mn-oxide compounds even in the ferromagnetic metallic state. The ferromagnetic state shows a very small spectral weight of the quasi-particle peak at the Fermi level in the photo-emission spectrum and also a minimal Drude weight in the optical conductivity spectrum [61]. Such orbital fluctuations may be a cause for the observed relatively incoherent charge dynamics in the ferromagnetic metallic state with minimum spin fluctuations. However, there are other possible explanations including that of microscopic phase separation [65] for such 'bad metal' features. For more details on the Mn-oxide compounds the readers are referred to the monograph by Dagatto [66] and review article by Goodenough [67].

6.3.3 Interesting properties of 1T-TaS$_2$

An interesting interplay between electron–electron and electron–lattice interaction gives rise to a very rich phase-diagram including metal insulator transition in the layered transition metal dichalcogenide compound 1T-TaS$_2$. The bulk 1T-TaS$_2$ has a layered structure. Each unit layer consists of a triangular lattice of Ta atoms, with each Ta-atom surrounded by S atoms in an octahedral coordination.

At high temperatures 1T-TaS$_2$ is a simple metal with a single Ta $5d$ electron band crossing the Fermi level [68, 69]. The Ta sublattice is prone to an in-plane deformation where 12 Ta atoms are displaced towards a 13th central Ta site [69]. Figure 6.15 shows this displacement of Ta atoms leading to the deformation, which is known as the 'David-star' deformation [70]. With lowering in temperature the compound undergoes a incommensurate (IC) charge density wave (CDW) transition at $T_{C0} = 543$ K (see second panel in figure 6.16). This IC-CDW structure tries to conform with the underlying crystal lattice structure, and on cooling below $T_{C1} = 350$ K the IC-CDW structure transforms into a nearly commensurate CDW (NC-CDW) state (see [70] and references therein). The NC-CDW state possess a regular domain-like structure where small commensurate (with underlying lattice structure) domains are separated from each other by regions of discommensuration (see third panel of figure 6.16). Further down in temperature below $T_{C2} \sim 180$ K 1T-TaS$_2$ enters a commensurate-CDW (C-CDW) state, where the David-stars are fully interlocked (see fourth panel of figure 6.16). The domain walls disappear and a $\sqrt{13} \times \sqrt{13}$ superlattice is formed [69]. This transition is of first order in nature and is associated with distinct thermal hysteresis.

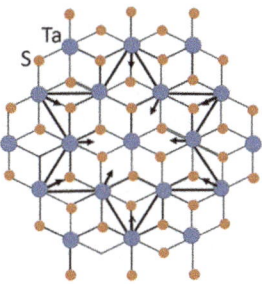

Figure 6.15. Schematic diagram showing the displacements of the Ta atoms and the formation of the David-star in the lattice structure of 1T-TaS$_2$.

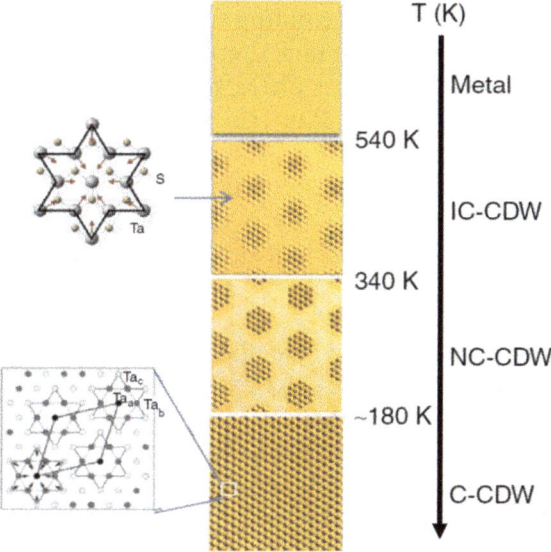

Figure 6.16. Schematic diagram showing the successive formations of IC-CDW, NC-CDW and C-CDW states with decrease in temperature, starting from the high temperature metallic state in 1T-TaS$_2$. (Reproduced from [70] with permission of Springer Nature.)

This commensurate lattice modulation is associated with reconstructions of electronic structure, which splits the Ta 5d-band into several submanifolds. This leaves with exactly one conduction electron per David-star, and the strong electron–electron interaction then localize these electrons leading to an insulating ground state [69]. This insulating CCDW is believed to be a Mott insulator with an energy gap of approximately 0.1 eV [71]. This Mott insulating state in 1T-TaS$_2$, differs from that of a conventional Mott-insulator that it resides inside a commensurate CCDW state. Instead of atomic sites in conventional Mott-insulators, the electron localization centres here are CDW superlattices and the electrons on each 13th Ta atom get localized. New equilibrium states can be obtained in 1T-TaS$_2$ on application of external pressure or doping, and both of those can make $1T$-TaS_2 superconducting [72, 73].

Coming back to the topic of Mott-MIT, in $1T$-TaS_2 the low temperature Mott-insulating state becomes metallic above 220 K while heating, where the CCDW order melts into small domains separated by nearly commensurate domain wall networks [74]. Apart from thermal excitations this NC-CDW metallic phase can also be induced by chemical doping [75, 76], pressure [72], photoexcitation [77, 78] and reduction of thickness [79]. There are quite a few suggestions on the metallic behaviour of the NC-CDW state: (i) metallic nature of the domain wall [72]; (ii) the metallic behaviour of the Mott-CDW domains due to the screening by free charge carriers in the domain walls [75]; (iii) the change in the interlayer stacking order [80]. Some of these questions were addressed by manipulating the insulator to metal transition of the Mott-insulating state at the nanometer scale by applying a voltage pulse from a scanning tunneling microscope (STM) tip [81]. This experiment showed that metallic domains could be formed and erased reversibly with an atomically abrupt phase boundary. This spectroscopic measurement with atomic resolution thus rules out the possibility of metallic behaviour originating from substantial free carriers along domain walls. It was concluded that the reduced CDW order led to an increase in the bandwidth W and at the same time reducing U, which in turn drove the Mott-CDW state into a metallic state. This metallic state is a correlated state induced by the moderate reduction of electron correlation due to the decoherence of the charge density [81].

$1T$-TaS_2 provides an example where the Mott state is clearly intercoupled with the CDW order. This provides the extra tunability of the Mott-insulating state by a distinct order. In fact there are reports of a hidden (H)-CDW state at low temperatures, which can play a role in the possible device applications of $1T$-TaS_2 [78]. We will discuss this topic more in section 8.3.1.

There is a puzzling and at the same time interesting aspect of $1T$-TaS_2, which needs some mentioning here. Within the standard framework of a Mott insulator the spins form local moments, and this may lead to an antiferromagnetic ground state due to exchange coupling. However, there is no report of any kind of magnetic ordering accompanying the insulating ground state in $1T$-TaS_2. Even the temperature dependence of magnetic susceptibility does not show the Curie–Weiss behaviour in the insulating state, thus questioning even the existence of local moment in the insulating state of $1T$-TaS_2. This serious deviation from the standard Mott picture was explained in terms of very small Lande g factor due to spin–orbit coupling [69]. This explanation, however, depends sensitively on the assumption of a cubic environment for the Ta atoms, which is not necessarily true [82]. Instead, it has been argued recently that the existing experimental data strongly point to the existence of a spin-liquid state in $1T$-TaS_2 [82]. Theoretically such a spin-liquid state may arise in a triangular lattice due to frustration [83, 84]. This frustration arises in a triangular lattice out of the inability to fulfill the requirement of antiferromagnetic interaction at all the lattice sites.

6.4 VO_2 and NbO_2: Peierls insulators or Mott insulators?

Vanadium dioxide (VO_2) is one of the most studied transition metal oxide systems, especially in the context of strong electron correlation and Mott physics [85]. Apart from the fundamental scientific interest, this material is drawing much attention in recent times from the applications point of view as electronic switch and memory devices, and also as sensing and cloaking materials. We will concentrate on the physics related aspects of VO_2 in this chapter, and the possibilities of technical applications will be covered in chapter 8.

VO_2 exists in monoclinic (M1) crystal structure in the temperature regime $T <$ 340 K with non-magnetic insulating (semiconducting) properties. At 340 K it undergoes a temperature driven displacive first order structural phase transition to tetragonal rutile (R) phase. In this high temperature phase, VO_2 shows paramagnetic metallic behaviour. This transition from the non-magnetic insulator state to paramagnetic metallic state was first reported by Morin [86], and a conductivity jump by a factor as large as 10^5 has been reported across this transition in various types of VO_2 samples [87]. Figure 6.17 shows such a conductivity jump as a function of temperature for a good quality single crystal sample of VO_2 [88]. The M1 phase has an optical gap of 0.6 eV. The vanadium atoms are arranged in one-dimensional chains in both the low temperature M1 and high temperature R phase [89] (see figure 6.18). In the monoclinic M1 phase the vanadium atoms dimerize and create long bonds of length 3.16 Å and short bonds of length 2.62 Å, while they remain equidistant in the tetragonal R phase. The dimerization in the M1 phase is also accompanied by zigzag deformation of chains and doubling of the unit cell in the chain direction [89]. There are also reports on two other insulating phases in VO_2 namely a monoclinic M2 phase where only one set of chains is dimerized, and a triclinic T phase, which is intermediate between M1 and M2 [90, 91].

The V-ion positions in various structural phases of VO_2 is shown schematically in figure 6.18 [92]. In all phases, the V-ions at the center of each unit cell (shown by the grid lines in figure 6.18) are offset from the others by $\frac{1}{2}$ unit cell (denoted by 1/2 in the rutile R panel). In the insulating phases, the V-ions are displaced from the corresponding positions in the high temperature metallic R phase, which are shown as open green circles in the panels of the insulating phases in figure 6.18. All of the V-ions dimerize in the M1 phase, and tilt in equivalent chains along the rutile c_R-axis. The M2 phase, in contrast, has two distinct types of V-ion chains. While in one type of chain in the M2 phase the V-ions form a pair but do not tilt, in the other type they tilt but do not pair. The vanadium ions in the latter type of chain are equidistant, and each carry a localized electron with a spin-1/2 magnetic moment and antiferromagnetic exchange coupling between nearest neighbors [90, 92]. The triclinic T phase, on the other hand, consists of two types of non-equivalent vanadium sublattices, where V-ions are paired and tilted to different degrees [90, 92]; the T phase is a kind of intermediate phase between the M2 and M1 phases.

There are mainly two approaches to understand the insulator–metal transition and origin of the insulating properties of the M1 phase in VO_2:

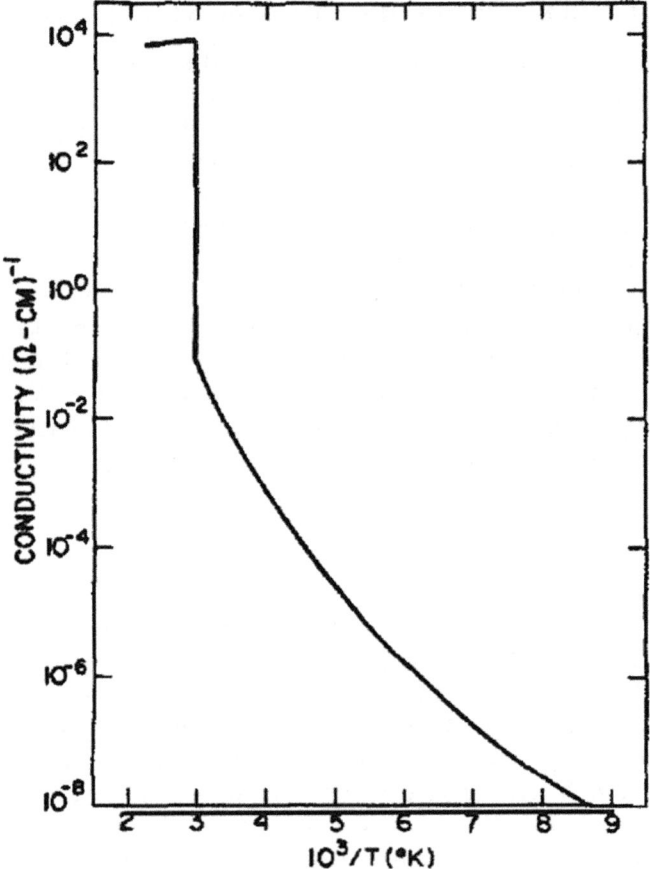

Figure 6.17. Temperature dependence of conductivity in a good quality single crystal sample of VO_2. (Reprinted from [88], 1969, with permission from Elsevier.)

1. In the first approach, the metal–insulator transition is understood within the standard one-electron band structure picture as being related to Peierls-type instability (see chapter 1, section 1.10) of the parent rutile R phase. It was suggested that the dimerization causes a splitting of the $d_{x^2-y^2}$ band, while the zigzag transverse displacements of vanadium atoms induced a shift of the d_{xz} and d_{yz} bands, with both the processes contributing to creation of the band gap [93, 94].

2. In the second approach, the metal–insulator transition was thought to be due to strong correlations of d electrons in vanadium atoms to the creation of Mott insulating state [85]. Using cluster dynamical mean-field theory in conjunction with density functional theory (DFT), Biermann *et al* [95] suggested that Coulomb correlations are necessary to open the Peierls gap in VO_2. The structural and electronic transitions appear to occur simultaneously, which raises the question, which is the primary one of these two transitions?

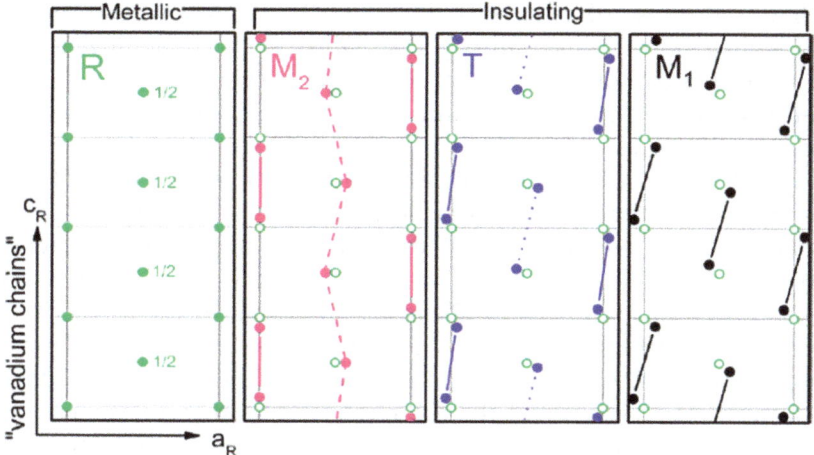

Figure 6.18. Schematic presentation of vanadium ion positions for the metallic rutile **R** and insulating **M2, T**, and **M1** phases of VO₂. (Reprinted figure with permission from [92]. Copyright (2017) by the American Physical Society.)

This debate on the origin of insulating gap in VO_2 between electronic band structure related Peierls instability [96, 97] and Mott physics [90, 98] has somewhat shifted in recent times towards Mott-insulator picture [93, 99–101]. This is mainly due to the refocus on the other two insulating phases, namely monoclinic M2 phase and triclinic T phase. It may be noted that in the theories involving Peierls instability, the metal–insulator transition involves only R and M1 phases. However, recent studies have revealed that M2 domains occur in many VO_2 samples near the metal–insulator transition, raising the question of its role in the transition (see [91] and references therein). Such studies refocus attention to the importance of measuring the electronic properties of the monoclinic M2 and triclinic T phases, in order to decouple the effects of the Peierls and Mott–Hubbard mechanisms [92]. To this end it may be instructive to summarize the arguments put forward by Pouget *et al* [90] in the mid 1970s, starting from a model of an isolated vanadium dimer in VO_2 with one electron per lattice site (see also [92]). In the limiting cases, the Hubbard model for a chain of such dimers will correspond to both the Peierls and Mott–Hubbard pictures, depending on whether the intradimer hopping parameter (t) or the intra-atomic Coulomb repulsion (U) is the dominant energy scale in the system. The qualitative description of the electronic structure will be the same in both the cases: an insulator with a bonded spin singlet on the dimer with the band gap resulting from the splitting of the bonding and antibonding a_{1g} bands (the lower and upper Hubbard bands in the Mott picture) [92]. The response of the energy gap to changes in the hopping parameter arising from the changes in lattice structure, can distinguish between these two cases. The bands broaden for the chain of dimers relative to the isolated dimer, thus decreasing the gap based on the interdimer hopping (t'). In the Peierls limit ($U \ll t, t'$), insulating behavior vanishes as t'

approaches t—the case of undimerized chains. On the other hand, in the Mott–Hubbard limit ($U \gg t, t'$) the gap is primarily set by U, hence it is insensitive to changes in the degree of dimerization. It is not quite possible to decouple the effect of dimerization from intra-atomic Coulomb correlations in the M1 phase, where all of the chains are dimerized and equivalent. This is, however, not the case with M2 and T phases. Huffman *et al* [92] have demonstrated in a recent study of broadband optical spectroscopy that the electronic structure of the VO_2 insulating phases is quite robust to the changes in lattice structure and vanadium–vanadium pairing. The energy gap is insensitive to the dimerization of the equally spaced vanadium ions with localized electrons in the M2 chains. This result rules out Peierls instability as the origin of the insulating gap in VO_2, and provides substantial evidence that this insulating gap is actually due to Mott–Hubbard type Coulomb correlations. Further support towards Mott physics of the M2 phase and its role in the metal–insulator transition in VO_2 came from the study of a granular VO_2 film using conductive atomic force microscopy and Raman scattering [102]. The recent dynamical mean-field theory with electronic structure calculations [100] also reported energy gaps for the M1 and M2 phases, which are consistent with these experimental results. In addition, the negative Knight shift is indicative of localized electrons on the equally spaced vanadium ions in the M2 chains [90]. The dimerized vanadium chains contain bonded spin singlets, which are localized on the vanadium dimers. This is in contrast to a more conventional Mott insulator, where valence electrons are localized on individual ions. In recent times within the framework of dynamical mean field theory (DMFT) Nájera *et al* [103] solved a Hubbard model with two orbital per unit cell, which captured the competition between Mott and singlet-dimer localization. The results of this theoretical study, which are in good agreement with the available experimental data on VO_2, show that the temperature-driven insulator-to-metal transition in VO_2 is compatible with Mott physics [103].

Like VO_2, the $4d$-electron system NbO_2 forms in distorted rutile-type crystal structure with Nb dimers, and undergoes an MIT at approximately 1083 K [104]. In $4d$ systems the d-orbitals are more extended in nature, hence Mott physics is expected to be less effective in this transition-metal oxide NbO_2. To this end the higher MIT transition temperature in NbO_2 is a bit surprising. Along with VO_2 this compound is considered to be a potential material for electronic switches and memory applications (see chapter 8) with an advantage of operation over a broader temperature range. Hence, the physical properties of NbO_2 remain a subject of continued interest.

The MIT in NbO_2 is accompanied by a structural transition from a rutile to a body-centered-tetragonal (bct) phase. The monoclinic (M1) structure of VO_2 and bct structures of NbO_2 have similar features. The pairs of transition-metal atoms in these phases dimerize and tilt with respect to the rutile c-axis. It was proposed by Goodenough [105] that such lattice distortion may cause opening of a band gap between the electronic states associated with the overlapping d-orbitals along the rutile c-axis and the remaining t_{2g} states. A possible explanation for the large difference of MIT temperature between VO_2 and NbO_2 is that the Peierls effect is stronger in NbO_2 due to larger overlap of Nb-$4d$ orbitals, thus leading to a greater orbital splitting between occupied and unoccupied d-states [106]. Consistent with this picture, an optical conductivity study has revealed that the orbital splitting is indeed approximately 0.3 eV larger in NbO_2 with splitting energy of ≈ 1.6 eV in NbO_2 and ≈ 1.3 eV in VO_2 [107].

However, as we have discussed above the insulating gap in VO_2 is due to the interplay between lattice distortions and electronic correlations, and not due to a purely Peierls-type transition alone. We saw that the Mott physics plays an important role in all phases of VO_2 [100]. Undimerized vanadium atoms undergo classical Mott transition through local moment formation in the M2 phase of VO_2, and strong super-exchange within V dimers introduces significant dynamic intersite correlations. The resulting metal–insulator transition in VO_2 is adiabatically connected to the Peierls-like transition, but it is better characterized as a Mott transition in the presence of strong intersite exchange [100]. In the same way the role of dynamic electronic correlations on the electronic structure of the metallic and insulating phases of NbO_2 has been investigated using a combination of density functional theory and cluster-dynamical mean-field theory calculations [108]. It is observed that electronic correlations lead to a strong renormalization of the t_{2g} sub-bands and there is an emergence of incoherent Hubbard sub-bands. This clearly indicates the presence of electronic correlations even in the metallic state of NbO_2. The effect of non-local dynamic correlations in the gap opening of bct phase of NbO_2 is less effective than in the monoclinic phases of VO_2. Nevertheless, the presence of electronic correlations in NbO_2 definitely indicates that it is not a purely Peierls-type insulator.

6.5 Sr_2IrO_4 and $Sr_3Ru_2O_7$: Mott insulators or Slater insulators?

We recall here the debate between Mott-type and Slater-type insulators discussed earlier in section 3.7. The Mott-type MIT is electronic-correlation (large value of U/t) driven with corresponding mass enhancement and formation of local moments. Long range magnetic order appears only subsequently. The Slater-type insulating behaviour, on the other hand, takes place for small value of U/t and is driven by the onset of long range antiferromagnetic ordering. The band-like Slater insulating state with an even number of electrons originates from the antiferromagnetic order induced modification of the Brillouin zone (see chapter 2, section 2.8). Here we revive this debate in real materials, and show the interplay between these two different types of mechanism of metal–insulator transition in two representative systems, namely a prototype relativistic Mott-insulator $5d$ transition metal oxide system $SrIr_2O_4$ and a $4d$ transition metal oxide system $Sr_3Ru_2O_7$.

6.5.1 Sr_2IrO_4

The relativistic Mott-insulator Sr_2IrO_4 forms in a layered cubic perovskite structure. We have learnt earlier in section 3.12 that the typical W values in $5d$ transition metal systems range from 3 to 4 eV, while the U value lies in the range 0.5–2 eV. In such systems Coulomb repulsion does not have the necessary strength to open a Mott–

Hubbard gap (since $U/W < 1$), and most $5d$ transition metal compounds are expected to be metallic within both the simple Mott–Hubbard picture and conventional one-electron band theory. The d orbitals in $5d$-compounds become more extended and tend to reduce the Coulomb repulsion U and hence diminish electron correlation effects. But the spin–orbit coupling energy ζ_{SO} in $5d$-transition metal systems is about 0.5 eV [109], which is quite comparable to U and W. This relatively large value of spin–orbit coupling (in comparison to transition metal compounds) leads to enhanced splittings between otherwise degenerate or almost degenerate orbitals and bands, and thus reduction of the kinetic energy gain. The last effect can offset the reduction in Coulomb electron repulsion energy U, and allow correlation physics back to the forefront. This combination of strong spin–orbit coupling and electron correlation opens up the insulating gap in Sr_2IrO_4 in spite of the rather moderate value of Coulomb electron interaction energy. It may be noted here that this spin–orbit coupling is a relativistic effect, which is quite different from the interactions between spin and orbital moments discussed above in section 6.3.2.

It is now understood that the ground-state of Sr_2IrO_4 comprises of a completely filled band with total angular momentum $J_{eff} = 3/2$ and a narrow half-filled $J_{eff} = 1/2$ band near the Fermi level E_F. This half-filled $J_{eff} = 1/2$ band is further split into an upper-Hubbard band (UHB) and lower-Hubbard band (LHB) due to on-site Coulomb interactions (see figure 3.11). Moreover this compound undergoes an antiferromagnetic ordering of the effective $J_{eff} = 1/2$ moments below $T_N = 240$ K [110]. The insulating ground state of Sr_2IrO_4 has been established through angle-resolved photoemission spectroscopy [111] and resonant x-ray scattering measurements [112]. However, absence of clear signature at the onset of antiferromagnetic ordering temperature T_N in transport [110, 113], thermodynamic [114] and optical conductivity [115] measurements raised the question of whether Sr_2IrO_4 is an usual Mott–Hubbard ($U > W$) insulator as that of 3d transition metal oxides or a Slater-insulator ($U \ll W$)? We may recall here again that a Mott–Hubbard MIT is usually a discontinuous transition taking place at temperatures greater or equal to the antiferromagnetic transition temperature, whereas Slater type MIT is a continuous transition driven by the onset of long range antiferromagnetic order. The results of time resolved optical spectroscopy measurements indicate that on cooling through the antiferromagnetic transition temperature the system evolves continuously from a metal-like phase to a gapped phase [116]. This is a typical signature of Slater MIT. On the other hand high energy reflectivity associated with optical transitions into the unoccupied $J_{eff} = 1/2$ band showed a sharp upturn at the antiferromagnetic transition temperature, which is consistent with a Mott–Hubbard MIT with spectral weight transfer into an upper Hubbard band [116]. A theoretical study with variational Monte Carlo method applied to large two-dimensional clusters showed that for a certain range of parameter U/t (where t is the hopping integral between neighbouring Ir $5d$ t_{2g} orbitals) lying between 3 and 7, the insulating antiferromagnetic ground state of Sr_2IrO_4 becomes a paramagnetic metal above the antiferromagnetic transition temperature [117]. This suggests a Slater MIT. In the regime with larger U/t the antiferromagnetic ground state becomes paramagnetic insulator above T_N, and thus the transition becomes Mott–Hubbard MIT. The insulating ground state thus changes from a weakly correlated Slater antiferromagnetic state to a strongly correlated Mott insulating antiferromagnetic state with increasing Coulomb electron interactions. It is suggested that Sr_2IrO_4 is located in the intermediate U/t regime between the weakly correlated and the strongly correlated antiferromagnetic insulators, and thus both the Slater-type and Mott-type characteristics can be observed in the experimental results on Sr_2IrO_4 [117]. This theoretical result got support from the bulk sensitive hard-x-ray photo-emission spectroscopy studies, which indicated that the insulating state of Sr_2IrO_4 was triggered by a combination of Slater-type antiferromagnetic correlation and Mott-type strong electron correlation [118].

6.5.2 $Sr_3Ru_2O_7$

The strength of electron correlation in $4d$ transition metal oxides is expected to be somewhere in between those $3d$ and $5d$ transition metal based systems. Competing phases and ordering phenomena have been observed in these materials, and in this respect ruthanate compounds have drawn considerable attention. We will take one member of this family $Sr_3Ru_2O_7$ as a platform to debate the issue: Mott insulator versus Slater insulator.

$Sr_3Ru_2O_7$ is known to be on the verge of being magnetic [119] and Mn has been doped at the Ru-sites to stabilize the magnetic order [120]. A metal to insulator transition has been observed in 5% Mn doped-$Sr_3Ru_2O_7$ samples around 50 K with the existence of long range antiferromagnetic order in the insulating state [120]. Subsequently it was reported that the antiferromagnetic phase transition takes place at a temperature below that of the metal–insulator transition temperature [121]. These observations naturally raise the question of whether the metal–insulator transition is strong electron correlation driven Mott–Hubbard transition or antiferromagnetic order driven Slater transition? This question was addressed using a temperature-dependent x-ray absorption (XAS) and resonant elastic x-ray scattering (REXS) study of the metal–insulator transition in $Sr_3(Ru_{1-x}Mn_x)_2O_7$ [122]. The XAS is sensitive to the nearest-neighbor spin correlation function, and hence to local antiferromagnetic correlations, whereas REXS is sensitive to long-range order in spin and charge. The combined results of XAS and REXS studies showed that the metal–insulator transition induced the onset of an exchange field at the Mn sites i.e. localized antiferromagnetic correlations. The long range antiferromagnetic order sets in at a distinctly lower temperature T_N. The MIT is thus of Mott–Hubbard type, where strong electron correlations drive both mass enhancement and associated electrical transport properties, and also the onset of local antiferromagnetic correlations. In the temperature region between T_{MIT} and T_N the magnetic correlations between separate Mn impurities are weak. Onset of true long-range antiferromagnetic order takes place only below T_N when the correlation length extends over a critical number of Mn impurities.

6.6 Mott insulating state in semiconductor surfaces

The surface atoms of covalently bonded semiconductors have reduced coordination, which can lead to the existence of a two-dimensional (2D) periodic array of partially filled dangling bonds. However, the reduced degree of freedom for electron delocalization decreases the kinetic energy gain in comparison to Coulomb repulsion. Thus the semiconductor surfaces provide an interesting platform for studying Mott physics. Indeed a Mott insulator was engineered on a silicon surface in an early experiment [123], where in the first step electrons were depleted in the Si surface by boron implantation and then electrons were introduced by adsorption of potassium just enough to take the system to the metallic limit. However, contrary to the expectation the photoemission spectroscopy indicated the presence of an energy band-gap [123], and the subsequent theoretical study [124], confirmed that this potassium-covered silicon surface was a Mott insulator.

The other interesting platforms to study Mott physics are the α phases of Sn and Pb on Si(111) or Ge(111) surface, which can be formed by adsorption of a 1/3 monolayer of Sn or Pb on the Si(111) or Ge(111) surface in an ordered $(\sqrt{3} \times \sqrt{3})R\,30°$ arrangement (see [125] and references within). The dangling bonds in the surface are passivated in this geometry, while each Sn or Pb adatom contributes one half filled dangling bond per unit cell (see figure 6.19). All these systems except Sn on Si(111) show a charge ordering transition at low temperature. Such instabilities arise due to the tendency of the system to saturate or eliminate partially filled dangling bonds via orbital rehybridization and/or the formation of new bonds. Angle-resolved photoemission experiments in Sn on Si(111) variously suggested this system to be an antiferromagnetic Mott insulator below 200 K [126] or a Slater insulator [127]. A subsequent study of scanning tunneling microscopy and spectroscopy, however, provided support towards the existence of Mott-insulating state. In this study using a modulation doping scheme, hole-doping up to 10% could

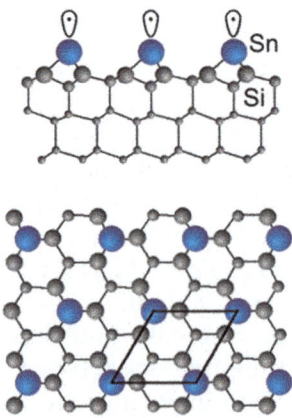

Figure 6.19. Side view (upper panel) and top view (lower panel) of the $\sqrt{3}$-Sn Mott insulating phase of Sn on Si(111). The Sn adatoms are shown in blue colour, each contributing a nominally half filled dangling bond orbital. (Reprinted figure with permission from [125]. Copyright (2017) by the American Physical Society.)

be achieved in $\sqrt{3}$-Sn dangling bond lattice without introducing structural and/or chemical disorder. Such modulation hole doping of these dangling bonds revealed clear signatures of Mott physics, such as spectral weight transfer and the formation of quasiparticle states at the Fermi level, well-defined Fermi contour segments, and a sharp singularity in the density of states [125]. All these studies open up the possibilities of realization of exotic quantum matter phases even on silicon-based materials platforms.

Self-assessment questions and exercises

1. What is a canonical Mott-insulator? The experimentally studied generic pressure–temperature phase diagram for some canonical Mott-insulators has revealed some striking differences with the corresponding pressure–temperature phase diagram for liquid to solid phase transition. What are these striking differences? What is another known physical system showing such behaviour?

2. What are the typical experimental signatures of filling control Mott transition?

3. What is the reason behind the asymmetry in filling control Mott transition with respect to electron and hole-doping in high T_C superconductor (HTSC) compounds?

4. What is the latest view on the mechanism of the Verwey transition in Fe_3O_4?

5. What is the David star structure in TaS_2 and how it is formed? What are the possible mechanisms of metallic behaviour in the non-commensurate CDW state of TaS_2?

6. The temperature dependence of magnetic susceptibility does not show the Curie–Weiss behaviour in the insulating state of TaS_2, which questions even the existence of local magnetic moment in the insulating state of $1T$-TaS_2. How is this serious deviation from the standard Mott picture explained in TaS_2?

7. What is the mechanism of metal–insulator transition in VO_2? Is it due to Peierls instability or is it a Mott-transition?

8. How can a Mott insulator be distinguished experimentally from a Slater insulator?

9. Why can band-semiconductor surfaces be considered as an interesting platform for studying Mott physics?

References

[1] Imada M, Fujimori A and Tokura Y 1998 *Rev. Mod. Phys.* **70** 1039
[2] Edwards P P 1998 *Phil. Trans. R. Soc. Lond.* A **356** 522
[3] Edwards P P and Rao C N R 2007 *Metal-insulator Transition Revisited* (London: Taylor and Francis)
[4] McWhan D B, Rice T M and Remeika J P 1969 *Phys. Rev. Lett.* **23** 1384
[5] McWhan D B, Remeika J P, Rice T M, Brinkman W F, Maita J P and Menth A 1971 *Phys. Rev. Lett.* **27** 941

[6] Kuwamoto H, Honig J M and Appel J 1980 *Phys. Rev.* B **22** 2626

[7] Limelette P, Georges A, Jérome D, Wzietek P, Metcalf P and Honig J M 2003 *Science* **89** 302

[8] Lupi S *et al* 2010 *Nat. Commun.* **1** 105

[9] Rodolakis F *et al* 2009 *Phys. Rev. Lett.* **102** 066805

[10] Janod E *et al* 2015 *Adv. Funct. Mater.* **25** 6287

[11] Kagawa F, Itou T, Miyagawa K and Kanoda K 2004 *Phys. Rev.* B **69** 064511

[12] Kanoda K 1997 *Physica* C **287** 299

[13] Ito H, Ishiguro T, Kubota M and Saito G 1996 *J. Phys. Soc. Jpn.* **65** 2987

[14] McKenzie R H 1997 *Science* **278** 820

[15] Lefebvre S, Wzietek P, Brown S, Bourbonnais C, Jérome D, Mézière C, Fourmigué M and Batail P 2000 *Phys. Rev. Lett.* **85** 5420

[16] Limelette P, Wzietek P, Florens S, Georges A, Costi T A, Pasquier C, Jérome D, Mézière C and Batail P 2003 *Phys. Rev. Lett.* **91** 016401

[17] Kagawa F, Miyagawa K and Kanoda K 2005 *Nature* **436** 534

[18] Bartosch L, de Souza M and Lang M 2010 *Phys. Rev. Lett.* **104** 245701

[19] Richardson R C 1997 *Rev. Mod. Phys.* **69** 683

[20] Pustogow A *et al* 2018 *Nat. Mater.* **17** 773

[21] Raju M, Banuti D T, Ma P C and Ihme M 2017 *Sci. Rep.* **7** 3027

[22] Sciortino F, Poole P H, Essmann U and Stanley H E 1997 *Phys. Rev.* E **55** 727

[23] Vaju C, Martial J, Janod E, Corraze B, Fernandez V and Cario L 2008 *Chem. Mater.* **20** 2382

[24] Pocha R, Johrendt D and Pettgen R 2000 *Chem. Mater.* **12** 2882

[25] Phuoc V T 2013 *Phys. Rev. Lett.* **110** 037401

[26] Camjayi A, Acha C, Weht R, Rodríguez M G, Corraze B, Janod E, Cario L and Rozenberg M J 2014 *Phys. Rev. Lett.* **113** 086404

[27] Katsufuji T, Okimoto Y and Tokura Y 1995 *Phys. Rev. Lett.* **75** 3497

[28] Katsufuji T and Tokura Y 1994 *Phys. Rev.* B **50** 2704

[29] Hays C C, Zhou J S, Markert J T and Goodenough J B 1999 *Phys. Rev.* B **60** 10367

[30] Kim H, Marshall P B, Ahadi K, Mates T E, Mikheev E and Stemmer S 2017 *Phys. Rev. Lett.* **119** 186803

[31] Mahajan A V, Johnston D C, Torgeson P R and Borsa F 1992 *Phys. Rev.* B **46** 10973

[32] Inaba F T, Arima T, Ishikawa T, Katsufuji T and Tokura Y 1995 *Phys. Rev.* B **52** R2221

[33] Dougier P and Hagenmuller P 1975 *J. Solid State Chem.* **15** 158

[34] Mott N F 1990 *Metal-Insulator Transitions* (London: Taylor and Francis)

[35] Sayer M, Chen P, Fletcher R and Mansingh A 1975 *J. Phys.* C **8** 2053

[36] Miyasaka S, Okuda T and Tokura Y 2000 *Phys. Rev. Lett.* **85** 5388

[37] Lee P A, Nagaosa N and Wen X-G 2006 *Rev. Mod. Phys.* **78** 17

[38] Peli S *et al* 2017 *Nat. Phys.* **13** 806

[39] Segawa K, Kofu M, Lee S-H, Tsukada I, Hiraka H, Fujita M, Chang S, Yamada K and Ando Y 2010 *Nat. Phys.* **6** 579

[40] Tokura Y, Takagi H and Uchida S 1989 *Nature* **337** 345

[41] Anderson P W 1987 *Science* **235** 1196

[42] Anderson P W 1997 *The Theory of Superconductivity in the High T_c Cuprates* (Princeton: Princeton University Press)

[43] Weng Z-Y 2011 *New J. Phys.* **13** 103039

[44] Walz F 2002 *J. Phys.: Condens. Matter.* **14** R285

[45] Lin J-F, Wu J, Zhu J, Mao Z, Said A H, Leu B M, Cheng J, Uwatoko Y, Jin C and Zhou J 2014 *Sci. Rep.* **4** 6282

[46] Verwey E J W 1939 *Nature* **3642** 327

[47] Miles P A, Westphal W B and von Hippel A 1957 *Rev. Mod. Phys.* **29** 279

[48] Verwey E J W and Haayman P W 1941 *Physica* **8** 979

[49] Verwey E J W, Haayman P W and Romeijn C W 1947 *J. Chem. Phys.* **15** 181

[50] Iizumi M and Shirane G 1975 *Solid State Commun.* **17** 433

[51] Zuo J M, Spence J C H and Petuskey W 1990 *Phys. Rev.* B **42** 8451

[52] Wright J P, Attfield J P and Radaelli P G 2001 *Phys. Rev. Lett.* **87** 266401

[53] Wright J P, Attfield J P and Radaelli P G 2002 *Phys. Rev.* B **66** 214422

[54] Jeng H-T, Guo G Y and Huang D J 2004 *Phys. Rev. Lett.* **93** 156403

[55] Lorenzo J E, Mazzoli C, Jaouen N, Detlefs C, Mannix D, Grenier S, Joly Y and Marin C 2008 *Phys. Rev. Lett.* **101** 226401

[56] Garcia J and Subias G 2004 *J. Phys.: Condens. Matter* **16** R145

[57] Senn M S, Wright J P and Attfield J P 2012 *Nature* **481** 173

[58] Ma Y, Johnson P D, Wassdahl N, Guo J, Skytt P, Nordgren J, Kevan S D, Rubensson J-E, Böske T and Eberhardt W 1993 *Phys. Rev.* B **48** 2109

[59] Seo H, Ogata M and Fukuyama H 2002 *Phys. Rev.* B **65** 085107

[60] Kim T *et al* 2018 *Sci. Rep.* **8** 5092

[61] Tokura Y and Nagaosa N 2000 *Science* **288** 400

[62] Maezono R, Ishihara S and Nagaosa N 1998 *Phys. Rev.* B **58** 11583

[63] Anisimov V I, Elfimov I S, Korotin M A and Terakura K 1997 *Phys. Rev.* B **55** 15494

[64] Damay F, Martin C, Maignan A and Raveau B 1997 *J. Appl. Phys.* **82** 6181

[65] Moreo A, Yunoki S and Dagotto E 1999 *Science* **283** 2034

[66] Dagotto E 2002 *Nanoscale Phase Separation and Colossal Magneto-Resistance* (Berlin: Springer)

[67] Goodenough J B 2003 *Handbook On the Physics and Chemistry of Rare Earths* ed K A Gschneidner Jr, J C Bunzli and V K Pecharsky (Amsterdam: Elsevier) p 249

[68] Wilson J A, Di Salvo F J and Mahajan S 1975 *Adv. Phys.* **24** 117

[69] Fazekas P and Tosatti E 1980 *Physica B+C* **99** 183

[70] Vaskivskyi I, Mihailovic I A, Brazovskii S, Gospodaric J, Mertelj T, Svetin D, Sutar P and Mihailovic D 2016 *Nat. Commun.* **7** 11442

[71] Rossnagel K 2011 *J. Phys. Condens. Matter* **23** 213001

[72] Sipos B, Kusmartseva A F, Akrap A, Berger H, Forró L and Tutiš E 2008 *Nat. Mater.* **7** 960

[73] Li L J, Lu W J, Zhu X D, Ling L S, Qu Z and Sun Y P 2012 *Eur. Phys. Lett.* **97** 67005

[74] Wu X L and Lieber C M 1990 *Phys. Rev. Lett.* **64** 1150

[75] Zwick F, Berger H, Vobornik I, Margaritondo G, Forró L, Beeli C, Onellion M, Panaccione G, Taleb-Ibrahimi A and Grioni M 1998 *Phys. Rev. Lett.* **81** 1058

[76] Ang R, Tanaka Y, Ieki E, Nakayama K, Sato T, Li L J, Lu W J, Sun Y P and Takahashi T 2012 *Phys. Rev. Lett.* **109** 176403

Ang R, Nakayama K, Yin W-G, Sato T, Lei H, Petrovic C and Takahashi T 2013 *Phys. Rev.* B **88** 115145

[77] Hellmann S *et al* 2010 *Phys. Rev. Lett.* **105** 187401

[78] Stojchevska L, Vaskivskyi I, Mertelj T, Kusar P, Svetin D, Brazovskii S and Mihailovic D 2014 *Science* **344** 177

[79] Yoshida M, Zhang Y, Ye J, Suzuki R, Imai Y, Kimura S, Fujiwara A and Iwasa Y 2014 *Sci. Rep.* **4** 7302

[80] Ritschel T, Trinckauf J, Koepernik K, Büchner B, Zimmermann M v, Berger H, Joe Y I, Abbamonte P and Geck J 2015 *Nat. Phys.* **11** 328

[81] Cho D, Cheon S, Kim K-S, Lee S-H, Cho Y-H, Cheong S-W and Yeom H W 2015 *Nat. Commun.* **7** 10453

[82] Law K T and Lee P A 2017 *Proc. Natl. Acad. Sci.* **114** 6996

[83] Anderson P W 1973 *Mater. Res. Bull.* **8** 153

[84] Fazekas P and Anderson P W 1974 *Philos. Mag.* **30** 423

[85] Zylbersztejn A and Mott N F 1975 *Phys. Rev.* **11** 4383

[86] Morin F J 1959 *Phys. Rev. Lett.* **3** 34

[87] Paul W 1970 *Mater. Res. Bull.* **5** 691

[88] Ladd L and Paul W 1969 *Solid State Commun.* **7** 425

[89] Plašienka D, Martoňák R and Newton M C 2017 *Phys. Rev.* B **96** 054111

[90] Pouget J P, Launois H, Rice T M, Dernier P, Gossard A, Villeneuve G and Hagenmuller P 1974 *Phys. Rev.* B **10** 1801

[91] Park J H, Coy J M, Kasirga T S, Huang C, Fei Z, Hunter S and Codben D H 2013 *Nature* **500** 431

[92] Huffman T J, Hendriks C, Walter E J, Yoon J, Ju H, Smith R, Carr G L, Krakauer H and Qazilbash M M 2017 *Phys. Rev.* B **95** 075125

[93] Goodenough J B 1971 *J. Solid State Chem.* **3** 490

[94] Eyert V 2002 *Ann. Phys.* **11** 650

[95] Biermann S, Poteryaev A, Lichtenstein A I and Georges A 2005 *Phys. Rev. Lett.* **94** 026404

[96] Wentzcovitch R M, Schulz W W and Allen P B 1994 *Phys. Rev. Lett.* **72** 3389

[97] Eyert V 2011 *Phys. Rev. Lett.* **107** 016401

[98] Rice T M, Launojs H and Pouget J P 1994 *Phys. Rev. Lett.* **73** 3042

[99] Weber C 2012 *Phys. Rev. Lett.* **108** 256402

[100] Brito W H, Aguiar M C O, Haule K and Kotliar G 2016 *Phys. Rev. Lett.* **117** 056402

[101] Gray A X *et al* 2016 *Phys. Rev. Lett.* **116** 116403

[102] Kim H, Slusar T V, Wulferding D, Yang I, Cho J-C, Lee M, Choi H C, Jeong Y H, Kim H-T and Kim J 2016 *Appl. Phys. Lett.* **109** 233104

[103] Nájera O, Civelli M, Dobrosavljević V and Rozenberg M J 2017 *Resolving the VO_2 Controversy: Mott Mechanism Dominates the Insulator-to-metal Transition* (arXiv: 1606.03157v4)

[104] Sakai Y, Tsuda N and Sakata T 1985 *J. Phys. Soc. Jpn.* **54** 1514

[105] Goodenough J B 1960 *Phys. Rev.* **117** 1442

[106] Eyert V 2002 *Eur. Phys. Lett.* **58** 851

[107] Wong F J, Hong N and Ramanathan S 2014 *Phys. Rev.* B **90** 115135

[108] Brito W H, Aguiar M C O, Haule K and Kotliar G 2017 *Phys. Rev.* B **96** 195102

[109] Kim S Y, Lee M-C, Han G, Kratochvilova M, Yun S, Moon S J, Sohn C, Park J-G, Kim C and Noh T W 2018 *Adv. Mater.* **30** e1704777

[110] Cao G, Bolivar J, McCall S, Crow J E and Guertin R P 1998 *Phys. Rev.* B **57** R11039

[111] Kim B J *et al* 2008 *Phys. Rev. Lett.* **101** 076402

[112] Kim B J, Ohsumi H, Komesu T, Sakai S, Morita T, Takagi H and Arima T 2009 *Science* **323** 1329

[113] Chikara S, Korneta O, Crummett W P, DeLong L E, Schlottmann P and Cao G 2009 *Phys. Rev. B* **80** R140407

[114] Ge M, Qi T F, Korneta O B, De Long D E, Schlottmann P, Crummett W P and Cao G 2011 *Phys. Rev. B* **84** 100402(R)

[115] Moon S J, Jin H, Choi W S, Lee J S, Seo S S A, Yu J, Cao G, Noh T W and Lee Y S 2009 *Phys. Rev. B* **80** 195110

[116] Hsieh D, Mahmood F, Torchinsky D H, Cao G and Gedik N 2012 *Phys. Rev. B* **86** 035128

[117] Watanabe H, Shirakawa T and Yunoki S 2014 *Phys. Rev. B* **89** 165115

[118] Yamasaki A *et al* 2014 *Phys. Rev. B* **89** 121111

[119] Ikeda S I, Maeno Y, Nakatsuji S, Kosaka M and Uwatoko Y 2002 *Phys. Rev. B* **62** R6089

[120] Mathieu R *et al* 2005 *Phys. Rev. B* **72** 092404

[121] Hu B, McCandless G T, Garlea V O, Stadler S, Xiong Y, Chan J Y, Plummer E W and Jin R 2011 *Phys. Rev. B* **84** 174411

[122] Hossain M A *et al* 2012 *Phys. Rev. B* **86** 041102(R)

[123] Weitering H H, Shi X, Johnson P D, Chen J, DiNardo N J and Kempa K 1997 *Phys. Rev. Lett.* **78** 1331

[124] Hellberg C S and Erwin S C 1999 *Phys. Rev. Lett.* **83** 1003

[125] Ming F, Johnston S, Mulugeta D, Smith T S, Vilmercati P, Lee G, Maier T A, Snijders P C and Weitering H H 2017 *Phys. Rev. Lett.* **119** 266802

[126] Li G, Höpfner P, Schäfer J, Blumenstein C, Meyer S, Bostwick A, Rotenberg E, Claessen R and Hanke W 2013 *Nat. Commun.* **4** 1620

[127] Lee J H, Ren X-Y, Jia Y and Cho J-H 2014 *Phys. Rev. B* **90** 125439

Part II

Applications of Mott insulators

The metal–oxide–semiconductor field-effect transistor (MOSFET) is the building block of microprocessors, memory chips and various micro-circuits, which in turn form the foundation of the complementary metal–oxide–semiconductor (CMOS) technologies in modern electronics industry. A modern CMOS device can contain billions of transistors, and the continued reduction in the size of CMOS transistors has caused enormous improvements in the density, switching speed, functionality and cost of the devices, say for example microprocessors. However, the advanced CMOS technology is facing problems, which include the rising difficulty in reduction of the supply voltage further, and prevention of the increasing leakage currents that degrade the switching ratio of 'ON' and 'OFF' currents. The problems are going to be more acute sooner or later, and there is definitely a need of at least some complementary technologies.

A field-effect transistor (FET) works by modulating the electrical charge carrier density of a thin semiconductor channel through the application of an electric field. The working principle is shown schematically in figure PII.1(a). The electric field in a FET structure is applied across a gate insulator using a gate electrode, which controls the charge carriers and creates a thin charge accumulation or depletion layer at the surface of the FET material. This later event in turn modifies the electrical conductivity between two contacts on the FET material namely source and drain. The creation of this accumulation or depletion layer by applied electric field is at the heart of FET operation. The characteristic width of this surface layer is represented by the electrostatic screening length, which in the metallic limit is the Thomas–Fermi length (see chapter 3). This screening length is extremely short in typical metallic systems, and this leads to negligible field effect. Larger screening

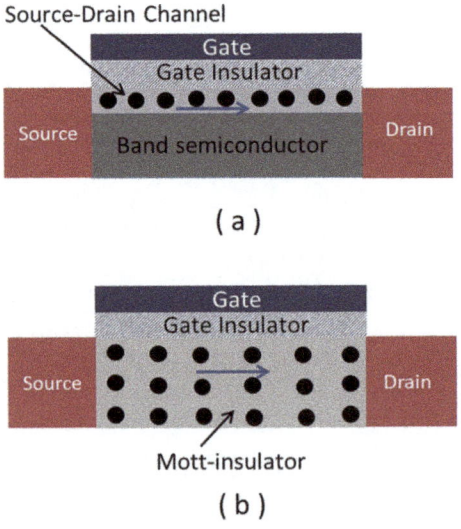

Figure PII.1. Schematic representation of the working of: (a) band-semiconductor based field effect transistor; (b) Mott-insulator based field effect transistor. The black dots represent charge carriers.

lengths and field effects are expected in low-carrier-density systems, and that is why standard transistors are fabricated with semiconductors. In addition to the sustainable charge carrier modulation, leakage currents in the FET channel through the gate insulator need to be much smaller than the channel current, and the density of localized interface states has to be small in comparison to the modulated carrier density. The classic MOSFET is based on silicon (Si) FET and a gate dielectric SiO_2, which is an excellent insulator with low leakage and almost free of interface states. The excellent properties of Si and SiO_2 are at the origin of the success of silicon-based electronics.

We have learnt in chapter 1 that within the independent electron approximation and one-electron band theory of solids, the Bragg scattering of the propagating electron wave in the Brillouin zone boundary of solid materials gives rise to an energy band gap. A band insulator is a solid with an even number of electrons per unit cell, and all the bands are either filled or empty at 0 K, with the highest filled band known as the valence band. Depending on the band gap, at finite temperatures there will be a small amount of charge carriers activated from the valence band to the next empty higher energy band i.e. conduction band. We also recall that band semiconductors belong to a special class of insulators with a relatively small energy band gap of the order of a few electron volt. The evolution of the semiconductor devices and the CMOS technology are indeed great success stories of electron band theory of solids, and this evolution will be briefly narrated below in chapter 7. The CMOS technology, however, is approaching a saturation level, and there is currently a need for going beyond the CMOS technology. This will be discussed in chapter 7.

We have learnt in chapters 5 and 6 that the electronic phase diagram in Mott-insulators is coupled with their carrier densities. An externally applied electric field can induce a certain amount of net carrier concentration in a Mott-insulator, and thus trigger a metal–insulator transition. This can lead to a sharp increase in the free carrier density in a Mott insulator, and in turn an enormous modulation of the electrical resistance, which is much more than that possible in a conventional FET. The working principle of a conceptual Mott-insulator based FET is shown schematically in figure PII.1(b). In this metal–insulator transition of Mott-insulators, the electrons are freed from their localized states near each atom without actually being moved through the bulk material. This is in contrast with the band semiconductor devices, where the charge carriers move through the material to a channel, thus enabling the devices to conduct current. Hence, Mott-insulator based devices have the potential to consume less power than band semiconductor devices. Furthermore, Mott metal–insulator phase transition can be triggered at sub-nanosecond timescales, the transition is robust and reversible in many cases, and the state of the device is accessed electrically. Thus, Mott-insulators are quite promising for ultrafast switches. These possibilities contributed to the very rapid growth of interest in Mott insulators since the early 2000 for logic and memory device applications. Beyond the logic and memory devices, the Mott insulators are also drawing considerable attention in recent times for various types of sensors and energy (both harvesting and storage) applications. In chapter 8 we will discuss the concepts and principles behind these emerging classes of Mott devices.

Chapter 7

Electron band semiconductor devices

The era of modern solid state semiconductor electronics really started in 1947 with the invention of point-contact transistor by John Bardeen and Walter Brattain [1] at Bell Laboratories, USA. This was then quickly followed by the development lead by Shockley of p–n junctions and bipolar transistors [2]. It may be noted here that I have used the term 'invention' instead of 'discovery' in the starting sentence! In this regard I would like to quote from pages 153–54 of the book by G Parker [3]: 'The work being carried out (at Bell Labs. in the USA) was a well-thought out, intense programme, trying to understand the mechanisms of solid-state physics at the atomic level. The scientists mentioned above were experts in the field of solid-state physics, and could apply that knowledge to eventually evolve the bipolar transistor (I don't even like the word invent here, although it is probably applicable). The point I am trying to make is that, true enough, the transistor's time had come, but it was due to the accumulation of a wealth of knowledge, not the fortuitous throwing together of semiconductors and dopants'. Two of the inventors of semiconductor transistor, Bardeen and Shockley, had worked before joining Bell Laboratories with Eugene Wigner and John Slater, respectively, who were the pioneers in the development of electron band theory of solids. In fact, both Bardeen and Shockley took active parts in the evolution of the one-electron band theory of solids. Even in those early days Bardeen had appreciated the importance of the electrostatic Coulomb repulsions between electrons i.e. electron–electron correlation in solids in addition to the exchange potential arising out of Pauli's exclusion principle [4]. Bardeen's theoretical approach of the late 1930s using energy-dependent potentials and relevant approximations is quite similar in its philosophy to the modern density-functional theory introduced subsequently in the mid 1960s [5]. On the other hand Shockley's early interest in the surface states of solids [6], played an influential role in the future development of the field effect transistors [7]. Bardeen and Shockley's continued deep interest in the electron theory of solids eventually lead to the key recognition of 'minority-carrier injection'—that an applied voltage can

doi:10.1088/2053-2563/ab16c9ch7

cause injection of valence band holes from the surface region of an n-type semiconductor near a metal-contact into the bulk of the material [8, 9]. For an early history of the solid state physics leading to the development of the field of solid state electronics, readers are referred to these interesting articles by Herring [5], Holonyak [8] and Weiner [10].

This early success story of semiconductor physics during 1930–50 made electron band theory with weakly interacting electron approximation very popular in the scientific community. In turn, the development of modern day solid state physics was greatly influenced by many exciting technological developments in the field of semiconductor electronics since the early 1950s. These devices based on band semiconductors eventually formed the basis of semiconductor electronics, which is now playing an enormous role in the present-day society and of course on the global economy. Presently there are 18 major semiconductor devices with over 140 device variations related to them [11, 12], and they form the foundation of the modern electronics industry. To highlight this success story of electron band theory, in this chapter we present a brief introduction to the semiconductor devices starting with the basic building blocks with which all the major semiconductor devices are made. A quite elaborate history of semiconductor devices (and the fabrication technologies) with some interesting historical photographs can be found at the website of the Computer History Museum [13]. I will then provide a basic introduction to the concepts and configurations of the band semiconductor devices important for logic and memory related applications. I will also discuss some applications beyond logic and memory, namely to photonic devices. During the course of this discussion we will find that the first integrated circuit containing one bipolar transistor, three resistors, and one capacitor was made in 1959, and then in 1965 Gordon Moore, co-founder of Intel Corporation (but at that time in Fairchild Semiconductor), predicted that the density of transistors on an integrated circuit would double every year [14]. We will see how science and technology working together in an intertwined manner sustained this so called 'more Moore' approach for the next five decades. Lastly, I will briefly narrate how this very robust electronics technology, which has grown very rapidly over the last sixty years will eventually reach a saturation level, and hence there is a need for an alternative technology. Presently there is indeed a necessity to go beyond the independent or weakly electron interaction driven so called 'solid state dream machine' [15]. The aim of this chapter is not to review the status of the semiconductor physics and technology, but to provide a basic framework to compare and contrast, with the emerging devices and technology based on strongly correlated electron systems, which will be the subject matter of the next chapter of this book. In the narratives on the semiconductor devices presented in this chapter, I have been considerably influenced by the excellent text books on semiconductor devices by Parker [3], Sze and Lee [11] and Neamen [16]. In the context of nanostructures and nanotechnology involving both electron band semiconductors and correlated electron materials, readers are referred to this fairly recent interesting textbook by Natelson [17].

7.1 Physics of intrinsic and extrinsic semiconductors

We have studied in chapter 1 that band semiconductors belong to a special class of insulators with a relatively small energy band gap of the order of a few electron volts. The energy band gap value E_g of typical semiconductor materials such as Si and GaAs at low temperatures are 1.12 eV and 1.42 eV, respectively. Thus, at a finite temperature the thermal energy k_BT can excite a number of electrons from the valence band to the conduction band, while an equal number of holes is left in the valence band. With the availability of many empty energy states in the conduction band, these electrons can be moved relatively easily by a small applied electric field resulting into a moderate current.

In an ideal intrinsic semiconductor there are no impurity atoms and no lattice defects in the semiconductor crystal. A practical intrinsic semiconductor is one like very pure silicon, which contains a relatively small amount of impurities compared with the thermally generated electrons and holes. The Fermi level E_F, lies in the band gap midway between the conduction and valence band. To get some idea on this total electron concentration, one first needs to know the electron distribution $n(E)$ in conduction band, which can be expressed in terms of Fermi–Dirac distribution function f (see appendix B) and density of states $g(E)$ introduced earlier in chapter 1:

$$n(E) = f(E)g(E). \tag{7.1}$$

Integration of this equation (7.1) over the entire conduction-band energy then yields the total electron concentration n_0 unit volume in the conduction band:

$$n_0 = \int_{E_C}^{E_{Top}} f(E)g(E)dE. \tag{7.2}$$

Here E_C and E_{Top} stands for the energy representing the bottom and top of the allowed conduction band energy. The Fermi–Dirac distribution function, however, approaches zero rapidly with increasing energy, hence E_{Top} can essentially be taken to be infinity. We rewrite the Fermi–Dirac distribution function here:

$$f(E) = \frac{1}{1 + e^{\frac{E-E_F}{k_BT}}}. \tag{7.3}$$

Here T stands for the absolute temperature in degrees Kelvin, k_B is the Boltzmann constant, and E_F is the Fermi level energy. It may be mentioned here that the Fermi–Dirac distribution function in general is described in terms of chemical potential μ, and strictly speaking $\mu = E_F$ only at $T = 0$. But conventionally in the semiconductor textbooks E_F is often used instead of μ.

The electrons within the conduction band have energy $E > E_C$. Now, if $(E_C - E_F) \gg k_BT$, then $(E - E_F) \gg k_BT$ and Fermi distribution function can be approximated as:

$$f(E) \cong e^{-\left(\frac{E-E_F}{k_BT}\right)}. \tag{7.4}$$

This is known as the Boltzmann approximation. However, it may be noted that the Maxwell–Boltzmann and Fermi–Dirac distribution functions are within 5% of each other when $E - E_F \approx 3kT$ [16]. So, while it is commonly used in the literature, the notation $\gg$ is then somewhat misleading to indicate when the Boltzmann approximation is valid [16].

With the help of equations (7.1), (7.4) and (1.76), the density of the electrons in thermal equilibrium can be expressed as:

$$n_0 = \int_{E_C}^{\infty} \frac{4\pi(2m_n^*)^{3/2}}{h^3} \sqrt{E - E_C} \exp\left[\frac{-(E - E_F)}{k_B T}\right] dE. \tag{7.5}$$

If we take:

$$x = \frac{E - E_C}{k_B T}. \tag{7.6}$$

Then equation (7.5) can be rewritten as:

$$n_0 = \frac{4\pi(2m_n^* k_B T)^{3/2}}{h^3} \exp\left[\frac{-(E_C - E_F)}{k_B T}\right] \int_0^{\infty} x^{1/2} e^{-x} dx. \tag{7.7}$$

The integral in equation (7.7) is the gamma function with a value of:

$$\int_0^{\infty} x^{1/2} e^{-x} dx = \frac{1}{2}\sqrt{\pi}. \tag{7.8}$$

Combining equations (7.7) and (7.8) we get:

$$n_0 = 2\left(\frac{2\pi m_n^* k_B T}{h^2}\right)^{3/2} exp\left[\frac{-(E_C - E_F)}{k_B T}\right]. \tag{7.9}$$

We introduce the parameter *effective density of states function in the conduction band N_C* as:

$$N_C = 2\left(\frac{2\pi m_n^* k_B T}{h^2}\right)^{3/2}. \tag{7.10}$$

Here m_n^* is the density of states effective mass of the electron. In terms of N_C, the thermal equilibrium electron concentration in the conduction band can be written as:

$$n_0 = N_C \exp\left[\frac{-(E_C - E_F)}{k_B T}\right]. \tag{7.11}$$

Assuming that m_n^* to be the same as the free electron mass m_0, the value of the effective density of states function at $T \sim 300$ K is $N_C = 2.5 \times 10^{19}$ cm^{-3}, which is the order of magnitude of N_C for most semiconductors [16].

The distribution of holes with respect to energy in the valence band is the density of allowed quantum states in the valence band multiplied by the probability that a state is not occupied by an electron [16]. This may be expressed as [16]:

$$p(E) = g(E)[1 - f(E)]. \tag{7.12}$$

Accordingly, the thermal-equilibrium concentration of holes in the valence band can be estimated by integrating equation (7.12) over the valence-band energy:

$$p_0 = \int g(E)[1 - f(E)]dE. \tag{7.13}$$

One can write:

$$1 - f(E) = \frac{1}{1 + e^{\frac{E_F - E}{k_B T}}}. \tag{7.14}$$

Now, $E < E_V$ energy states in the valence band, where E_V stands for the energy at the top of the valence band. If $E_F - E_V \gg k_B T$, then the equation (7.14) can be rewritten as:

$$1 - f(E) = \exp\left[\frac{-(E_F - E)}{k_B T}\right]. \tag{7.15}$$

Combining equations (7.13) and (7.14) one can obtain the thermal-equilibrium concentration of holes in the valence band as:

$$p_0 = \int_{-\infty}^{E_V} \frac{4\pi(2m_p^*)^{3/2}}{h^3} \sqrt{E_V - E} \exp\left[\frac{-(E_F - E)}{k_B T}\right]dE. \tag{7.16}$$

Since the exponential term in equation (7.15) decays fast enough, it is possible to approximate the lower limit of integration as minus infinity instead of the bottom of the valence band.

If we take:

$$y = \frac{E_V - E}{k_B T}. \tag{7.17}$$

Then equation (7.16) can be rewritten as [16]:

$$p_0 = -\frac{4\pi(2m_p^* k_B T)^{3/2}}{h^3} \exp\left[\frac{-(E_F - E_v)}{k_B T}\right]\int_{\infty}^{0} y^{1/2}e^{-y}dy. \tag{7.18}$$

It may be noted that the lower limit of integration in y becomes ∞ when $E = -\infty$. If the order of integration is changed then one needs to introduce another minus sign. Combining equations (7.8) and (7.18) we get:

$$p_0 = 2\left(\frac{2\pi m_p^* k_B T}{h^2}\right)^{3/2} \exp\left[\frac{-(E_F - E_V)}{k_B T}\right]. \tag{7.19}$$

We introduce the parameter *effective density of states function in the valence band* N_V as:

$$N_V = 2\left(\frac{2\pi m_p^* k_B T}{h^2}\right)^{3/2}. \tag{7.20}$$

Here m_p^* is the density of states effective mass of the hole. In terms of N_V, the thermal equilibrium hole concentration in the valence band can be written as:

$$p_0 = N_V \exp\left[\frac{-(E_F - E_V)}{k_B T}\right]. \tag{7.21}$$

At $T \sim 300$ K, N_V for most semiconductors is also of the order of 10^{19} cm^{-3} [16].

At a constant temperature the effective density of states functions, N_C and N_V, are constant for a given semiconductor material [16]. From the discussions above it is clear that the thermal-equilibrium concentration of electrons (holes) in the conduction (valence) band is directly related to the effective density of states function and to the Fermi energy level.

The number of electrons per unit volume n_i in the conduction band is equal to the number of holes per unit volume p_i in the valence band for an intrinsic semiconductor, and these parameters n_i and p_i are usually referred to as the intrinsic electron concentration and intrinsic hole concentration, respectively. However, since $n_i = p_i$, in the literature one normally uses the parameter n_i as the intrinsic charge carrier concentration. The Fermi energy level E_F for the intrinsic semiconductor is designated as the intrinsic Fermi energy E_{Fi}. From equations (7.11) and (7.21), we can write for intrinsic semiconductors:

$$n_i = N_C \exp\left[\frac{-(E_C - E_{Fi})}{k_B T}\right] \tag{7.22}$$

and

$$p_i = N_V \exp\left[\frac{-(E_{Fi} - E_V)}{k_B T}\right]. \tag{7.23}$$

Taking the product of the equations (7.22) and (7.23) we can write:

$$n_i^2 = N_C N_V \exp\left[\frac{-(E_C - E_{Fi})}{k_B T}\right] \cdot \exp\left[\frac{-(E_{Fi} - E_V)}{k_B T}\right] \tag{7.24}$$

or,

$$n_i^2 = N_C N_V \exp\left[\frac{-(E_C - E_V)}{k_B T}\right] = N_C N_V \exp\left[\frac{-E_g}{k_B T}\right]. \tag{7.25}$$

Here E_g stands for the band gap energy of the semiconductor. Equation (7.25) indicates that the larger the bandgap energy is, the smaller will be the intrinsic carrier density of the semiconductor. This equation also indicates that at a constant temperature, the value of n_i is a constant for a given semiconductor material, and it is independent of the Fermi energy. However, n_i will vary appreciably with temperature. Figure 7.1 shows schematically the band diagram, the density of states $g(E)$ and the carrier concentrations for an intrinsic semiconductor. At room temperature (300 K), n_i is 9.65×10^9 cm^{-3} for Si and 2.25×10^6 cm^{-3} for GaAs [11]. These numbers are quite small in comparison to the total number of electrons available in the valence band, hence the intrinsic semiconductors are not quite suitable for semiconductor electronic device applications. We will see below that required electron/hole concentration level for device applications can be achieved by selective doping of intrinsic semiconductor materials.

On doping with impurities, impurity energy levels are introduced in a semi-conductor. Such semiconductors are known as extrinsic semiconductors. When a pentavalent arsenic (As) atom is introduced in Si, it forms covalent bonding with its four neighboring Si atoms. The fifth arsenic electron has a relatively small binding energy, and can be transferred to the conduction band at a moderate temperature. Thus, an electron is donated to the conduction band, and the As atom is known as a donor. The silicon becomes an n-type semiconductor because of this addition of the negative charge carriers. In a similar way, a positively charged hole is created in the valence band of Si on doping with trivalent boron (B) atom, and Si becomes a p-type semiconductor. The boron in silicon is termed as an acceptor. The perfect periodicity of the Si lattice is disrupted by the impurity atoms, and multiple energy levels are introduced in the energy band gap.

In a semiconductor the addition of donor or acceptor impurity atoms changes the distribution of electrons and holes in the semiconductor. The Fermi energy is related to the carrier distribution function, hence the Fermi energy will change with the addition of dopant atoms. The density of electrons (holes) in the conduction (valence) band will change when the Fermi energy changes from near the midgap value. Using equation (7.11) we can write the thermal-equilibrium concentrations of electrons in an extrinsic semiconductor as:

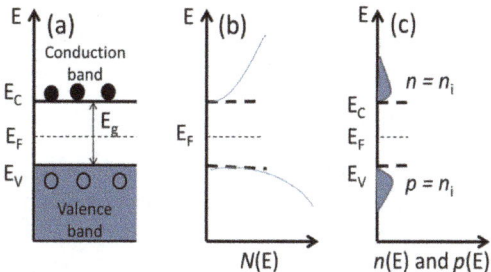

Figure 7.1. Schematic representation of an intrinsic semiconductor: (a) band diagram, (b) density of states, and (c) carrier concentration.

$$n_0 = N_C \exp\left[\frac{-(E_C - E_F)}{k_B T}\right]. \tag{7.26}$$

Denoting the Fermi energy of an intrinsic semiconductor as E_{Fi} equation (7.26) can be rewritten as:

$$n_0 = N_C \exp\left[\frac{-(E_C - E_{Fi}) + (E_F - E_{Fi})}{k_B T}\right] \tag{7.27}$$

or,

$$n_0 = N_C \exp\left[\frac{-(E_C - E_{Fi})}{k_B T}\right] \cdot \exp\left[\frac{(E_F - E_{Fi})}{k_B T}\right]. \tag{7.28}$$

Using the expression for the intrinsic carrier concentration n_i from equation (7.22) one can further rewrite equation (7.28) as:

$$n_0 = n_i \exp\left[\frac{(E_F - E_{Fi})}{k_B T}\right]. \tag{7.29}$$

In a similar manner we can write the thermal-equilibrium concentrations of holes in an extrinsic semiconductor as [16]:

$$p_0 = n_i \exp\left[\frac{-(E_F - E_{Fi})}{k_B T}\right]. \tag{7.30}$$

It is evident from equations (7.29) and (7.30) that with the change of Fermi level from the intrinsic Fermi level, the thermal-equilibrium concentrations of electrons and holes n_0 and p_0 in an extrinsic semiconductor change from the n_i value of an intrinsic semiconductor. In a semiconductor where $E_F > E_{Fi}$, there will be $n_0 > n_i$ and $p_0 < n_i$, thus $n_0 > p_0$. Such a semiconductor is characterized as n-type semiconductor. In a similar way, in a p-type semiconductor, $E_F < E_{Fi}$. This leads to the situation with $p_0 > n_i$ and $n_0 < n_i$, and thus $p_0 > n_0$. In an n-type (p-type) semiconductor, electrons (holes) are referred to as the majority carrier and holes (electrons) as the minority carrier.

Taking the product of the general expressions for n_0 and p_0, one can write:

$$n_0 p_0 = N_C N_V \exp\left[\frac{-(E_C - E_F)}{k_B T}\right] \cdot \exp\left[\frac{-(E_F - E_V)}{k_B T}\right] \tag{7.31}$$

and this equation can be further rewritten as:

$$n_0 p_0 = N_C N_V \exp\left[\frac{-(E_C - E_V)}{k_B T}\right] = N_C N_V \exp\left[\frac{-E_g}{k_B T}\right]. \tag{7.32}$$

It is to be noted that equation (7.32) has been derived for a general value of E_F, hence the values of n_0 and p_0 are not necessarily equal. This equation, however, is exactly the same as equation (7.25) obtained earlier for an intrinsic semiconductor. Hence for a semiconductor in thermal equilibrium one can write:

$$n_0 p_0 = n_i^2 \qquad (7.33)$$

which states that, for a given semiconductor the product of n_0 and p_0 is always a constant at a given temperature. This is one of the fundamental principles of semiconductors in thermal equilibrium [16].

In n-type Si and GaAs, for shallow donors there usually is enough thermal energy to ionize all donor impurities at room temperature, and hence to supply the same number of electrons in the conduction band. This is known as complete ionization, and in this situation the electron density is practically the same as the donor concentration. The larger the donor concentration, the smaller will be the energy difference $E_C - E_F$, and the Fermi level will get closer to the bottom of the conduction band [11]. In a similar way, the Fermi level will get closer to the top of the valence band for larger acceptor concentration. Figure 7.2 shows schematically the band diagram, the density of states and the carrier concentrations for an extrinsic n-type semiconductor. The Fermi level in this figure is marked E_D. From these discussions it is now clear that the number of charge carriers can be increased significantly in the extrinsic semiconductors. A very specific number of charge carriers can be generated with the choice of suitable dopants, which may required a particular type of device functioning.

7.2 Building blocks for semiconductor device

There are four basic building blocks with which different types of semiconductor devices are constructed:

1. First one is the metal–semiconductor interface building block, where these two different types of materials are in intimate contact (see figure 7.3(a)). Such metal–semiconductor interface found early usage as a semiconductor device in the form of a Schottky barrier diode. This building block can be used as a basic device allowing easy flow of electrical current only in one direction, i.e. a rectifying contact or as a device where electrical current can flow in either direction with a very small voltage drop i.e. an ohmic contact.

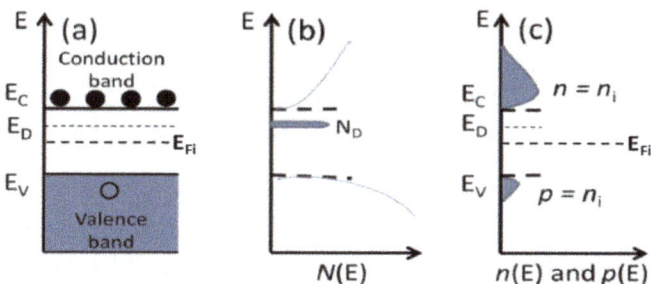

Figure 7.2. Schematic representation of an n-type extrinsic semiconductor: (a) band diagram, (b) density of states, and (c) carrier concentration.

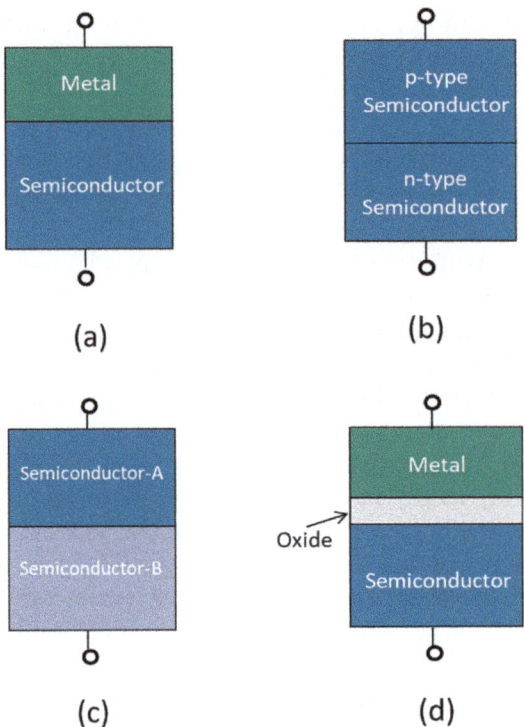

Figure 7.3. Four basic building blocks for semiconductor devices: (a) interface between metal and semi-conductor, (b) p–n junction, (c) interface between two dissimilar semiconductors-heterojunction interface; and (d) metal–oxide–semiconductor structure.

These can be combined to form a metal–semiconductor field-effect transistor, where a rectifying contact acts as the gate and two ohmic contacts act as the source and drain. It may be noted here that vacuum tube based rectifiers also perform quite well and they were extensively used from the mid-1920s in radio applications. However they are quite fragile and burn out easily. In comparison, the solid devices are relatively robust and of long durability.

2. Second building block is the junction formed between two semiconductor materials, one with positively charged carriers (i.e. p-type semiconductor) and the other with negatively charged carriers (i.e. n-type semiconductor). This p–n junction (see figure 7.3(b)) serves as a key building block for many devices. In fact the device constructed in 1947 by combining two p–n junctions was the first p–n–p bipolar transistor for which Bardeen–Brattain and Shockley received the Nobel prize in 1956.

3. Third building block is constituted by the interface between two dissimilar semiconductors, i.e. heterojunction interface (figure 7.3(c)). Gallium arsenide (GaAs) and aluminium arsenide (AlAs) are typical examples of dissimilar semiconductors used to form such heterojunction, which are actually key components for high-speed electronics and photonic devices.

4. Fourth building block is the metal–oxide–semiconductor (MOS) structure (figure 7.3(d)), which is a combination of a metal–oxide interface and an oxide–semiconductor interface. A metal–oxide–semiconductor field-effect transistor (MOSFET) is constructed by using the MOS structure as the gate and two *p–n* junctions as the source and drain. The MOSFET is the key element in advanced integrated circuits (ICs), and there are typically tens of thousands of MOSFETs in an integrated circuit chip.

7.3 A brief history of major semiconductor devices and technologies

The first documented account of the semiconducting property of a material was by Michael Faraday, when he had observed that the electrical conductivity of silver sulfide increased with increasing temperature [13]. This is in contrast with typical metals such as copper where the conductivity decreases with increasing temperature. The metal–semiconductor interface was the first semiconductor device ever studied, when in 1874 it was discovered by Karl Ferdinand Braun that the resistance of contacts between metals and metal sulfides depended on the magnitude and polarity of the applied voltage (see [11]). In the mid 1890s Jagadish Chandra Bose, an Indian scientist, demonstrated the use of lead sulfide (Galena) crystals in contact with a metal point to detect millimeter electro-magnetic waves [18], and a US patent was filed in 1901 for a point-contact semiconductor rectifier for detecting radio signals [19]. In the early 1900s electro-luminescence phenomenon was discovered by H J Round, with the generation of yellowish light from a crystal of carborundum by applying a potential of 10 V between two points on the crystal [11]. Subsequently various materials were tested in the USA, Europe and Japan to assess their rectification properties. Among these, silicon crystals showed some of the best result, and in 1906 G W Pickard filed a US patent for a silicon point-contact detector [20]. These point-contact detectors were commonly known as 'cat's-whisker' detectors. This name originated from the fine metallic probe used to make electrical contact with the crystal surface, and these detectors are now considered to be the primitive form of Schottky diodes. Such 'cat's-whisker' crystal radio detectors were used in the operation of simple radio sets, but by the mid-1920s they were replaced with the vacuum-tubes with better performance in most radio applications.

In the early 1930s Alan Wilson [21] used the evolving quantum theory of solids to propose the idea of electronic semiconductors, and associated their many interesting properties to the presence of impurity atoms in otherwise pure crystals of such materials. A satisfactory explanation of rectification in terms of an asymmetric barrier created by a concentration of electrons on the semiconductor surface to current flow, was provided by the late 1930s independently by Boris Davydov [22] in Russia, Nevill Mott [23, 24] at England and Walter Schottky [25] at Germany. However, the name Schottky barrier is more common in the literature. In fact all these theoretical developments along with the experimental works on copper-oxide rectifiers in Bell Laboratories provided an early impetus towards the quest for a semiconductor based amplifier as a viable and robust alternative for vacuum tubes [26].

As mentioned in the introduction of this chapter, the era of modern solid state semiconductor electronics really started with the invention of point-contact transistor by Bardeen and Brattain [1]. The semiconductor material used in that first transistor was germanium (Ge). The two point contacts were made from two strips of gold foil separated by about 50 micron and pressed onto the Ge surface [11]. The transistor action in the form of amplified input signal was observed with one gold contact having positive voltage with respect to the third terminal i.e. forward biased and the other one reverse biased. The birth of bipolar transistor was followed by a spurt of scientific activities leading to the inventions of various key semiconductor devices in the next three decades. A list of these key inventions is provided below [11, 13]:

1. Solar cell by Chapin, Fuller and Pearson in 1954 using a silicon p–n junction.
2. Heterojunction bipolar transistor to improve transistor performance by Kromer in 1957.
3. Heavily doped p–n junction based tunnel diode by Esaki in 1958.
4. MOSFET by Kahng and Atalla in 1960.
5. GaAs p–n junction based semiconductor laser by Hall *et al* in 1962.
6. Heterostructure laser by Kroemer, Alferov and Kazarinov in 1963.
7. Three important microwave devices: Gunn-diode invented by Gunn in 1963, IMPact ionization Avalanche Transit-Time diode (IMPATT) diode by Johnston, DeLoach and Cohen in 1965, and metal–semiconductor field-effect transistor (MESFET) by Mead in 1966.
8. Nonvolatile semiconductor memory (NVSM) by Kahng and Sze in 1967.
9. The charge-coupled device (CCD) by Boyle and Smith in 1970.
10. The resonant tunneling diode (RTD) by Chang, Esaki and Tsu in 1974.
11. Modulation-doped field-effect transistor (MODFET) by Mimura *et al* in 1980.

The first known integrated circuit (IC) was made by Kilby in 1959 (see [11]), which contained one bipolar transistor, three resistors, and one capacitor. All these device elements were made by using germanium and then connected by wire bonding. Subsequently Noyce created the monolithic IC by fabricating all devices in a single semiconductor substrate and connecting the devices by aluminum metallization. These inventions were the starting point of the rapid growth of the modern electronics industry. With the increase in the complexity of the integrated circuit, one moved from the n-channel MOSFET (NMOS) to the complementary MOSFET (CMOS) technology. The CMOS technology incorporates both the n and p-channel MOSFETs (NMOS and PMOS) to form the logic elements, which draw significant current only during the transition from one state to another (e.g. from 0 to 1) but otherwise very little current between transitions. This low power consumption is a distinct advantage of the CMOS technology. The CMOS concept was proposed by Wanlass and Sah in 1963 and CMOS technology is presently the dominant technology for electronics industry. In a related development the dynamic random-access memory (DRAM), was invented by Dennard in 1967. The memory cell contains one MOSFET and one charge-storage capacitor, and the MOSFET acts as a switch to charge or discharge the capacitor. DRAM is volatile and is used in most electronic systems as an important working memory where information is held temporarily before being saved for long-term storage (e.g. in NVSM) [11].

The first microprocessor was made in the early 1970s with the entire central processing unit (CPU) of a simple computer on a single chip. This was a four-bit microprocessor (Intel 4004), with a chip size of 3 mm × 4 mm, and it contained 2300 MOSFETs and operated at 0.1 MIPS (million instructions per second) [11]. This microprocessor performed in the similar level as those in IBM computers of the early 1960s, which needed a CPU the size of a large desk. This event was another major milestone in the history of the electronics industry, as we know the microprocessors constitute the largest segment of the industry today. The interconnect material in ICs since the early 1960s has been aluminium, which suffers from electromigration at high electrical current. In early 1990s copper interconnect was introduced to replace aluminium for minimum feature lengths below 100 nm. Over the period increased component density and improved fabrication technology led towards the system-on-a-chip (SOC), which is an IC chip containing a complete electronic system, and by the early 2000 the SOC was integrated into a three-dimensional system with improved performance.

Apart from electronics device applications, the semiconductor materials also drew considerable attention in the 1950s for their thermoelectric properties. Thermoelectric generators transform heat into usable electrical energy, and the conversion efficiency is defined by the dimensionless figure of merit of the thermo-electric materials $zT = \frac{S^2\sigma}{k}T$, where S, σ and k are the Seebeck coefficient, electrical conductivity and thermal conductivity, respectively, of the material. A high electrical conductivity is required for a small internal resistance of the generator. On the other hand, thermal conduction through the thermoelectric generator is one major loss mechanism, hence the thermal conductivity needs to be low. The Seebeck coefficient for semiconductors and insulators can have relatively high values than the metals. This is particularly so when the Fermi level lies deep within the forbidden gap [27]. This fact along with reasonable electrical conductivity and low thermal conductivity, was the reason for the exploration of semiconducting materials for thermoelectric generators. The best figures of merit of thermoelectric are mostly measured at around $zT \cong 1$, and for most technical applications zT needs to be around 1.5 [28]. Some of the commercial thermoelectric materials are: Bi_2Te_3, Sb_2Te_3, $PbTe$, $(GeTe)_{0.85}(AgSbTe_2)_{0.15}$ and $SiGe$ [29].

We will now briefly discuss below the working principle of some key semi-conductor devices.

7.4 Single-junction semiconductor devices

A single-junction semiconductor device consists of a single-crystal semiconductor material containing both p- and n-type regions that form a p–n junction. In this single-crystal semiconductor, one region is doped with acceptor impurity atoms to form the positively charged or p-region and the adjacent region is doped with donor impurity atoms to form the negatively charged or n-region. To understand the functional properties of a *p–n* junction, we will first consider *p*-doped and *n*-doped semiconductors separately (see figure 7.4(a)). The *p*-doped system has free positively charged holes and the *n*-doped system has free negatively charged electrons. Both systems, however, are electrically neutral as a whole since the charges of the mobile carriers are compensated by charged ions. The Fermi energy E_F in the *n*-doped semiconductor is in the band gap region just below the conduction band, whereas in the *p*-doped semiconductor the E_F is just above the valence band. When the two materials are brought into contact to form the *p–n* junction, the electrons in the conduction band will go to the valence band and fill the empty hole states (see figure 7.4(a)). This leads to pair-annihilation of both the electron and the hole, and after that there will be a region near the interface devoid of free charge carriers. This is known as the depletion region or space charge region (see figure 7.4(b)). This space charge region, however, is electrically charged, as there are charged ions but without

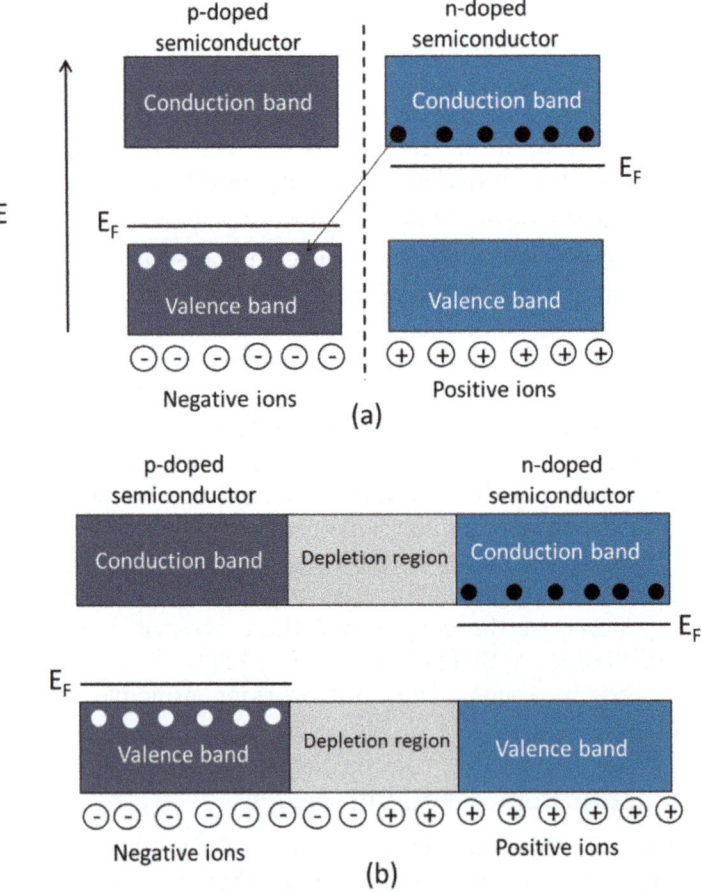

Figure 7.4. Schematic representation of a single-junction semiconductor device consisting of a *p–n* junction. (a) shows the *p*-type and *n*-type region separately, and the electron–hole pair annihilation process. (b) shows the expanded view of the junction highlighting the depletion or space charge region.

any charge carriers to neutralize them. As a result there is a net electric field pointing from the n-doped semiconductor to the p-doped semiconductor region. This acts like a capacitor with spatially separated positive and negative charges and an electric field in between. If an additional electron now moves across the space charge region in order to annihilate another hole, there will be a gain in energy E_G but at the same time there will be also a cost in energy of $-e\Delta\phi$, where ϕ is the electrostatic potential. With the increase in depletion region the last effect will be sufficiently large so that it is no longer energetically favorable for further electron-hole annihilation process. As a result the space charge region will grow only to a width where these two energies are in the same scale. In thermal equilibrium, the diffusion force on the majority carriers and the force from electric field will exactly balance each other [16].

There is, however, a problem in the representation of the p–n junction in figure 7.4(b) that the E_F of the n-type semiconductor is plotted higher than the E_F of the p-type semiconductor, and it does not reflect the significance of the electrostatic potential generated by the charges in the junction region. Hence, the more realistic picture is shown in figure 7.5 where the E_F on two sides are now at the same level, and that represents the fact that the drop in band energy is compensated by the change in electrostatic potential.

A very well known electrical property of the p–n junction is that it allows current to flow easily in one direction only i.e. rectification. It is a key building block for the bipolar transistor, as well as for MOSFETs. With proper biasing conditions or when exposed to electromagnetic radiation, the p–n junction can also function as micro-wave and photonic devices. We will briefly narrate here how a p–n junction acts as a rectifier. Figure 7.6(a) represents a p–n junction with some voltage V applied to it. This figure is similar to figure 7.5 but with potential barrier (height of the energy gap E_G) reduced by the applied voltage V. As a result the E_F on the p-doped and n-doped

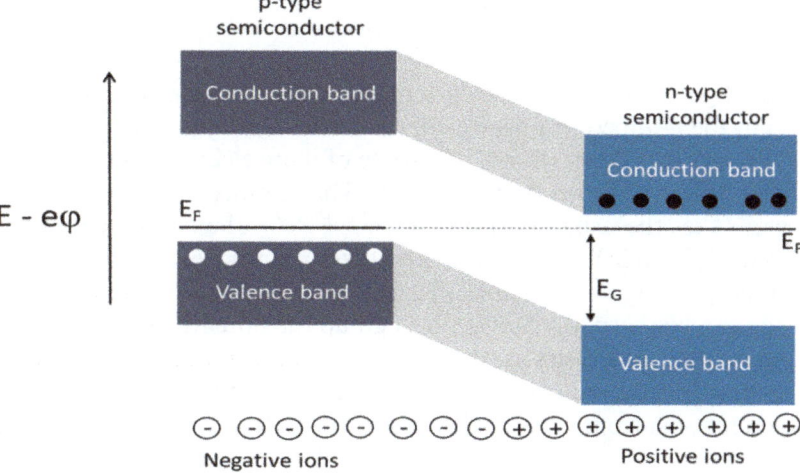

Figure 7.5. Schematic representation of the energy band diagram of a p–n junction device without any external bias after adding the electrostatic potential to the band energy.

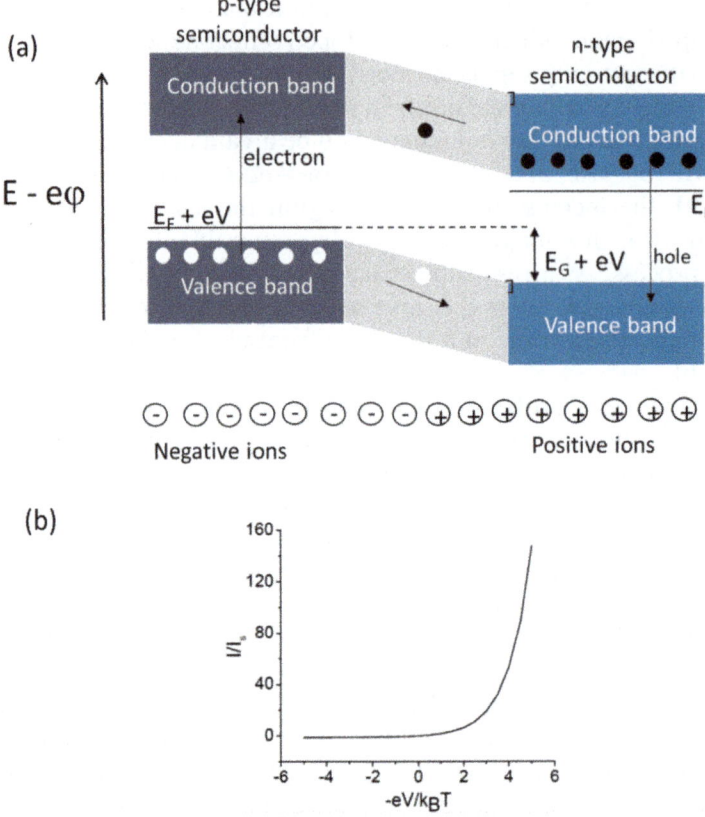

Figure 7.6. (a) Schematic representation of the energy band diagram of a p–n junction device with an external bias. The n-type semiconductor side of the diagram is pulled downwards by an applied voltage V. It is to be noted that $+eV$ is negative in this figure. The arrows show the various processes that can create a current in the devise (see text for details). (b) The characteristic I–V curve of a p–n junction diode represented by diode equation. The current is normalized with respect to the saturation current I_S.

sides are not aligned any more. The current can be generated in this device by several processes. On the p-doped (n-doped) side, electrons (holes) can be thermally excited into the conduction (valence) band, and some of these electrons (holes) will go down (up) the slope to the n-doped (p-doped) side. The number of excited charge carriers in both cases takes the usual activated form $e^{-E_G/k_B T}$. The charge current (electron) will flow to the left (right) in both cases and takes a form $I \propto e^{-E_G/k_B T}$. This current is designated as I_L. It is also possible that electrons in the conduction band on the n-doped side can be thermally activated to go up the potential slope in the depletion layer and annihilate with holes on their arrival at the p-doped side. In the absence of applied voltage the potential barrier that the electrons need to overcome is E_G, and thus the amount of current generated is proportional to $I \propto e^{-E_G/k_B T}$. In a similar way the holes in the valence band in the p-doped side can be thermally activated to go across the potential barrier towards the n-doped side and annihilate with electrons. Again in this process a current $I \propto e^{-E_G/k_B T}$ can be generated in the absence of an applied voltage. The charge current will be flowing to the right in these two cases and is designated as I_R. In the presence of an applied voltage V the height of the potential barrier is modified from E_G to $E_G + eV$, and accordingly the current I_R is modified as $I_R \propto e^{-(E_G+eV)/k_BT}$. The bias voltage V, however, will not change the number of excited carriers in the processes generating the current I_L, hence I_L is independent of bias voltage. The total current flow I_{Total} in this p–n junction device is the sum of I_L and I_R. In the absence of an applied voltage there cannot be any net current flow in the system, and the total current flow in general is expressed as:

$$I_{Total} = I_S(T)(e^{-eV/k_BT} - 1) \tag{7.34}$$

where $I_S(T) \propto e^{-E_G/k_BT}$ is known as the saturation current. In the literature p–n junction devices are commonly termed as diodes, and the equation (7.34) is known as the diode equation. The characteristic features of a diode as represented by equation (7.34) are displayed in figure 7.6(b), which shows that the current flows easily in one direction i.e. forward bias direction, and rather poorly in the opposite direction i.e. reverse bias direction.

7.5 Bipolar transistor

The transistor is a multi-junction three terminal semiconductor device, where the basic action involves the control of current at one terminal of the device by the voltage applied across the other two terminals. The bipolar transistor, which is also known as bipolar junction transistor (BJT), is one of two major types of transistors; the other being the field effect transistor (FET). As mentioned earlier in this chapter, the first bipolar transistor invented in 1947 had two metal wires with sharp points making contact with a germanium substrate. In the modern bipolar transistors germanium is replaced with silicon and the point contacts by two closely coupled p–n junctions forming two complementary structures n–p–n or p–n–p. Usage of two types of devices in the same circuit enables very versatile electronic circuit design.

Figure 7.7 shows the schematic representation of an n–p–n bipolar transistor and a p–n–p bipolar transistor, each consisting of three separately doped regions and two p–n junctions. The three terminal connections are termed as emitter, base and collector. The base region has a smaller width in comparison to the minority carrier diffusion length. The region marked (++) in figure 7.7 indicates the heavily doped region and the region marked with (+) is moderately doped. The emitter region has the largest doping concentration, and the collector region has the smallest. The actual structure of the bipolar transistor in reality is more involved [16], and figure 7.8 shows a cross section of a classic n–p–n bipolar transistor fabricated in an integrated circuit configuration. The terminal connections in actual devices are made at the surface. Heavily doped n^+ buried layers are included in order to minimize semiconductor resistances. Secondly, multiple bipolar transistors are fabricated on a single piece of semiconductor material. The individual transistors are isolated from each other, and this isolation is accomplished by adding p^+ regions so that the transistors are separated by reverse-biased p–n junctions as shown in figure 7.8. It is

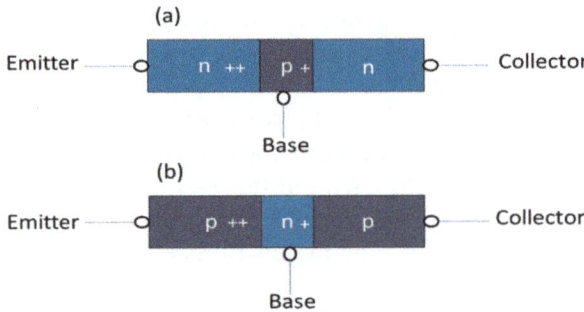

Figure 7.7. Schematic representation of (a) *n–p–n* and (b) *p–n–p* bipolar semiconductor devices.

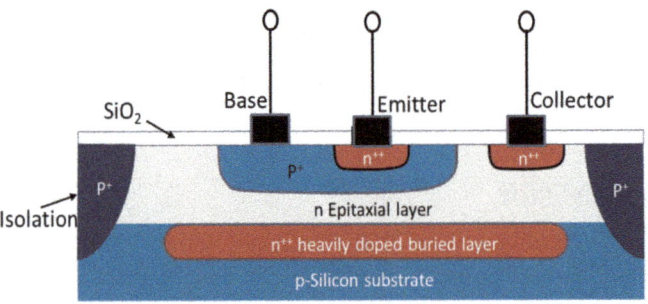

Figure 7.8. Cross section of a conventional integrated circuit *n–p–n* bipolar transistor.

to be noted that the bipolar transistor is not a symmetrical device; the impurity doping concentrations in the emitter and collector as well as the geometry of these regions are quite different.

The bipolar transistor is integrated with other circuit elements for voltage gain, current gain or signal-power gain. This is one of the most important semiconductor devices, which finds extensive usage in high-speed circuits, analog circuits, and power applications [11]. In contrast with a bipolar transistor, a heterojunction bipolar transistor is a transistor where one or both *p–n* junctions are formed between dissimilar semiconductors. The applications of the heterojunction bipolar transistor are essentially the same, but it has higher-speed and higher-frequency capability in circuit operation. For these reasons heterojunction bipolar transistors are preferred in photonic, microwave, and digital applications.

7.6 Metal–oxide–semiconductor field-effect transistor (MOSFET) and CMOS technology

A field effect transistor (FET) is a device whose electrical resistance is controlled by an external bias voltage. An FET is comprised of a conducting channel with two contacts termed as the source and the drain. There is also a third contact to this conducting channel: the gate, which is separated from the conducting channel by an insulating layer. In an operating FET, this gate contact remains biased with respect to the conducting channel. Different amounts of charge can be induced in the channel depending on the magnitude and sign of the gate voltage. The so called 'field-effect' involves a change in conductance of a solid induced by the application of transverse electric field, and this was subject of intensive research in the 1920–30s [30]. In fact, the first FET was invented by Julius Lilienfeld in 1925 [30], but there is no clear evidence that the conceptual FET actually worked. The first unequivocal demonstration of field-effect in the form of appreciable modulation of conductance in the surface region of a semiconductor was made by Shockley and Pearson in 1948 [7]. In the early 1950s this field-effect was applied to many prototype devices, in which the transverse electric field caused the majority carrier density to be modulated in a semiconductor leading to a conductance changes between two suitably placed ohmic contacts on the semiconductor surface. The useful devices based on this principle require that the ratio of surface area perpendicular to the applied electric field to volume (where the conductance is modulated) must be sufficiently large. Thin film technology was needed to meet this geometrical requirement and the first thin film field-effect transistors were eventually reported in 1961. Such devices usually lead to high impedances in the amplifier circuits, thus limiting operation to low frequencies [30].

In between, in 1960 a remarkably simple electronic device based on silicon technologies, and without the geometrical constraints mentioned above, was introduced by Kahng and Atilla of Bell Laboratories [31]. This is the metal–oxide–semiconductor (MOSFET) which subsequently became the backbone of the modern semiconductor electronics industry. This device is based on the idea of inversion layer conductance modulation, which was envisaged earlier during the studies on bipolar transistors (see [30] and references within). It was also realized that if the space between the control gate electrode and the semiconductor surface was filled with a dielectric material, the channel could be maintained in a controlled manner until the gate was suitably biased to reverse the polarization, and hence erase the channel. Over the period SiO_2 turned out to be the universal gate dielectric because of its excellent insulating properties with low leakage. It is also chemically quite stable and almost free of interface states. Figure 7.9(a) shows the schematic diagram of a planar MOSFET structure.

MOSFETs are primarily used as logic switches, but they can also be used in various other applications involving low-voltage, low-power and high-speed operation. This is a device of relatively small size, and is used extensively in digital circuit applications where millions of devices are included in a single integrated circuit. At the heart of the MOSFET technology is a metal–oxide–semiconductor structure known as the MOS capacitor [11, 16]. While the metal in a MOS capacitor may be aluminium or some other type of metal, in many real cases a high-conductivity polycrystalline silicon (deposited on the oxide) can be used. The source and drain contacts to the channel should ideally be of low resistance ohmic contact. The channel could be a thin film semiconducting material, or a two-dimensional electron gas in a quantum well. Depending on the nature of charge carrier, there can be n-channel MOSFET (NMOS), where the majority carrier is electron, and the p-channel MOSFET (PMOS), where the majority carrier type is hole. Most

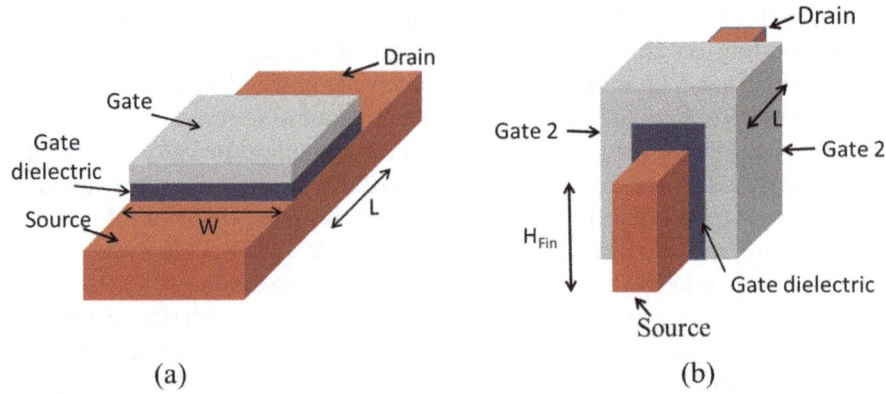

Figure 7.9. Schematic diagrams showing the comparison between (a) planar MOSFET structure and (b) FinFET structure.

common MOSFETs are Si-based, but other semiconductor materials are also being used. In contrast with bipolar junction transistor (BJT), which is a current controlled device, MOSFET is actually a voltage controlled device. In a MOSFET switching or amplification takes place by application of the gate voltage, and practically no current is required. This leads to the rather low power consumption in MOSFETs. Of course, there is another profound difference between these two types of transistors: the MOSFET is a unipolar device transporting one type of carrier (electrons or holes), whereas BJT is bipolar device needing both electrons and holes to operate. We will discuss more some basic concepts and device properties of MOSFET in chapter 8, while comparing those with similar three-terminal devices based on Mott-insulators.

7.6.1 CMOS technology

The present information technology is based on the integrated circuits (ICs), which are used for data processing i.e. logic, data storage i.e. memory, and also in the data transfer networks involving opto-electronic devices like light emitting diodes (LEDs) and semiconductor lasers. This leads to the demand for more energy efficient and economic devices. During the evolution of MOSFET it was realized that electronic circuit design becomes more versatile when the two complementary configurations of MOSFET, i.e. *n*-channel MOSFET and *p*-channel MOSFET, are used in the same circuit. These electronic circuits are known as complementary metal oxide semiconductor (CMOS) circuits, and they involve complementary *p*-channel and *n*-channel MOSFET pairs. With the increase in the complexity of the integrated circuit (IC), the technology gradually moved from MOSFET to CMOS technology, which has the advantage in that the logic elements draw significant current only during the transition from one state to another (e.g. from 0 to 1) and draw very little current between transitions. This low power consumption makes the CMOS technology the dominant technology for advanced ICs. Reduction of the MOSFET dimension i.e. scaling down has become a continuous trend in the electronic industry, since smaller size devices enable higher device density in an IC. In addition, a smaller channel length improves the driving current, and in turn performance of the device.

7.6.2 Semiconductor memory device

There are two types of semiconductor memories: (i) volatile memories such as dynamic random-access memories (DRAMs) and static random-access memories (SRAMs), in which the stored information is lost on switching off the power supply; (ii) nonvolatile memories, where the stored information can be retained on a long-term basis. DRAM with its high density and relatively low cost and SRAM with high speed are used extensively in personal computers and workstations. The nonvolatile memory, on the other hand finds extensive usage in portable electronics systems such as the cell phone, digital camera, and smart IC cards because of low-power consumption and nonvolatility [11].

DRAM is a two element circuit incorporating a MOSFET and a MOS capacitor. In a DRAM cell, the MOSFET acts as a switch to control the writing, refreshing, and read-out actions, and the MOS capacitor is used for charge storage (see figure 7.10). The MOSFET is switched ON during a write cycle, and the logic state in the bit line is transferred to the storage capacitor. The operation of DRAM in practical devices is dynamic, because the charges stored in the MOS capacitor will disappear gradually due to the small but non-negligible leakage current of the storage node, and as a result the data in DRAM need to be refreshed periodically within a fixed interval, typically 2–50 ms. Aggressive scaling down of the DRAM cell size is necessary to increase the storage density of a chip. The scaling, however, will also decrease the capacitor electrode area, which will result in the degradation of the storage capability of the MOS capacitor. Three-dimensional capacitor structures and high dielectric-constant materials are possible remedies of this problem.

In contrast to DRAM with the requirement of periodic refreshing, in a SRAM device memory can be retained as long as the power is ON. The typical structure of a 6-MOSFET SRAM cell is shown in figure 7.11. The core of the SRAM cell is built

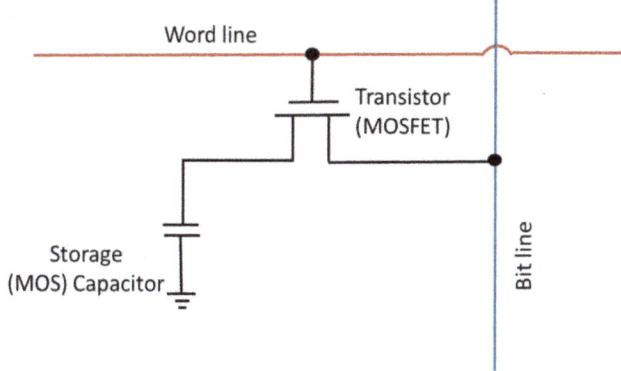

Figure 7.10. Schematic diagram of a CMOS DRAM cell.

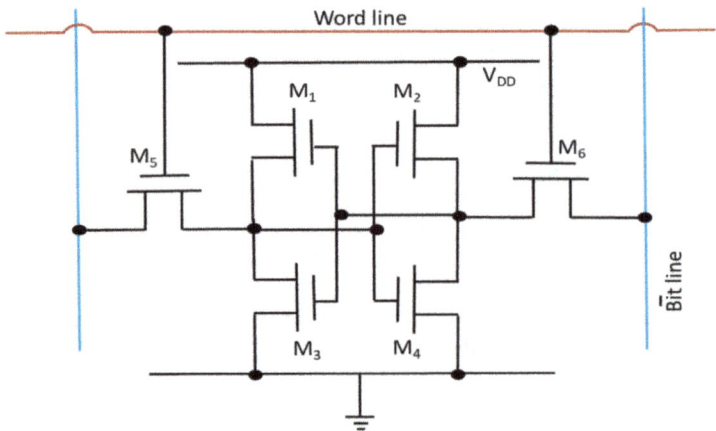

Figure 7.11. Schematic diagram of a CMOS SRAM cell.

with two cross-coupled CMOS inverters to store the logic state (see figure 7.11). In this matrix, M1 and M2 are *p*-channel MOSFETS serving as the load transistors, whereas the *n*-channel MOSFETS M3 and M4 are the drive transistors. The output potential of each inverter becomes the input into the other, and the inverters are stabilized to their respective state by this feedback loop. Two additional *n*-channel MOSFETs M5 and M6 are called access transistors, and they along with the word line (WL) and bit line (BL) are used to access (read and write) the SRAM cell. Each cell in an array of SRAM cells, can be read or written in a random order, irrespective of which cell was accessed last.

A nonvolatile memory device is achieved by modifying the gate electrode of a conventional MOSFET so that semi-permanent charge storage inside the gate is possible. Such nonvolatile memory devices have been used extensively in ICs as the erasable-programmable read-only memory (EPROM), electrically erasable pro-grammable read-only memory (EEPROM), and flash memory. There are two groups of nonvolatile memory devices, (i) floating-gate devices and (ii) charge-trapping devices.

In floating-gate memory devices charges are injected from the silicon substrate across the first insulator and stored in the floating gate. These stored charges cause a shift in the threshold-voltage and the device is switched to a high-threshold state, i.e. programmed or logical 1 [11]. A gate voltage or other external sources, such as ultra-violet (UV) light, can be used to erase the stored charges, and the device returns to the low-threshold state, i.e. erased or logical 0. In a well-designed memory device, the charge storage time can be over 100 years. The first EPROM was developed using heavily doped poly-silicon as the floating-gate material. UV light or x-ray can be used to erase the memory. These radiations can excite the stored charges into the conduction band of the gate oxide and back to the substrate. In contrast, the EEPROM uses the tunneling process to erase the stored charges, and cells in an EEPROM can be erased on a selection basis.

In the early 1980s Flash memory devices were developed from EEPROM. Three layers of poly-silicon constitutes the cell structure of flash memory, and the cell is programmed by a channel hot carrier injection mechanism similar to EPROM [11]. The erasing process is controlled by the field emission of electrons from the floating gate to an erase gate. In a flash memory device the storage cells are divided into several blocks, and erasing takes place on one selected block with the tunneling process. The name flash is attributed to the erasing speed in such devices, which is much faster than that of EPROM. Flash memory is categorized into NAND Flash and NOR Flash depending on the cell structure. NAND Flash is suitable for a high storage capacity and it supports a fast writing/erasing rate. They are widely used for storage applications like memory cards, USB drives, and solid state drives. On the other hand, NOR Flash provides a faster readout rate, but this comes at the expense of storage density. NOR Flash finds usage where fast code execution is involved, and in various hand-held devices including cell-phones [32].

7.7 Beyond logic and memory: opto-electronic devices

Semiconductor devices can also be designed to convert optical energy or light into electrical energy, and vice versa. In general terms these devices are called opto-electronic or photonic devices. There are four groups of photonic devices: (i) light-emitting diodes (LEDs), (ii) light amplification by stimulated emission of radiation (lasers), (iii) photodetectors, and (iv) solar cells. Of these, LEDS and lasers convert electrical energy to optical energy, photodetectors electrically detect optical signals, and solar cells convert optical energy into electrical energy.

In a solid, interaction between an electron and photon can take place via three processes: (i) absorption, (ii) spontaneous emission, and (iii) stimulated emission. At room temperature most of the atoms in a solid are in the ground state. Assuming two energy levels E_1 and E_2 representing ground state and excited state, respectively, of an atom, any transition between these states involves the emission or absorption of a photon with frequency ν_{12} given by $h\nu_{12} = E_2 - E_1$. Figure 7.12 presents a schematic diagram showing (a) induced absorption, (b) spontaneous emission, and (c) stimulated emission processes. The process of the absorption of radiation is shown in figure 7.12(a), where an incident photon is absorbed and an electron is transferred from the energy level E_1 to the energy level E_2. In the event of the electron returning to the ground state E_1 after sometime without any external stimulus and giving off a photon of energy $h\nu_{12}$, the process is called spontaneous emission. This process is represented in figure 7.12(b). On the other hand, when a photon of energy $h\nu_{12}$ falls on an atom while it is in the excited state E_2 (see figure 7.12(c)) the atom can be stimulated to make a transition to the ground state E_1. The transition of the electron from the excited state E_2 to the ground state E_1 produces a photon of energy $h\nu_{12}$, which is in phase with the incident radiation. The process is called stimulated emission, which involves two photons, and hence can lead to optical gain or amplification. In LEDS the dominant operating process is the spontaneous emission, whereas stimulated emission is the necessary condition for the lasing action. The functioning of photodetectors and the solar cells on the other

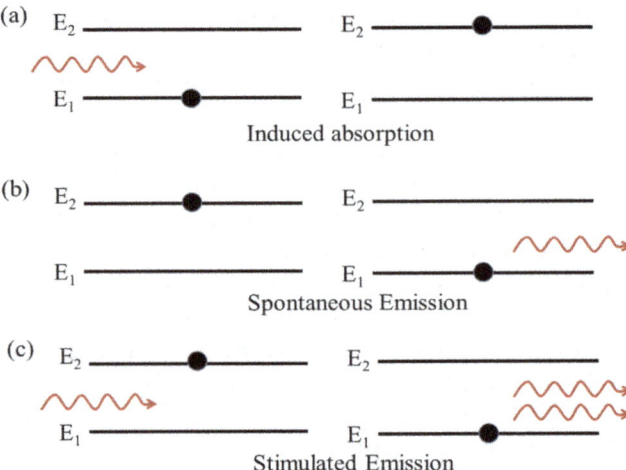

Figure 7.12. Schematic diagram showing (a) induced absorption, (b) spontaneous emission, and (c) stimulated emission processes.

hand involves the absorption of the external radiation. We will discuss the function of a conventional semiconductor based solar cell in chapter 8 while comparing the same with a solar cell based on Mott-insulators.

7.8 Scaling of MOSFET

The reduction in the size of semiconductor devices would allow an increase in their density on an integrated circuit (IC), which in turn would increase the functionality of the circuit for a constant IC size. Gordon Moore in 1965 published a seminal paper, where he predicted that the density of transistors on a chip would double every year [14]. This is now known as the celebrated Moore's law. The most complex chip in 1965 had about 64 transistors, but true to the prediction of Moore the number of transistors in a chip went on increasing, but at the same time the chip making process also became quite complex. Realizing this increasing complexity, Moore in the mid 1970s slightly modified his law, and from then onwards it was said that the number of devices on a chip would double every two years [33]. This difference with the original prediction of Moore is often rationalized by saying that the microchip complexity would double every 18 months [34]. Moore's law also gave rise to the concept of 'More Moore', which represents the technological approach that would advance Moore's law with continued scaling of the physical feature sizes of silicon based CMOS transistors [35]. The semiconductor industry has remarkably kept pace with this exponential growth, which provided a great impetus in the evolution of sophisticated technologies involving electronics devices. The starting point of this success story is the scaling theory proposed in another seminal paper in 1974 by Dennard *et al* [36]. Increasing the density of transistors on a chip by a factor of 2 leads to the reduction of the chip's linear dimensions (length and width) by a scaling factor equal to $\sqrt{2}$ [37], which is generally termed as κ. With an assumption of maintaining a constant electric field inside the transistor, Dennard *et al* [36] showed that scaling the device by a factor κ increased the switching speed by κ, reduced the power dissipation by κ^2 and improved the power-delay product by κ^3. This scaling law would also indicate a reduction in the supply voltage and the threshold voltage by the same factor κ.

Dennards scaling law was followed in the semiconductor industry until approximately 2005 [37, 38]. The silicon processing length scale has reduced tenfold every 15 years since 1971 and around 2010 it reached to 22 nm [39]. Extrapolation of this scaling trends for CMOS devices, however, indicates that the transistor power consumption in modern integrated circuits actually reduces more slowly than their size [40]. In practice the effective transistor channel length is even smaller, about 15 nm. At such small dimensions, there is a discrete, countable number of donors (10–100) in the transistor channel, and the electrical characteristics of individual dopant atoms are expected to be prominent in the commercial MOSFETs, especially at low temperatures. In addition the width of the gate dielectric has now reached the size of several atoms, which can create several problems. A few missing atoms can alter transistor performance and quantum tunneling of electrons through potential barriers would limit the ability to confine charges to a densely packed array [39]. To circumvent these problems the transistors are redesigned with wider dielectric layers. In this structure known as FinFET there is a thin vertical silicon body on a substrate (see figure 7.9(b)). The structure got its name FinFET because of this thin Si body, which resembles the back fin of a fish. Since the fin Si body is very thin, there is no leakage path which can be far from the gate. And then the wrapping of the gate around the channel provides excellent control from three sides of the channel. This configuration leads to some improvement in the control of the electric field and reduction in current densities and leakage. In contrast with a planar MOSFET structure where the channel is horizontal, the channel in FinFET is vertical and the height of the channel H_{FIN} (see figure 7.9(b)) determines the width of the FinFET. Each transistor can have several fins, and thus extend transistor scaling for some more years.

7.9 Limit of existing semiconductor technologies

It is quite clear now that there is a problem with almost exponential growth of transistor density in semiconductor chips as predicted by Moore's law. In fact Moore himself pointed out in 2003—'no exponential change of a physical quantity can, however, continue forever' [41]. There are various key factors, which limit continued scaling of CMOS devices [42]:

1. Physical limit: This is due to the increase in tunneling and leakage currents as the CMOS devices become smaller, which influence the performance and functionality of the devices. The transistor dimensions eventually will approach the size of the atom and molecule. It is clear that the devices smaller than the dimension of a single molecule may not be possible, and some dimensions of the device are more than a molecule wide. Even if the technology constraints are sorted out, the transistors are unlikely to reach single molecule gate length. Various non-ideal effects will start to dominate,

and the performance gains from each successive generations of devices become less than the gains from the last generation. One of such non-ideal effects is the off state power consumption, which needs to be addressed for current scaling of CMOS devices to continue. In the context of digital logic, the sources that contribute to off-state power consumption are: junction leakage, gate induced drain leakage, subthreshold channel current, and gate tunnel currents. The leakage currents grow exponentially as gate length decreases. Thus scaling CMOS devices all the way down to the molecular level may not produce increasingly better devices.

2. Materials limit: This limit arises from the inability of the dielectric and wiring materials to provide reliable insulation and conduction, respectively, with the reduction in the dimensions of devices. Performance of the concerned materials such as silicon, silicon dioxide, aluminium and copper are limited by their physical properties such as dielectric constant, carrier mobility, carrier saturation velocity, conductivity and breakdown field. The reliability of SiO_2 deteriorates as it becomes thinner, resulting ultimately in breakdown. Copper is less sensitive to electromigration than aluminium, but the material is more susceptible to the defects when used as interconnect wires. Low-k materials are utilized as insulator between adjacent interconnect wires, suffer from the high mechanical and thermal stress during packaging of the device. High-permittivity materials have been used to replace silicon SiO_2 as gate dielectric to address the current leakage problem. However, these materials are prone to change in their physical properties in high temperature, which needs to be addressed carefully. Strained silicon has been introduced to improve electron and hole mobility. This technique has certain disadvantages from the fabrication point of view, since it is not compatible with many of the evolving technologies.

3. Power and thermal limit: This limit arises from the ever increasing number of transistors integrated per unit-area, demanding larger power consumption with higher thermal dissipation. There are two types of power density dissipation per unit area of IC chips namely dynamic power density and static power density [43]. Dynamic power is dissipated when the transistor is switched ON. On the other hand, static power density dissipation originates from the source–drain leakage current when the transistor is switched OFF. The power density in CMOS devices has in fact been growing, because the supply voltage is not scaling as fast as channel length [44]. While scaling makes it possible to have one billion of CMOS transistors in a single chip, the actual augmentation and integration process contributes to the power and thermal problems. This influences the performance and reliability of the device.

4. Technological limit: CMOS transistors are patterned on wafer by means of lithography technique, which needs to meet the demand of the continuous scaling of CMOS devices. Trade-off between complex and costly lithography techniques/masks and evolving newer design features of a device is required for patterning smaller features than wavelength of light. The minimum feature size that can be provided by the conventional photolithography

system is confined to approximately 40 nm due to the diffraction limit [45]. Patterning of features to 20 nm and below has been demonstrated by a variety of techniques, but their economic viability for industrial production needs to be assessed.

5. Economic limit: The cost of production, and testing has increased exponentially with time with the scaling down of CMOS devices. The cost explosion is primarily due to the equipment cost, clean room facilities, and the complexity of the lithography process. The smaller size circuits are also vulnerable to various kinds of defects, and they need to be tested thoroughly for the required quality. This results in the additional testing steps and time consuming sophisticated test methods, and hence adding to the costs. This continued rise in investment may eventually offset the profit margin. In Moore's own language, which is sometime termed as Moore's second law: '... the rate of technological progress is going to be controlled by functional realities' [46].

As an example, we will now discuss the possible limitations of the CMOS technology in the context of Flash memory devices [47], which have dominated the semiconductor storage market so far because of its ultra-high density, low cost, short data latency, and nonvolatility.

1. The physical scaling size of Flash memory depends on the limit of the available photolithography techniques.
2. To avoid possible electron leakage the tunnel oxide layer inside Flash memory needs to be thicker than 8 nm in order.
3. The gate coupling ratio has to be at a value greater than 0.6 to control the conductive channel and prevent gate electron injection. This can be achieved by wrapping the control gate around the floating gate to geometrically increase the gate coupling ratio. The required space to have such a wrapping structure may be unavailable with the continuation of the down-scaling process
4. A relatively long distance between two adjacent cells inside Flash memory is required to suppress the crosstalk effect of the electrons, which adversely affect the performance of scaled devices.

Due to these limitations, the state-of-the-art NAND and NOR Flash devices are restricted to 20 nm node and 45 nm node sizes, respectively [47].

7.10 Beyond CMOS technology

In recent times the concept 'more than Moore' is becoming more and more familiar, which encompasses the recent technology developments within the semiconductor industry that achieve low power consumption using new materials and structures rather than just by miniaturization within 'more Moore' approach. In this 'more than Moore' approach, the development of semiconductor electronic devices takes place by the diversification of functions and also the improvement of the performance of systems by the introduction of new technologies like micro-electromechanical systems (MEMS) technology [48]. This is particularly important, when the difficulties in further scaling within CMOS technology as well as the need for the alternative technologies are now well recognized in the International Technology Roadmap for Semiconductors (ITRS) 2015 [35]. In this backdrop 'beyond CMOS' refers to the technologies used to create devices that exhibit performance exceeding that of CMOSs on the basis of principles different from those of CMOSs [48]. The cartoon figure 7.13 obtained from the International Roadmap for Semiconductor Materials (ITRS15) indicates that 'beyond CMOS' technologies are actually being developed in parallel with 'more Moore' and 'more than Moore' technologies, rather than as a successor of Moore-CMOS technologies [35]. This may facilitate compensation to the drawbacks of existing CMOS-technologies and ultimately lead to the newer kind of semiconductor electronic devices and systems with better performance [48].

During the last decade or so significant efforts have been made towards exploring non-charge based semiconductor memories that are intrinsically immune to the issues that are limiting the Flash memory. These new emerging memory classes are (i) ferroelectric random access memory (FeRAM), (ii) magnetic RAM (MRAM), (iii) phase change RAM (PCRAM), and (iv) resistive RAM (ReRAM). We provide below a brief introduction to each of these memory classes.

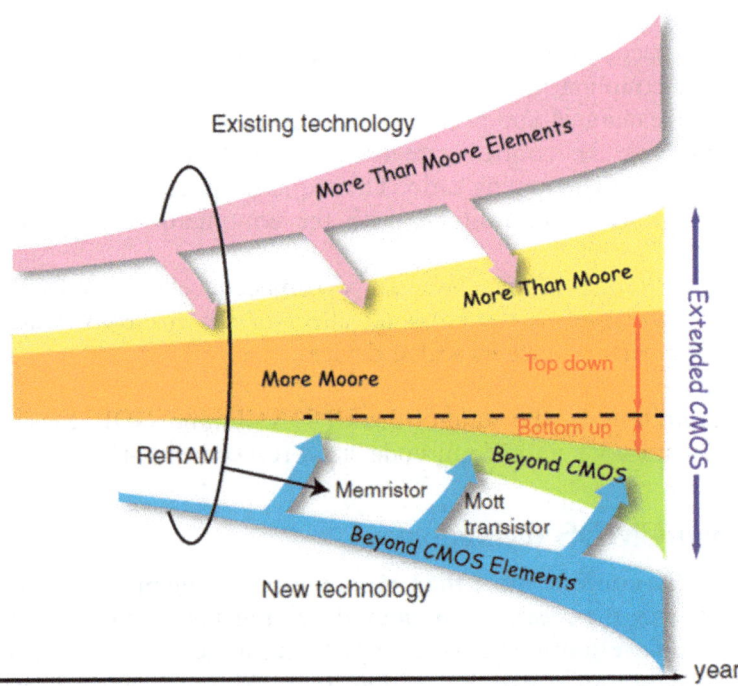

Figure 7.13. Relationship among three research directions of 'more Moore', 'more than Moore', and 'beyond CMOS'. (Reproduced from [35] with the permission of Semiconductor Industry Association.)

7.10.1 Ferroelectric random access memory and logic

The core of a ferroelectric random access memory (FeRAM) device consists of a capacitor of ferroelectric materials such as lead zirconate titanate, also known as PZT, sandwiched between two metallic electrodes. The principle of functioning of a FeRAM device is based on the electric field induced switching of the polarization state in the ferroelectric layer. The FeRAM has been considered as one of the candidates to be the next nonvolatile RAM because of its inherent speed, low-power consumption, low-voltage operation, and the possibility of straightforward CMOS integration [49]. However, FeRAM devices have problems, like the decrease in remanent polarization with cycling and loss of stored polarization over time [50]. Also ferroelectric material in FeRAM will lose its essential ferroelectric properties at very small thicknesses. Some of these issues have been addressed either by changing electrode metals such as Pt and Ir with metal oxides such as RuO_2, IrO_2 and $SrRuO_3$ or by changing ferroelectric material PZT to strontium bismuth tantalite or lanthanum-doped bismuth titanate. While the potential of FeRAM as a suitable alternative of CMOS Flash memory for storage class memory devices is well recognized, a significant breakthrough is required in the integration of ultra small cells using 3D ferroelectric capacitors without sacrificing reliability or memory performance [50]. There is also a concept of a field effect transistor with a ferroelectric capacitor as the gate electrode for creating a non volatile memory device [51, 52]. In fact, alternative device concepts involving ferroelectric field effect transistor (FeFET) and ferroelectric tunnel junction may enable a non-destructive detection of the memory state with improved scalability of the memory cell [35]. However, there are a number of problems on the way to make an actual device, which include the necessity of integrating ferroelectric materials directly on silicon [52].

7.10.2 Phase change random access memory

The function of a phase change random access memory (PCRAM) device is based on the phase transition from the amorphous state to the crystalline state (or vice versa) of the phase-change material. The phase transition is induced by an applied current pulse. The crystalline phase of the phase change materials has low electrical resistivity, whereas the amorphous phase exhibits a four to five orders of magnitude higher electrical resistivity. The PCRAM device exploits the large resistance contrast between the amorphous and crystalline states in phase-change materials [53]. The future promise of PCRAM arises from its simple cell structure and low voltage operation, and makes it attractive for embedded memory applications. However, there are concerns that the interaction of phase change materials with electrodes may give rise to long-term reliability issues and limit the cycling endurance [47]. There is some recent progress in PCRAM technology, which highlights its excellent scaling potential down to the 5 nm scale and beyond [54].

7.10.3 Magnetic random access memory and spintronics

The memory element in a magnetic random access memory (MRAM) device is a magnetic tunnel junction (MTJ). Two ferromagnetic layers separated by a thin layer of insulator material form such a MTJ, where the insulating layer acts like a tunnel barrier between the ferromagnetic materials. The effective resistance to the current flow through the barrier depends on the spin alignments of the ferromagnetic layer, and the magnitude of the tunneling current can be used to indicate the stored memory. Magnetic field switching MRAM is fast, non-volatile and can be cycled indefinitely. Also MRAM is probably closest to the concept of a universal memory [35]. However, electromigration limits the current density for producing the required value of switching magnetic field in scaled MRAM devices. As a result it is expected that the field switching MTJ-MRAM is unlikely to scale beyond 65 nm node [35, 55]. To this end there is some potential solution in the form of spin-transfer torque (STT) approach, where a spin-polarized current transfers its angular momentum to the ferromagnetic layer and switches its magnetization. Note that there is no external magnetic field involved here. However, substantial current passes through MTJ during spin transfer process, which effects the writing endurance. In addition, the stability of the storage element with scaling is subject to thermal noise.

In recent times newer mechanisms, including voltage-induced magnetization switching and spin-Hall effect, have been discovered for STT-RAM [35]. The voltage writing technique is based on the electric field (applied at the interface between a magnetic metal and insulator) induced change in magnetic anisotropy, which in turn switches magnetization from the perpendicular to in-plane direction [56]. The usage of high resistance MTJs are allowed due to the voltage driven nature of the technology, and it is compatible with the existing CMOS technology [35]. Voltage torque magnetic random access memory (MRAM) has the potential to offer ultra low power, high speed and long endurance memory cells beyond conventional STT-MRAM technology [35]. The spin Hall effect (SHE) originates from asymmetric charge scattering via spin–orbit coupling, and can convert charge current to a transverse pure-spin current in non-magnetic materials [57]. This spin current can transfer spin angular momentum to an adjacent magnetic layer via the STT effect, and thus enables manipulation of magnetization [58]. In contrast to the conventional STT where the charge current is passed through the ferromagnetic layer, here a current in the proximity does the job. A closely related effect is the Rashba effect originating from the interfacial electric field between dissimilar metals [59, 60]. This interfacial electric field appears as magnetic field in the reference frame of a moving electron (assuming that the electron is moving at least with some fraction of relativistic speed) and results into a reorientation of the spin of electron. The STT originating from the spin-Hall effect and Rashba effect is now commonly termed as spin orbit torques [35, 61]. They have actually been used as the writing mechanism is MTJ devices (see [35] and references within).

There are conceptual spin-field effect transistors, which are divided into two categories: (1) spin-FET proposed by Datta and Das [62], and (2) spin-MOSFET proposed by Sugahara and Tanaka [63]. Together they are termed as non-conventional charge-based CMOS devices [35]. Both of these spin-transistors consist of a ferromagnetic source functioning as spin-injector and a ferromagnetic drain acting as spin-detector, but their operating principles are quite different. In spin-FETS the spin-direction of the charge carriers is controlled by the applied gate voltage through Rashba effect, whereas in spin-MOSFETS the gate has the same current switching function as that of a standard MOSFET but the variable current is controlled by the configuration of magnetization of the ferromagnetic electrodes. Spin injection, manipulation and detection are the core ingredients for a fully operational spin-transistor. The read and write operations of a spin-FET and spin-MOSFET are yet to be established completely, but significant progresses have been made in the concerned technologies (see [35] and references therein).

7.10.4 Resistive random access memory

Resistive random access memory (ReRAM) devices function by voltage or current induced switching of the resistive material inside ReRAM between a high-resistance state and a low-resistance state. The materials class for ReRAM is comprised of oxides, chalcogenides (including glasses), semiconductors and organic compounds (including polymers), and ReRAM can be classified into several groups in terms of the resistive materials used [35]. Several recent developments, which include the suitable switching and electrode materials and the introduction of the crossbar structure give rise to the promise of ReRAM devices with good stability, ultra-high density and fast switching speed [64]. In most of the cases the enhanced conductivity is correlated to the formation of defect-related conductive filament, and a one time formation process is required before the bistable switching can be started [35]. The most discussed models for the ReRAM mechanism are based either on the growth and rupture of a metal filament inside the oxide materials under the action of externally applied electric current, or on the redox processes responsible for the formation of some high-conductivity or low-conductivity local inclusions corresponding to a particular oxygen stoichiometry [65]. However, a complete understanding of the physical mechanism of the resistive switching is still in progress, and this remains a key challenge to this evolving technology.

In spite of the promise of all these evolving technologies discussed above, the Flash memory and other CMOS devices are still dominating the semiconductor/electronics market. In the backdrop of this present scenario, to achieve further dimensional and functional scaling beyond the CMOS technology there have been considerable efforts in recent times on some novel concepts of nonvolatile memory including carbon memory, molecular memory and Mott memory. In fact IBM has identified three main approaches to the development of a so-called 'beyond silicon' electronics [65]: (1) quantum computers, (2) molecular electronics, and (3) Mott transition field effect transistor (Mott-FET). The thin film Mott-FET is considered to be the most promising because it is closest to existing CMOS technologies. Mott-FET and Mott memory devices are also being considered as prominent emerging devices in the recent International Road Map for Semiconductors 2015 [35]. With some exposure in the earlier chapters of this book on the science of Mott insulators and metal–insulator transition, in the next chapter we will now explore how these strong electron correlation based novel phenomena can be translated into various interesting applications.

Further reading

1. Parker G 2003 *Introductory Semiconductor Device Physics* (London: Taylor and Francis).
2. Sze S M and Lee M 2012 *Semiconductor Devices: Physics and Technology* 3rd edn (New York: Wiley).
3. Neamen D A 2012 *Semiconductor Devices and Physics: Basic Principles* 4th edn (New York: McGraw-Hill).
4. Natelson D 2015 *Nanostructures and Nanotechnology* (Cambridge: Cambridge University Press).

Self-assessment questions and exercises

1. A sample of pure Si is doped with 10^{15} As atoms/cc. What will be the carrier concentrations and the Fermi level in the resultant extrinsic Si sample at 300 K?
2. What are the basic building blocks with which different types of semiconductor devices are constructed?
3. Electron–hole pairs can be excited by exposing a semiconductor to an electromagnetic radiation if the energy of a photon is greater than the bandgap energy. Show that a voltage can be generated spontaneously in a *p–n* junction by exposing it to light.
4. What are the basic building blocks in dynamic random-access memories (DRAMs)? Why does the data in DRAM need to be refreshed periodically within a fixed interval?
5. What is the basic reason of low power consumption in CMOS technology?
6. What is Denard scaling? How such scaling is expected to improve switching speed and reduce power dissipation in a CMOS device?
7. What are the limitations with Denard scaling, and how are these presently circumvented in MOSFETs?
8. What are the physical and power limitations of present CMOS technologies?
9. What is a conceptual spin-MOSFET and what are its possible advantages over a conventional MOSFET?

References

[1] Bardeen J and Brattain W H 1948 *Phys. Rev.* **71** 230
[2] Shockley W 1949 *Bell Syst. Tech. J.* **28** 435
[3] Parker G 2003 *Introductory Semiconductor Device Physics* (London: Taylor and Francis)
[4] Wigner E and Bardeen J 1935 *Phys. Rev.* **48** 84
    Bardeen J 1936 *Phys. Rev.* **49** 653
[5] Herring C 1992 *Phys. Today* 26
[6] Shockley W 1939 *Phys. Rev.* **56** 317

[7] Shockley W and Pearson G L 1948 *Phys. Rev.* **74** 232

[8] Holonyak N Jr 1992 *Phys. Today* 36

[9] Bandopadhyay P K 1998 *Proc. IEEE* **86** 63

[10] Weiner C 1973 *IEEE Spectr.* p 24

[11] Sze S M and Lee M 2012 *Semiconductor Devices: Physics and Technology* 3rd edn (New York: Wiley)

[12] Ng K K 2002 *Complete Guide to Semiconductor Devices* 2nd edn (New York: Wiley)

[13] http://www.computerhistory.org/siliconengine/timeline/

[14] Moore G E 1965 *Electronics* **8** 114

[15] Anderson P W 2011 *More and Different: Notes from a Thoughtful Curmudgeon* (Singapore: World Scientific)

[16] Neamen D A 2012 *Semiconductor Devices and Physics: Basic Principles* 4th edn (New York: McGraw-Hill)

[17] Natelson D 2015 *Nanostructures and Nanotechnology* (Cambridge: Cambridge University Press)

[18] Sengupta D L, Sarkar T K and Sen D 1998 *Proc. IEEE* **86** 235

[19] Bose J C 1904 *US Patent* 755840

[20] Pickard G W 1906 *US Patent* 836531

[21] Wilson A H 1931 *Proc. R. Soc. Lond.* A **133** 458
Wilson A H 1931 *Proc. R. Soc. Lond.* A **134** 277

[22] Davydov B 1938 *C. R. Dokl. Acad. Sci.* **20** 279

[23] Mott N F 1938 *Proc. Cambridge Philos. Soc.* **34** 568

[24] Mott N F 1939 *Proc. R. Soc. Lond.* A **171** 27

[25] Schottky W 1939 *Naturwissenschaften* **26** 843

[26] Shockley W 1976 *IEEE Trans. Electron Devices* **23** 597

[27] Wood C 1988 *Rep. Prog. Phys.* **51** 459

[28] Bell L E 2008 *Science* **321** 1457

[29] Snyder G J and Toberer E S 2008 *Nat. Mater.* **7** 105

[30] Kahng D 1976 *IEEE Trans. Electron Devices* **23** 655

[31] Kahng D and Atalla M M 1960 *IRE-AIEE Solid-state Device Res. Conf.* (Pittsburgh, PA: Carnegie Inst. of Technol.)

[32] Lai S K 2008 *IBM J. Res. Dev.* **52** 529

[33] Moore G 1975 *IEEE Int. Electron Devices Meeting Tech. Dig.* p 11

[34] Bandopadhyay P K 1998 *Proc. IEEE* **86** 78

[35] International Technology Roadmap for Semiconductors 2.0: 2015 Edition (Semiconductor Industries Association) https://www.semiconductors.org/wp-content/uploads/2018/06/0_2015-ITRS-2.0-Executive-Report-1.pdf

[36] Dennard R H, Gaensslen F H, Rideout V L, Bassous E and LeBlanc A R 1974 *IEEE J. Solid-State Circuits* **9** 256

[37] Ferain I, Colinge C A and Colinge J-P 2011 *Nature* **479** 310

[38] Bohr M A 2007 *IEEE Solid-State Circuits. Soc. Newslett.* **12** 11

[39] Morton J J L, McCamey D R, Eriksson M A and Lyon S A 2011 *Nature* **479** 345

[40] Esmaeilzadeh H, Blem E, St Amant R, Sankaralingam K and Burger D 2013 *Commun. ACM* **56** 93

[41] Moore G E 2003 *Digest of Technical Papers. 2003 IEEE Int. Solid-State Circuits Conf. (ISSCC)* vol 1 p 20

[42] Haron N Z and Hamdioui S 2010 *IEEE Xplore*

[43] Frank D J 2002 *IBM J. Res. Dev.* **46** 235

[44] Nowak E J 2002 *IBM J. Res. Dev.* **46** 169

[45] Lan H and Liu H 2013 *J. Nanosci. Nanotech.* **13** 3145

[46] Mann C C 2000 The end of Moore's law? *MIT Technology Review* https://www.technologyreview.com/s/400710/the-end-of-moores-law/

[47] Wang L, Yang C H, Wen J and Gai S 2014 *J. Nanomater.* **2014** 927696

[48] Akinaga H 2013 *Jpn. J. Appl. Phys.* **52** 100001

[49] Kato Y, Yamada T and Shimada Y 2005 *IEEE Trans. Electron Devices* **52** 2616

[50] Burr G W, Kurdi B N, Scott J C, Lam C H, Gopalakrishnan K and Shenoy R S 2008 *IBM J. Res. Dev.* **52** 449

[51] Wu S Y 1974 *IEEE Trans. Electron Devices* **21** 499

[52] Ma T P and Han J P 2002 *IEEE Electron Device Lett.* **23** 386

[53] Raoux S *et al* 2008 *IBM J. Res. Dev.* **52** 465

[54] Kolobov A and Popescu M 2012 *Phys. Status Solidi* B **249** 1824

[55] Chun K C, Zhao H, Harms J D, Kim T-H, Wang J-P and Kim C H 2013 *IEEE J. Solid-State Circuits* **48** 598

[56] Shiota Y 2012 *Nat. Mater.* **11** 39

[57] Hoffmann A 2013 *IEEE Trans. Magn.* **49** 5172

[58] Liu L 2012 *Science* **336** 555

[59] Miron I M, Gaudin G, Auffret S, Rodmacq B, Schuhl A, Pizzini S, Vogel J and Gambardella P 2010 *Nat. Mater.* **9** 230

[60] Miron I M, Garello K, Gaudin G, Zermatten P-J, Costache M V, Auffret S, Bandiera S, Rodmacq B, Schuhl A and Gambardella P 2011 *Nature* **476** 189

[61] Gamberdella P and Miron I M 2011 *Philos. Trans. R. Soc. Lond.* A **369** 3175

[62] Datta S and Das B 1990 *Appl. Phys. Lett.* **56** 665

[63] Sugahara S and Tanaka M 2004 *Appl. Phys. Lett.* **84** 2307

[64] Jeong D S, Thomas R, Katiyar R S, Scott J F, Kohlstedt H, Petraru A and Hwang C S 2012 *Rep. Prog. Phys.* **75** 076502

[65] Pergament A L, Stefanovich G B and Velichko A A 2013 *J. Sel. Topics Nanoelectr. Comput.* **1** 24

IOP Publishing

Mott Insulators
Physics and applications
Sindhunil Barman Roy

Chapter 8

Mott insulator and strongly correlated electron materials based devices

The idea of a Mott insulator based device like the Mott field effect transistor (Mott-FET) dates back to the mid 1990s, which originally involved a hypothetical molecular layer undergoing a Mott metal–insulator transition at sub-nanosecond timescales [1, 2]. This phase transition can be triggered by thermal energy, electric field, and optical stimuli, and can be studied via electrical or magnetic signals. From the device points of view it is also advantageous that the electronic-structural phase transitions are robust and reversible for many cycles in the thin film form of Mott insulators [3]. During the last two decades there have been significant advances in the synthesis of thin-film Mott-insulators by a variety of techniques, such as molecular beam epitaxy, sputtering, chemical vapor deposition, pulsed laser deposition and atomic layer deposition with emphasis on the device applications. Apart from the Mott-FET, there are many other potential applications of Mott-insulators, including two-terminal electronic switches, memristive devices, electronic oscillators, optical devices, thermal and chemical sensors, photovoltaic cells and battery materials.

In this chapter we shall discuss the working principles of such Mott-insulator based devices. We will focus particularly on the device concepts behind the Mott-FET and discuss the pros and cons of such devices in comparison to band semiconductor based MOSFETs. We will also discuss a few other potential applications taking the example of some well known real materials. This is not a review article, and hence the examples of materials and possible devices will by no means be exhaustive. In fact the field is developing quite rapidly and new device concepts involving even newer materials are regularly appearing in the literature. The purpose of this chapter is to highlight how an understanding of the basic science (including physics, chemistry and materials science) enables one to engineer new and novel devices with Mott-insulators, which can complement and may even compete with the existing CMOS technology.

doi:10.1088/2053-2563/ab16c9ch8

8.1 Mott metal–insulator transition: phase-coexistence and hysteresis

The first order nature of the metal–insulator phase transition in Mott insulators plays an influential role in the device applications. The salient features and characteristics of a first order phase transition (FOPT) are highlighted here, which will be helpful in the subsequent discussions on Mott devices. Within the realm of the Ehrenfest classification of phase transitions, a first order phase transition is marked by a discontinuity in the first derivative of the free energy, taken with respect to the control variable (say temperature) that causes the phase transition. Experimentally, the entropy and the order parameter distinguishing two phases change discontinuously across the first order phase transition point. The identification of a FOPT is complete if the change or jump in the order parameter and the latent heat satisfy the Clausius–Clapeyron relation. The supercooling (superheating) of the high (low) temperature phase across a FOPT arises naturally within the Ehrenfest scheme, but the limit or extent of such supercooling (or superheating) remains undefined. Hence, the hysteresis often observed across the metal–insulator phase transition in Mott insulators remains somewhat unexplained within the Ehrenfest scheme. (We shall see later that this hysteresis across a FOPT plays a very important role in the workings of Mott devices.) In this regard the relatively modern Landau theory of phase transition provides a better platform to understand the phenomenology of FOPT and its characteristic features [4, 5]. Assuming temperature T to be the control parameter, the free-energy density can be expressed within Landau theory in terms of the order parameter S as [4]:

$$f(T, S) = \left(\frac{r(T)}{2}\right)S^2 - wS^3 + uS^4. \tag{8.1}$$

Here w and u are positive and temperature independent and it is assumed that symmetry does not prohibit odd order terms [4]. In case the odd order terms are prohibited, then the free energy would be expressed as:

$$f(T, S) = \left(\frac{r(T)}{2}\right)S^2 - wS^4 + uS^6 \tag{8.2}$$

and it is easy to carry through all subsequent arguments, and the assumption of the form of equation (8.1) is thus made without loss of generality. A schematic representation of the free energy density (f) as expressed in equation (8.1) is shown in figure 8.1. At the transition temperature $T = T_N$, the high- and low-temperature phases coexist with $f = 0$ at $S = 0$ and at $S = w/2u$, separated by an energy barrier (see the curve (a) in figure 8.1). These results are independent of any assumption about the detailed temperature dependence of the term $r(T)$. The standard treatment assumes that $r(T) = a[T - T^*]$, where a is positive and independent of temperature. The parametric temperature T^* is the limit of metastability of the high temperature phase i.e. supercooling temperature, where $\frac{d^2f}{dS^2}$ at $S = 0$ goes to zero. This limit of

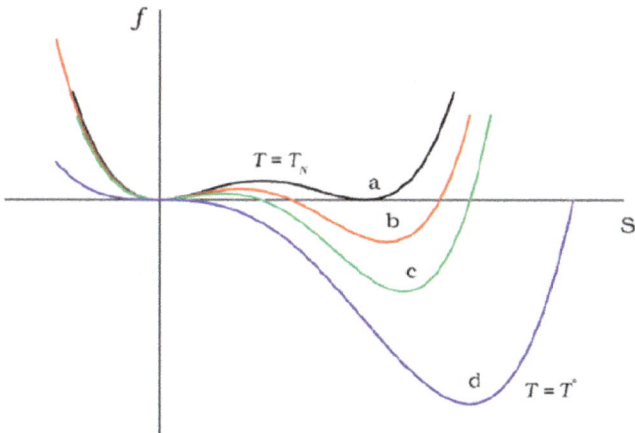

Figure 8.1. Schematic representation of free energy across a first order phase transition (Reproduced from [5]. (c) IOP Publishing. Reproduced with permission. All rights reserved.).

supercooling is reached at $T^* = T_N - \frac{w^2}{2ua}$ [4]. The barrier height between the stable low temperature phase and metastable high temperature phase at the transition temperature (see curve (a) in figure 8.1) decreases with the reduction in temperature (see curves (b) and (c) in figure 8.1) and finally disappears at the supercooling temperature T^* (see curve (d) in figure 8.1). Similarly, in the increasing temperature cycle the limit of the superheating is reached at $T^{**} = T_N + w^2/16ua$ [4]. It may be noted here that the supercooling (or superheating) can persist till T^* (or T^{**}) only in the limit of infinitesimal energy fluctuations, as finite energy fluctuations can destroy the supercooled (superheated) state in the temperature regime $T^* < T < T_N$ ($T_N < T < T^{**}$). Designating the stable state as the '0-state' and the metastable state as '1-state', it is clear that the relaxation to 0-state is dictated by the kinetics of the phase transition. If the barrier height between 0 and 1 states is much smaller than the dissipative energy associated with the driving force, the metastable 1-state cannot be sustained for a long period of time and the memory of the 1-state will be volatile. If on the other hand the energy barrier height is large the 1-state will be nonvolatile, and another external stimulus will be required to switch the 1-state to 0-state. The existence of supercooling and superheating in the FOPT gives rise to hysteresis in an experimental observable as a function of the control parameter and enables a nonvolatile operation.

It is now instructive to investigate the Mott metal–insulator phase transition (triggered by thermal energy, or electric field, or optical stimuli) in the above-mentioned framework of FOPT. Figure 8.2 represents some of the generic features of how the electrical conductivity of a classical Mott insulator responds to external stimulation, say thermal energy:

1. The material is in the insulating phase at low temperature. As the temperature increases and cross a threshold value, a sharp phase transition causes the conductivity to increase by several orders of magnitude and the material goes to the metallic phase. In this case this threshold temperature is the Mott

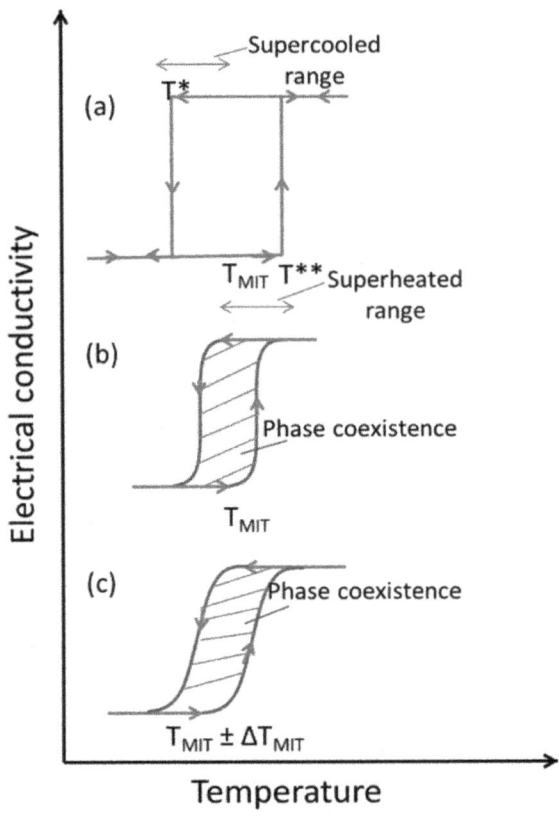

Figure 8.2. Generic features of electrical conductivity across a first order metal–insulator phase transition.

metal–insulator transition temperature T_{MIT}. On decreasing temperature from the high temperature metallic phase, the material eventually goes back to the insulating phase.

2. The superheating (supercooling) of the low (high) temperature insulating (metallic) state is possible only in ultra-pure single crystal samples (see figure 8.2(a)). The transition is very sharp and takes place not at the metal–insulator transition temperature T_N, but at the limit of superheating T^{**} (supercooling T^{*}) and without any phase-coexistence.

3. In real materials a finite amount of defects will give rise to nucleation and growth of the emerging equilibrium state across the phase transition temperature. The phase transition process starts at T_{MIT} but is completed only at T^{**} (or T^{*}). This is reflected by the somewhat smooth change in conductivity as a function of temperature with the phase-coexistence of metal and insulator phases in the finite temperature regime between T_{MIT} and T^{**} (or T^{*}). This process gives rise to the distinct hysteresis across the phase transition (see figure 8.2(b)). Thus, the magnitude of conductivity in the transition region depends not only on the temperature but also the history of temperature excursion.

4. The quenched disorder in the materials will also cause a local variation of the phase transition temperature giving rise to disorder influenced transition temperature landscape in the sample [6–8]. Thus, in real materials instead of a single T_{MIT} there will be a small but finite band of T_{MIT}. This gives rise to a somewhat tilted hysteresis loop with the impression that the onset of phase transition while increasing the temperature takes place at a lower temperature than that in the decreasing temperature part of the temperature cycle (see figure 8.2(c)). This behaviour has been seen across the first order phase transition in many classes of magnetic and ferroelectric materials [5].

All these features of the temperature-induced first order Mott metal–insulator transition is expected to be observed with other external stimuli too. It may also be noted that nucleation and growth of the equilibrium phase across a first order phase transition following the Kolmogorov–Johnson–Mehl–Avrami rule, is very much possible also with driving parameters other than temperature [9].

In a canonical Mott-insulator metal–insulator transition is purely electronic in origin and is not assisted by a structural transition. However, in practice many Mott insulators have close coupling between charge, spin, and lattice degrees of freedom, so that the metal–insulator transition many a times is accompanied by a structural transition. In addition, the devices are expected to be made from thin-film or nano form of materials, and surfaces of Mott insulator materials may display properties distinct from bulk. However, both insulating and metallic phases in the potential Mott insulator device materials, like VO_2, remain the same as in the bulk for thin films down to ~4 nm. This property suggests that VO_2 can be scaled down to 4 nm sizes without degradation of the sought after functional properties [3]. Also in systems like layered ruthenate Mott insulator $Ca_{1.9}Sr_{0.1}RuO_4$ the surface layer shows a purely electronic Mott metal–insulator transition without any structural transition [10]. Further, it has been argued that some of the concepts useful in designing the conventional semiconductor devices are still valid in such strongly correlated electron system devices [11]. It may be noted here that in heavily doped silicon a Mott insulator–metal transition takes place at a free charge carrier density of $n_c \sim 3.5 \times 10^{18}$ cm^{-3} [12]. This is, however, a second order phase transition and not accompanied by a large conductivity jump. In contrast the conductivity changes by 4–5 orders of magnitude across the metal–insulator transition in a typical Mott insulator like VO_2 (see figure 6.17 in chapter 6). However, it may be mentioned here that the temperature induced Mott metal–insulator transition as such will not be quite suitable for thermally induced switching in electronics devices, since the process can be slow and difficult to reverse.

Significant developments have taken place in the field during the last two decades, and the terminology 'Mottronics' [13] has been coined in the literature to represent the concept of electronics technology utilizing the Mott transition. In the original proposal for Mott-FET [2], exotic materials like the quasi-monomer organic conductor (K^+TCNQ^-) or the doped fullerene (KC_{60}) were proposed. Over the years several types of Mott insulators have been investigated for various kinds of device applications. A detailed review of all these materials and devices is beyond the

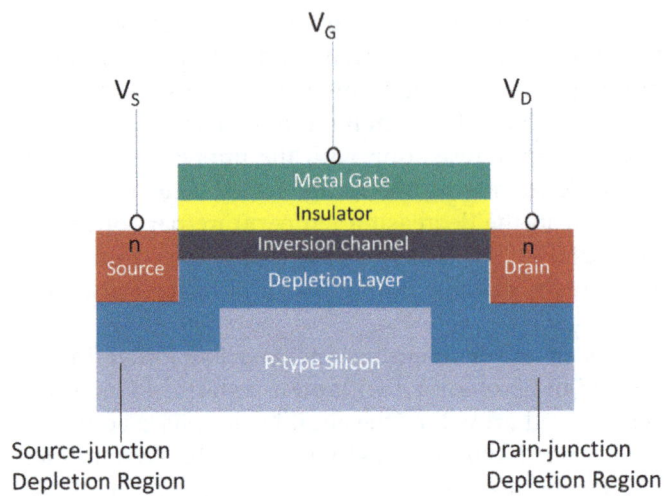

Figure 8.3. Schematic representation of an *n*-channel MOSFET.

scope of this book, and we shall rather focus here on the concepts of different types of Mott devices (giving more emphasis on memory and logic devices) with the help of a few examples of well studied Mott insulator oxide compounds including VO_2, NbO_2 and one layered dichalcogenide compound TaS_2.

8.2 Mott devices and mechanism of operation

8.2.1 Mott-field effect transistor (MOTT-FET) and a comparison with MOSFET

We will start here with the discussion on the functioning of a band semiconductor based MOSFET. Figure 8.3 shows the schematic presentation of a Si-based *n*-channel MOSFET (NMOS) device. The device has two regions of *n*-type Si called the source and the drain, and they are separated by a region of *p*-type Si known as the substrate. The region between the source and the drain is covered by a thin layer of silicon dioxide. This insulating SiO_2 layer, often referred to as gate-oxide, in turn is covered by a metal electrode called the gate. The source and the *p*-type substrate are grounded in normal bias conditions, with a positive voltage applied to the drain. The drain *p–n* junction is reverse biased in this condition with no current flowing between the substrate and the drain. There is also no current flowing between the substrate and the source since the bias across the source *p–n* junction is zero. As a result, there is no current flow between the source and the drain; the transistor is turned OFF and acts as an open switch.

If a positive voltage is applied to the metal gate electrode with respect to the semiconductor ground, the gate acts as one plate of a capacitor and the semiconductor forms the other plate of the capacitor. As a result a positive voltage on the gate attracts negative charge to the region just under the oxide insulator. This attraction of charge is termed as a 'field effect', because this is originated due to an electric field caused by the gate. Thus on the application of a large enough positive voltage to the gate, if the gate voltage is larger than some particular threshold voltage, an electron-rich layer is formed in the region of the semiconductor underneath the gate oxide. This electron-rich layer is called the *n*-channel, which builds a continuous electron bridge between the source and the drain, and enables a current flow between these two electrodes. In this situation the transistor is in a turned ON state, and behaves as a closed switch. A relatively small applied gate voltage can control a large current between the source and drain.

It may be noted that a *p*-channel MOSFET is also possible, where the regions around the source and drain would be *p*-doped Si and the rest of the semiconductor would be *n*-doped. For a *p*-channel MOSFET conductance between source and drain would be turned on when the applied gate voltage is sufficiently negative so that holes are attracted to the region underneath the gate oxide layer, and thus producing a conductive *p*-channel between source and drain.

The first MOSFET fabricated in the early 1960s had a thermally oxidized silicon substrate with a channel length of 20 μm and a gate oxide thickness of 100 nm [14]. MOSFETs have been scaled down considerably, but the silicon and thermally grown silicon dioxide remain the most important combination.

A voltage V_G applied across the MOS capacitor induces charge into the MOSFET channel near the semiconductor–dielectric interface. The electric field penetrates the semiconductor, which causes the energy spectrum of the electron states near the interface to shift. This later effect is known as band bending. The position of the Fermi level with respect to the delocalized bands shifts as a result. This changes the density of mobile charge carriers and the conductance of the interfacial region. Depending on the direction and amount of the shift in the energy levels induced by the gate voltage, the carrier density of the interfacial region is enhanced, reduced, or even reversed in sign compared to the carrier density in the bulk. Figure 8.4 presents the energy-band diagrams of the MOS capacitor with a *p*-type substrate for various gate bias voltages. Figure 8.4(a) shows the case when there is no bias across the MOS device. Since the substrate is a *p*-type semi-conductor, the intrinsic Fermi level E_{Fi} (i.e. the Fermi level of the undoped pure semiconductor) is shifted down towards the valence band and is denoted by E_F. The energy bands in the semiconductor are flat, which indicate that there is no net charge in the semiconductor. If a negative voltage $(-V_G)$ is applied to the gate electrode, the bands near the semiconductor surface are bent upward (see figure 8.4(b)). The Fermi level in the semiconductor remains constant in an ideal MOS capacitor, since no current flows in the device irrespective of the value of the applied gate voltage. The valence-band edge is closer to the E_F at the oxide–semiconductor interface than in the bulk material. This upward bending of the energy band at the semiconductor surface causes an increase in the energy $E_{Fi} - E_F$ near the oxide–semiconductor interface, and this gives rise to an enhanced concentration or accumulation of holes in that region (see figure 8.4(b)). On the other hand, when a small positive gate voltage $(+V_G)$ is applied to an ideal MOS capacitor, the energy bands near the semiconductor surface are bent downward (see figure 8.4(c)). The conduction band and intrinsic Fermi level move closer to the Fermi level E_F of the *p*-type semi-conducting substrate. This gives rise to a region of width x_d, where the majority carriers, i.e. holes, are depleted (see figure 8.4(c)). With the application of a larger

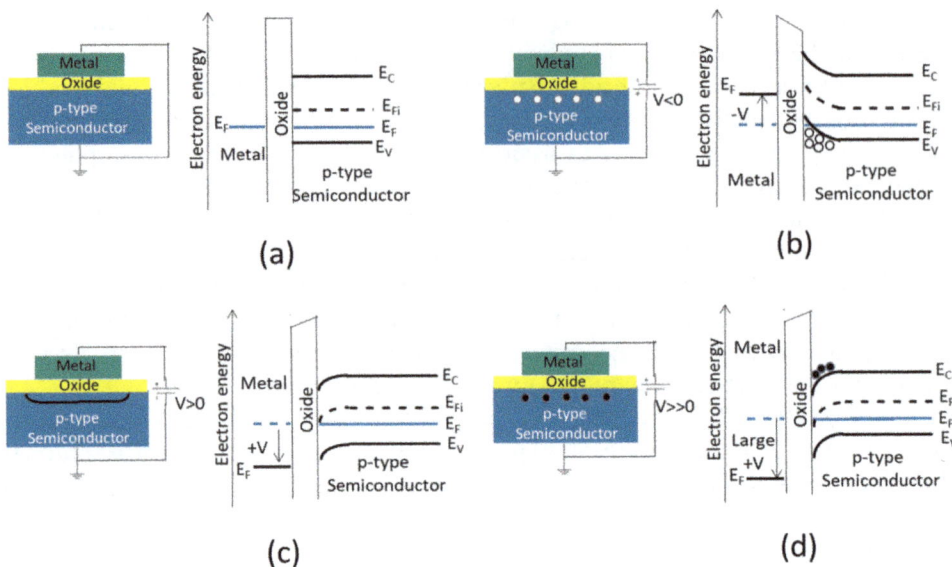

Figure 8.4. The energy-band diagram of a MOS capacitor with a *p*-type substrate for (a) a zero applied gate voltage, (b) a negative gate voltage, (c) a small positive gate voltage, and (d) a large positive gate voltage.

positive V_G, the energy bands bend downward even more, and eventually the intrinsic Fermi level E_{Fi} at the surface crosses over the Fermi level E_F (see figure 8.4(d)). In this situation the gate voltage starts to induce excess negative carriers i.e. electrons at the oxide–semiconductor interface, and the surface of the semiconductor gets inverted from a *p*-type to an *n*-type semiconductor. An inversion layer of electrons is thus created at the oxide–semiconductor interface. This is the mechanism of the formation of the electron-rich channel layer, which forms a continuous electron bridge between the source and the drain in the MOSFET device shown in figure 8.4. The carrier density in the bulk of a doped semiconductor is typically 10^{17}–10^{18} cm^{-3}, which translates to an aerial carrier density of 10^{12}–10^{13} cm^{-2} for a channel thickness of 10 nm [15]. Gate voltage can induce changes in carrier density of the order of 10^{12}–10^{13} cm^{-2} in the MOSFET channel before dielectric breakdown occurs in typical dielectrics like SiO$_2$. Even higher densities in excess of 10^{14} cm^{-2} can be obtained with the usage of SrTiO$_3$ or ferroelectric materials like lead zirconate titanate (PZT). Thus, the application of gate voltage leads to a large change in the channel conductance. The ratio of currents in the presence and absence of a gate voltage is often termed to as the ON–OFF ratio.

In practical cases the induced charge in the MOSFET channel initially populates the localized states, and no change in the channel conductance is observed until the gate voltage reaches threshold voltage V_T. Several factors, such as the density of localized states in the semiconductor, traps at the oxide–semiconductor interface, and immobile charges or defects in the dielectric materials, can influence the magnitude of this threshold voltage V_T. When a bias voltage V_D is applied between the source and drain of the MOSFET device, the charge density induced by the gate at a given point in the channel will depend on the local channel potential $V_C(x)$ ($0 < V_C(x) < V_D$). With the assumption that the potential in the channel does not vary appreciably over lengths of the order of the insulator thickness, the current at a point in the channel may be expressed as [16]:

$$I_x = -\mu C_I[(V_G - V_T) - V_C(x)]W\frac{dV_C(x)}{dx} \tag{8.3}$$

where x is a coordinate along the length of the channel, μ is the carrier mobility, C_I is the capacitance per unit area of the gate insulator, and W is the width of the channel. We will now discuss the operation of a MOSFET taking the example of a MOSFET on a p-type semiconductor substrate. The conducting channel here is the inversion layer of electrons. Heavily electron-doped i.e. n-type regions form source and drain contacts, which ensure very thin ohmic contact via a very thin Schottky barrier. The drain current in this n-channel MOSFET enters the drain electrode and is a constant along the entire channel length. Letting $I_D = -I_x$, an expression for the current–voltage characteristic of the MOSFET channel can be obtained by integrating the equation (8.3) over the length of the channel L [16]:

$$I_D = \mu C_I[(V_G - V_T)V_D - V_D^2/2]\frac{W}{L}. \tag{8.4}$$

This equation (8.4) is valid for $V_G \geqslant V_T$ and $0 \leqslant V_D \leqslant V_{DSat}$. Here V_{DSat} is the value of V_D above which the drain current I_D saturates.

Figure 8.5(a) shows schematically the MOSFET structure with $V_G > V_T$ and a small applied source to drain bias voltage V_D. In this case the thickness of inversion channel layer (which qualitatively indicates the relative charge density) is essentially the same across the entire channel length. The corresponding drain current varies linearly with the bias voltage V_D. This can be seen in figure 8.6, which displays representative I_D versus V_D curve for various values of gate voltage V_G. With the increase in V_D the voltage drop across the gate oxide layer near the drain terminal decreases, and in turn the charge density in inversion channel also decreases (see figure 8.5(b)). This will lead to a decrease in the slope of I_D versus V_D curve (see figure 8.6). With further increase in V_D when the potential drop across the gate oxide layer near the drain terminal is equal to V_G, the charge density in the inversion channel becomes zero at the drain terminal. This effect is shown schematically in figure 8.5(c). This means at this point the incremental channel conductance at the drain is zero, which in turn makes the slope of the I_D versus V_D curve zero. This condition is known as 'pinch off'. We can obtain the peak current value I_D from equation (8.4) with the condition $\partial I/\partial V_D = 0$, and the peak current occurs when,

$$V_D = V_G - V_T. \tag{8.5}$$

This is the value of bias voltage V_{DSat} at which the drain current I_D reaches value $I_D = \mu C_I\left[\frac{(V_G - V_T)^2}{2}\right]\frac{W}{L}$. With further increase in V_D the length of undepleted channel

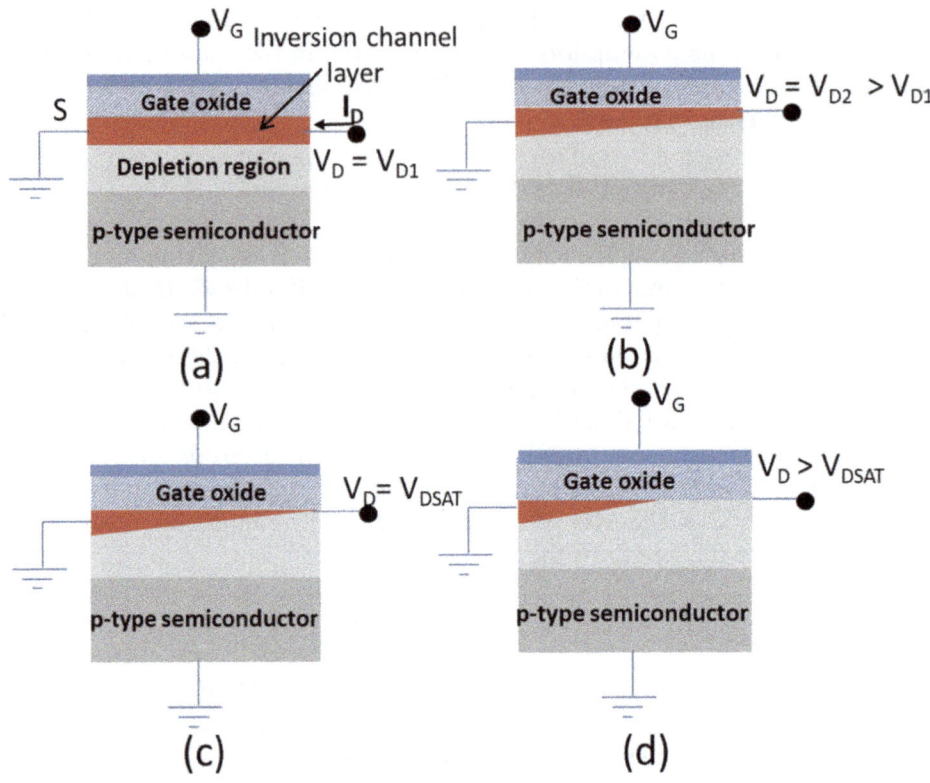

Figure 8.5. Schematic cross-section of MOSFET with $V_G > V_T$ for: (a) a small value of V_D, (b) a larger value of V_D, (c) $V_D = V_D(Sat)$ and (d) $V_D > V_D(Sat)$.

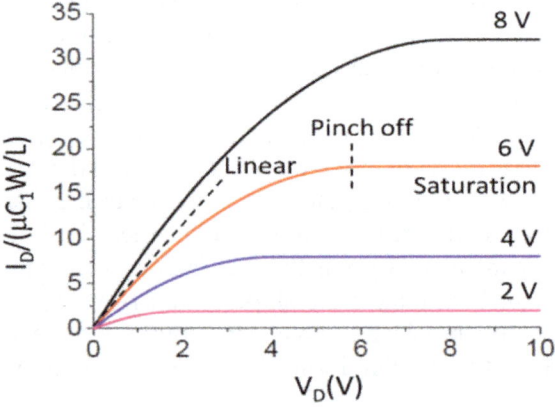

Figure 8.6. Representative I_D versus V_D curves for various values of gate voltage marking the low V_D linear regime, and with increase in V_D the pinch-off and saturation regime.

with free charge carriers becomes shorter (see figure 8.5(d)). However, the accompanying carrier density is also lower, and these effects together keep the integrated resistance of this region unchanged. The energy barrier seen by the carriers in the depleted region is compensated by a longitudinal electric field. The charge carriers in the depleted region of the channel are driven to the drain by this electric field. The I_D versus V_D curve saturates and becomes flat above a value V_{DSat} (see figure 8.6), with the saturation value of drain current $I_D = \mu C_I \left[\frac{(V_G - V_T)^2}{2} \right] \frac{W}{L}$ [16].

In a MOSFET the gate voltage only penetrates up to a thin layer of semiconductor and modulates the carrier density within a characteristic length: the Debye screening length L_D [17]. This is because the external field is screened by free charge carriers. We had talked about this screening length earlier in the context of degenerate electron gas obeying Fermi–Dirac statistics i.e. the Thomas–Fermi screening length in chapter 2. However, in the present case of non-degenerate electron gas obeying Maxwell–Boltzmann statistics we are dealing with the Debye screening length L_D, which is given by [17]:

$$L_D = \left(\frac{\epsilon_r \epsilon_0 k_B T}{e^2 (p + n)} \right) \tag{8.6}$$

where ϵ_r is the dielectric constant of the channel material, ϵ_0 the vacuum permittivity, k_B is the Boltzmann constant, T is temperature, p and n are hole and electron density, respectively. Gate voltage in a MOSFET acts on both the gate oxide and the screening layer. Hence, the differential capacitance of the MOS capacitor system can be treated as gate oxide capacitance C_{ox} in series with surface charge capacitance C_{Si}, and the potential drop will be divided between gate oxide and channel in proportion to the inverse of their capacitance [17].

A perfect switch has zero current flow when it is open and zero resistance when it is closed. In addition, it should be capable of switching instantly from the 'OFF' state to the 'ON' state, and vice versa [18]. In MOSFETs, however, the off current is not really zero, the on current is limited, and the switching requires finite time. Hence, they are not perfect switches. In addition, the switching action in a MOSFET does not start at a specific value of gate voltage. The switching takes place gradually over a range of gate voltage. Figure 8.7 presents a plot showing the drain current as a function of gate voltage in a MOSFET. There are two curves representing the same data: (i) blue curve plotted using a linear scale with the y-axis on the right, and (ii) a red curve plotted in the logarithmic scale with y-axis on the left. When the drain current is plotted on a linear scale, there is no current below a threshold gate voltage. Above this threshold voltage, the increase in the current is linear with the applied gate bias. On the other hand, in the logarithmic scale, an exponential variation in the drain current with the gate voltage is clearly seen below the threshold gate voltage, and the off current is clearly not equal to zero. A parameter termed as the 'subthreshold slope (SS)' characterizes the rate of increase of the drain current (I_D) with the gate voltage V_G below the threshold voltage. The SS is defined as the change in gate voltage V_G necessary for one decade change in the drain current I_D [14]:

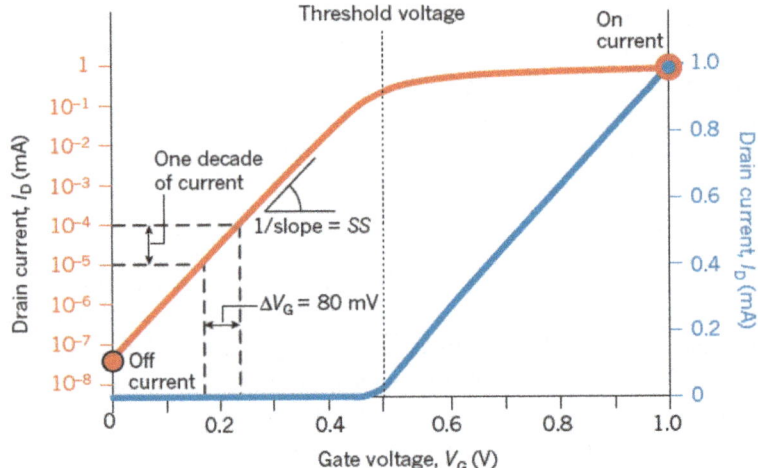

Figure 8.7. Variation of drain current in a MOSFET as a function of gate voltage. (Reproduced from [18] by permission from Springer Nature.)

$$SS \equiv \frac{dV_G}{(d \log_{10}(I_D))} = \frac{dV_G}{d\Psi_S} \frac{d\Psi_S}{(d \log_{10}(I_D))} \cong \left(1 + \frac{C_D}{C_{ox}}\right)\frac{k_B T}{q}(\ln(10)) \qquad (8.7)$$

where k_B is the Boltzmann constant, T is the temperature in Kelvin, q is the absolute value of the charge of an electron, $\ln(10)$ is the natural logarithm of 10, Ψ_S is the semiconductor surface potential and C_D and C_{ox} are the capacitances of the depletion layer in the subthreshold region and gate oxide, respectively. The SS is expressed in millivolts per decade of current.

We have mentioned above that the gate voltage V_G will be distributed between the gate oxide and channel. The factor $\left(1 + \frac{C_D}{C_{ox}}\right)$ in the above equation (8.7) is termed as the body factor η, and it says how efficient the semiconductor surface potential or the channel region is coupled to the gate voltage. The minimum subthreshold slope of a conventional MOSFET can be found by putting $C_D \rightarrow 0$ and/or $C_{ox} \rightarrow \infty$. In this case $\eta = 1$ and from equation (8.7):

$$SS_{min} = \ln(10)\frac{k_B T}{q}. \qquad (8.8)$$

Here $\frac{k_B T}{q}$ is thermal voltage and the SS is at the thermoionic limit. At a room temperature of $T = 300$ K, $SS_{min} = 59.6$ mV decade^{-1}. In reality, the control of the channel region by gate voltage is not perfect because of the electrostatic coupling between the channel and the substrate through the depletion layer. This leads to a typical η value in bulk MOSFETs between 1.2 and 1.5, which in turn results in SS values in the range of 70–90 mV decade^{-1} (see figure 8.7). So in conventional MOSFETs there is a fundamental barrier due to Boltzmann statistics, hence it is impossible to reduce the value of subthreshold slope to below 59.6 mV decade^{-1} at

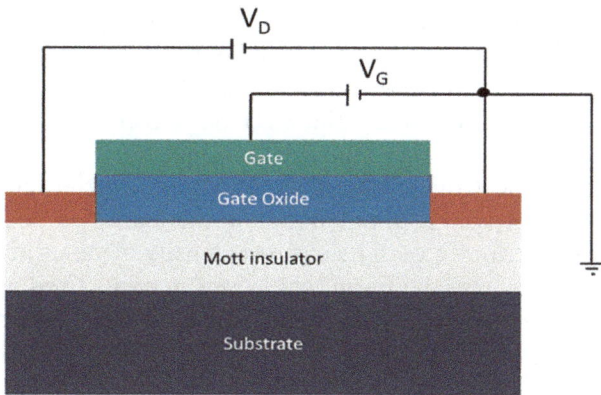

Figure 8.8. Schematic diagram of a Mott field-effect transistor.

room temperature because of thermodynamic reasons, and the best one can hope for is to approach this limit as close as possible [18]. It may, however, be possible to avoid this limit by using quantum-mechanical band-to-band tunneling instead of thermal injection to inject charge carriers into the device channel. This can possibly be achieved by using tunnel FETs based on ultra-thin semiconducting films or nanowires [19]. The other possibility is the negative capacitance FET (NC-FET), where the negative capacitance effect of certain ferroelectric materials is utilized to achieve an energy gain by changing the polarization [20–22].

The n-channel and p-channel MOSFET device structure can be further modified to allow two modes namely accumulation mode and depletion mode. A channel region of a material type corresponding to the majority carrier (i.e. n-type material for NMOS and p-type material for PMOS) is physically implanted at the time of fabrication of the device to connect the source and drain for enabling the device to work in the depletion mode. In this mode MOSFET is normally in an ON condition until an appropriate bias condition is generated to deplete the channel region of charge carriers, and remove the conduction path between the source and the drain. On the other hand in the accumulation or enhancement mode, the MOSFET normally is in a OFF state until the conduction channel is created, so that the current can flow between the source and drain.

With this background information on electron band semiconductor based MOSFET, we shall now study the possibility of such a device based on Mott-insulators. In a three-terminal Mott device or Mott transistor a change in charge carrier density is induced by electrostatic carrier doping through field effect. This process does not usually give rise to Joule heating because of low gate–drain current. Such a Mott field-effect transistor (Mott-FET) can be fabricated on a homogeneous material without spatially varying dopant profiles. This is in contrast with the conventional MOSFET devices where doping profile variation is an essential ingredient for functioning.

Figure 8.8 presents a schematic diagram of a Mott-FET. There will be carrier accumulation in the interface of the Mott insulator material at a finite gate voltage.

At a threshold voltage this electrostatic modification of charge density will reach the necessary level to induce a Mott insulator to metal phase transition. This will cause a large increase in the electrical conductance of the channel, and the device is turned ON. In this ON state of Mott-FET the electron density of states becomes finite around the Fermi level of the Mott insulator, while the same is zero in the OFF state. The evolution of electron density of states with the variation of doping level in Mott insulators is a nontrivial phenomenon, and there are two possible ways how it takes place [3]. The appearance of finite electron density of states at the Fermi level is one possible way. In this case the Fermi level itself remains fixed and the developed states at the Fermi level are called midgap states [23]. The second way involves a doping-induced spectral weight transfer between the Hubbard bands, which we have discussed in section 3.10. The phenomenon of spectral weight transfer is an experimental signature of electron–electron correlation, and a Mott-insulator can be differentiated from a band insulator through such spectral weight transfer. In a Mott-FET the upper and lower Hubbard bands merge at the Mott insulator-metal transition, and that activates the Mott-FET channel. In contrast in a conventional MOSFET the electronic bands in the band-semiconductor are rigid, and there is no change in band gap on application of the external electric field. It is the inversion due to band bending above a threshold voltage, which triggers switching in conventional MOSFETS. This contrasting behaviour of Mott-FET and MOSFET is shown schematically in figure 8.9.

When the gate dielectric (which is often a band insulator) and the Mott insulating channel of Mott-FET are brought into contact, a redistribution of electron takes

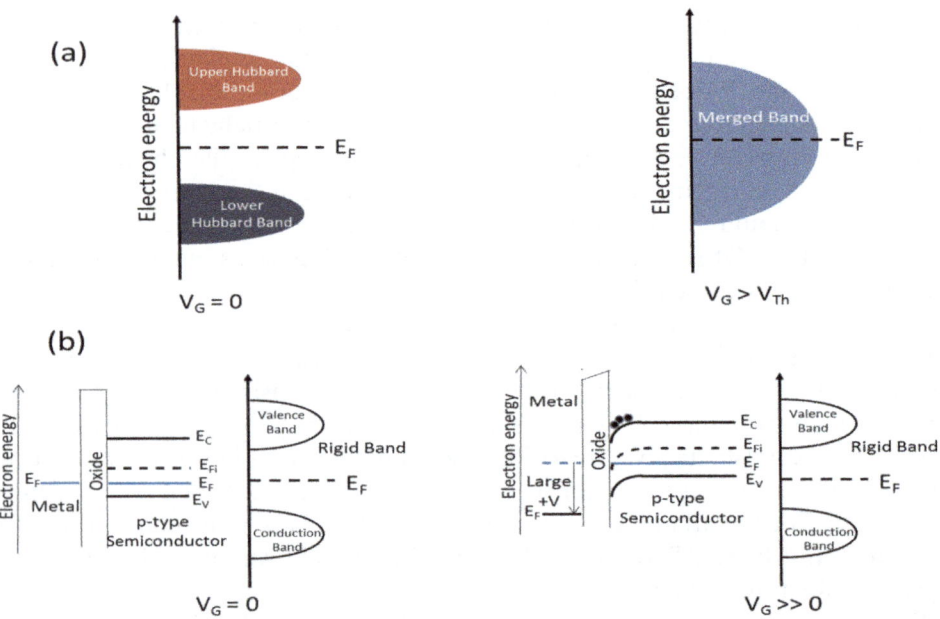

Figure 8.9. Schematic diagram showing: (a) collapse of lower and upper Hubbard bands above a threshold voltage in a Mott-FET, (b) band bending in conventional n-channel MOSFET.

place due to the difference in work function. This in turn will build up a potential at the dielectric/channel interface, which will transfer charge carriers in the Mott insulator channel. This eventually may lead to a band filling-control Mott-insulator to metal phase transition in the channel. In addition, there may also be a change of parameters such as bandwidth W and the on-site electron–electron repulsion energy or Hubbard U near the interface in Mott insulator channel with fixed band-filling. When a positive charge is created in a Mott insulator in contact with a metal, this event will simultaneously induce a negative image charge in the metal. The attraction between this image charge in metal and the net charge in the Mott-insulator will result in a reduction in the energy cost for electron hopping by $2U_1 = e^2/2\pi\epsilon D$, where D is the distance between the electron and image charge, and ϵ is the dielectric constant of the Mott-insulator [17]. This reduction in U does not involve charge redistribution between metal and the Mott-insulator; it is a purely surface phenomenon. The effective value of Hubbard U is $U' = U_0 - 2U_1$, where U_0 is the bulk value. In a similar manner the difference in the dielectric constant between a band-insulator and Mott-insulator will also induce an image charge across the band-insulator/Mott-insulator interface, leading to a change in Hubbard U in the surface region.

We have discussed above that the gate voltage in a conventional MOSFET penetrates only up to a thin layer and modulates the carrier density within a characteristic length i.e. Debye screening length. We need to find out this screening length in a Mott-FET. The band structure of a Mott insulator is not rigid upon carrier doping, and the electron density profile is related to the location of Fermi level and therefore its surface potential [3]. The free electron/hole density in the channel is described by the general relation $n = n(E_F)$ or $p = p(E_F)$. The next step now is to study how the surface potential modifies the carrier density in the Mott-insulator, which would provide information on screening length and performance of Mott-FET. There are some theoretical studies reporting that the interface electron density is well explained by the Poisson equation considering the lower (upper) Hubbard band as valence (conduction) band in conventional band bending picture [11]. Starting from the Poisson equation and keeping the first order term, the Thomas–Fermi screening length in a Mott insulator can be expressed as [15, 17]:

$$\lambda_{TF} = \left(\frac{\epsilon_r\epsilon_0}{e^2\chi_C}\right) \tag{8.9}$$

where ϵ_r is the dielectric constant of the channel Mott-insulator material, ϵ_0 the vacuum permittivity, and $\chi_C = \frac{\partial n}{\partial E_F}$ is the charge compressibility in Mott insulators with electrons as majority carrier. This charge compressibility vanishes in the Mott insulating state at $T = 0$ K [24], and the screening length is very large. The charge compressibility in the Mott insulating state, however, becomes non-zero at finite temperatures due to the thermal excitations of the free carriers, and hence the Mott-FET channel acquires a finite screening length at finite temperatures. Assuming that the electron density of states do not change upon doping, to the zeroth order one can use the formula for conventional MOSFET (see equation (8.6)) for the estimation of the screening length. Lying between the semiconductors like Si and GaAs and conventional metal, the carrier concentration is often relatively large in Mott-FET channel, hence the screening length is expected to be quite small i.e. a few nm. This assumption demands the Mott-insulating material that could be metalized to be quite thin, thus making the Mott metal–insulator transition characteristics almost two dimensional [17].

Another important parameter is the SS, which is the rate of increase of the drain current (I_D) with the gate voltage V_G below the threshold voltage. The free carrier density in the conventional MOSFET channel is the same as the net carrier induced by the gate oxide. This is not the case in Mott-FET, where the free carrier density in Mott-FET channel is different from the net carrier density injection due to possible electron band structure changing or even band closing. We have earlier discussed in section 5.2 that Mott metal–insulator transitions can be either the mass diverging type as in the Brinkman–Rice picture, which mainly influences mobility, or the carrier density vanishing type as in the Mott–Hubbard picture, which involves change in free carrier density. However, in many classical Mott insulator systems like VO_2, $NdNiO_3$ and $Ni_{1-x}S$ it has been observed experimentally that both for thermally induced and electrostatically induced Mott metal–insulator transitions, the mobility does not change as much as the carrier density across the phase transition (see [17] and references within). The channel conductance under a certain gate voltage is determined by the free carrier density in the channel, which is enhanced in a Mott-FET channel. A new parameter K is introduced for Mott-FET to represent the charge enhancement, which is the ratio of free electrons to the net electrons in the channel. With this information, the SS for Mott-FET is expressed as [17]

$$\text{SS} = \frac{dV_G}{d\Psi_S}\frac{d\Psi_S}{(d\ln_{10}(I_D))} = \left(1 + \frac{C_{Mott}}{C_{ox}}\right)\frac{d\Psi_S}{d\ln_{10}n} \tag{8.10}$$

where Ψ_S is the surface potential, n is the free carrier density, C_{ox} is the capacitance of the gate dielectric and C_{Mott} is the effective capacitance of the Mott insulating channel.

The gate voltage drop in the Mott-FET channel will be determined by the relative magnitude of C_{ox} and C_{Mott}. The effective capacitance of the correlated electron systems is related again to the charge compressibility of the material due to the screening of electric field [3]. The free carrier density (n) in the OFF state of Mott-FET is equal to the net carrier density (n_{net}). Above the threshold voltage n is enhanced over n_{net} by factor represented by the parameter K. Assuming that the n_{net} changes with surface potential in a similar way as in the conventional MOSFET, the average subthreshold slope SS in Mott FET will decrease by $1/\ln_{10}(K)$.

8.2.2 Boltzmann switch versus Landau switch

In band semiconductors, the transport factor $\frac{d\Psi_S}{(d\ln_{10}(I_D))}$ is intrinsically limited by the carrier statistics through the Boltzmann factor $10^{-\Delta E/k_B T}$. Here, ΔE is barrier height relative to the Fermi level. Modulation of the drain current I_D by the gate voltage V_G requires the SS as conversion factor $I_D \sim 10^{-\Delta E/k_B T} \sim 10^{-\Psi_S/SS}$ with SS $\geqslant (q/K_B T)$, where q is the elemental charge. This tells that a minimum $\sim$60 mV change in Ψ_S is required to modulate the drain current I_D at room temperature by one order of magnitude. This is the well known 60 mV decade^{-1} limit of band semiconductor FETs. This implies that the supply voltage needs to be well beyond this limit for suitably distinguishing the OFF-state and the ON-state current. So, it is quite clear that the Boltzmann factor puts a constraint on the scaling of supply voltage, and in turn on the switching energy of a band semiconductor FET.

In a Mott-insulator charge carriers can change their character from localized to itinerant behaviour and the entire band structure gets modified in the vicinity of the Fermi level (see figure 8.9(a)). This is in striking contrast to the band semi-conductors, where an electric field only shifts electronic bands with respect to the Fermi level (see figure 8.9(b)). Switching characteristics of Mott-FETs are governed by the free energy of the competing metal and insulator phases, and not by the statistics of charge carriers. Mott-insulators are therefore well suited for going beyond the intrinsic 60 mV decade^{-1} limit of conventional semiconductor based *Boltzmann switch*. The underlying principle behind this very sharp switching in Mott insulator is its characteristic electronic phase transition, which may be understood within the framework of Landau theory of phase transition. This has led to the coining of the term *Landau switch*. There is added advantage with Landau switches that the switching can be triggered by various external stimuli like strain, magnetic field, optical excitation, temperature or pressure apart from electric field. Figure 8.10 presents a schematic diagram showing a comparison between the Boltzmann switch and Landau switch.

Like conventional bipolar FET with homogeneous doped channel, a Mott-FET can be turned ON both with positive and negative gate bias; either hole doping or electron doping can induce a Mott insulator to metal transition. However, the critical density for hole and electron doping may be asymmetric, which results in different threshold voltages for positive or negative gate bias. We will discuss more on Mott-FET with some actual examples later in this chapter.

8.2.3 Two-terminal Mott devices

A two-terminal Mott metal–oxide–metal device (see inset of figure 8.11) works on the principle of electrically induced metal–insulator transition (E-MIT). This E-MIT can be triggered both by current and voltage source. The main frame of figure 8.11 shows typical current (I)–voltage (V) curve arising from E-MIT in a two-terminal Mott device driven by a voltage source. In the low voltage region the Mott-insulator material of a Mott device remains in its insulating phase, and the device response is linear. A deviation from this linear behaviour in the I–V curve appears with the

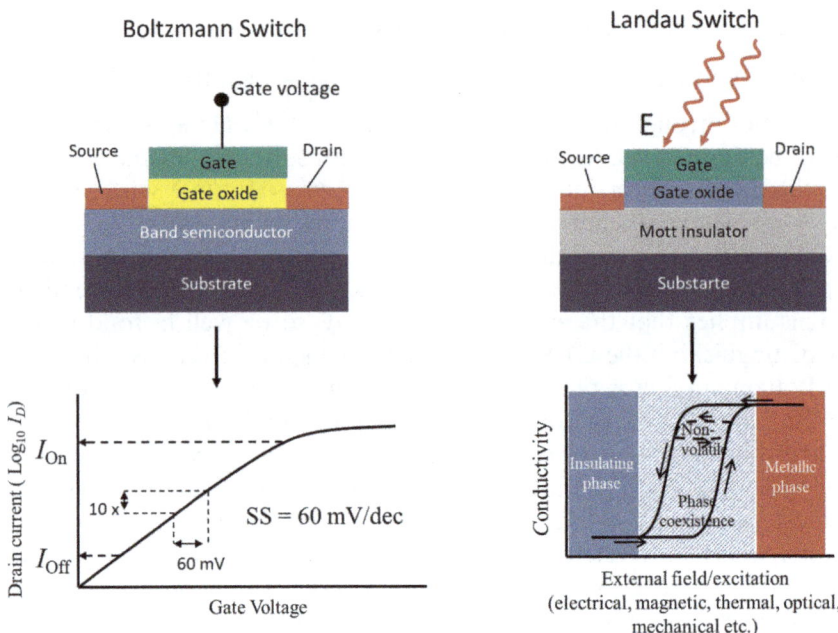

Figure 8.10. Schematic diagram showing a comparison between the Boltzmann switch and Landau switch.

application of a large voltage, which is indicative of the nucleation of the low resistance metallic phase. Above a threshold voltage (V_{Th}^{**}) the resistance of the device decreases drastically. This is akin to the limit of superheating T^{**} in figure 8.2 above which the whole sample is in the metallic state. The value of electric field corresponding to the threshold voltage can be 10^4–10^5 V cm^{-1} depending on the Mott-insulator materials. In the voltage region above the threshold voltage V_{Th} the I–V curve and the device is considered to be in the switched ON state. The nature of this I–V curve does not depend on the voltage polarity in a symmetric device. The memory state of the device can be switched by successive application of electric field of either the same or opposite polarities, and this is known as unipolar switching. This kind of switching is essentially volatile, since the low resistance metallic state is maintained only with an applied electric field, which takes the device above the threshold voltage V_{Th}. Such volatile threshold switching devices can be used as selectors in resistive random access memory (ReRAM) crossbar arrays (to be discussed below in section 8.4.1).

This two-terminal device is turned OFF at a threshold voltage V_{Th}^* in the decreasing voltage cycle with a sharp increase in the resistivity in the Mott-insulator material (see figure 8.11). It is to be noted that akin to the limit of supercooling T^* in figure 8.2, this threshold voltage V_{Th}^* is clearly lower than the threshold voltage in the increasing voltage cycle V_{Th}, and there is a distinct hysteresis in the I–V curve (see figure 8.11). This hysteresis can be exploited for device functionality such as memory. Under some circumstances multiple steps can occur during the Mott

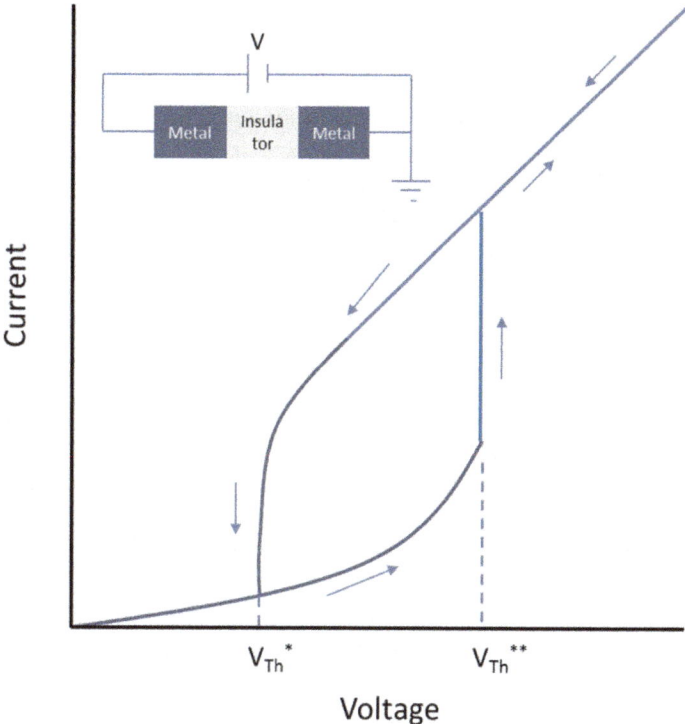

Figure 8.11. Schematic presentation of a typical current–voltage curve arising from a metal–insulator phase transition in a two-terminal Mott device.

metal–insulator transition, which lead to newer possibilities such as the creation of nonlinear resistor circuits.

Multi leveled functionality can be achieved in the two-terminal Mott switches, where memories can be toggled between the memory states by application of successive electric field of alternate polarity. Local switching mediated by point defects in materials leading to bipolar switching behavior has been reported in Mott-insulators with structural and compositional disorder [3]. Under certain circum-stances a first order phase transition can be kinetically arrested [5, 25] giving rise to further possibilities of multiple memory states for device applications. Such kinetic arrest of a disorder influenced first order phase transition has actually been observed across the first order phase transition in many classes of magnetic and ferroelectric materials including various manganese oxide based charge ordered systems [5, 25].

The switching mechanism in two-terminal Mott devices is quite distinct from resistive random access memory (ReRAM) and phase change random access memory (PCRAM) discussed in chapter 7:

1. In contrast with the defect influenced conductive filament formation in ReRAM, the switching here is due to the electrically induced metal–insulator transition (E-MIT).

2. The magnitude of the thermally induced metal–insulator transition in the core Mott-insulator material, matches with the ON/OFF ratio of E-MIT in the device.
3. There is a scaling of ON/OFF resistance with the device area.
4. Sometimes there can be multiple jumps in the $I-V$ curves due to partial switching in local regions of the sample. However, in contrast with nano scale conducting filaments in ReRAM, the conductive path in Mott devices is usually in micrometers to millimeters and comparable with the device area.
5. In PCRAM the resistive switching involves a structural transition from the crystalline to amorphous phase, whereas in Mott devices involving intrinsic Mott-insulators the E-MIT is purely electronic in origin.

There are various mechanisms, which can cause E-MIT in a two-terminal Mott device. Application of an electric field to a Mott device will cause a current flow, which will result in the electric field-induced generation of carriers and also thermal generation of additional carriers due to the Joule heating. In chapter 5 (see section 5.1) we have learnt about the band filling-control (FC) Mott metal–insulator transition. The electric field induced processes can induce a FC metal–insulator transition in the Mott insulator core material of the device. Such processes include [3]: (i) Fowler–Nordheim tunneling through a Schottky barrier or thermionic emission across a Schottky barrier when the metal/insulator contact is non-Ohmic, (ii) trap-to-band tunneling or Poole–Frenkel emission in the presence of defect states, and (iii) band-to-band tunneling.

A certain critical carrier density is required before Mott insulators become metallic. In fact we have seen in section 3.2 that a correlated electron gas becomes an insulator when the average electron separation given by $n^{-1/3}$ becomes much greater than four times Bohr radius a_0. This leads to the famous Mott criterion for insulator-metal transition: $n_c^{1/3}a_0 \sim 0.25$. Here in a two-terminal Mott device, n_c is the critical carrier density at the metal–insulator transition at $T = 0$ K, and a_0 is the Bohr radius of electrons orbiting around the dopant center. Then there is of course the possibility of Joule heating by the current passing through the Mott device, thus elevating the device temperature and inducing a Mott MIT. The nonlinearity of $I-V$ curves just below V_{Th} can be taken as a signature of Joule heating. However, Poole–Frenkel emission and/or disorder broadening of the first order phase transition can also cause such nonlinearity in the $I-V$ curves. High electric fields applied in short pulses helps in minimizing the dissipation of heat energy in the device.

8.3 Theoretical models of resistive switching

The electric field controlled insulator to metal transition in a Mott device leading to resistive switching is actually an out-of-equilibrium phenomenon, which goes beyond the well established bandwidth- and band filling-controlled insulator to metal transition (see [26] and references within). The microscopic mechanism involves nontrivial out-of-equilibrium effects in correlated electron systems, and can be classified into three categories: (i) thermal effects, (ii) electric-field-assisted ionic migration, and (iii) pure electronic effects. In the first category it is envisaged that in insulator–metal transition induced by a dc voltage or dc current, the local temperature can reach the transition temperature, and thus Joule heating plays a dominat role. The sharp resistive transition arises from the sudden formation of a hot and highly conductive filament (along the electric field) induced by Joule heating [26]. Some classical Mott-insulators like VO_2 and NbO_2 belong to this class where the insulator to metal transition is thought to be essentially temperature driven. In both of these compounds, a change in the lattice structure accompanies the drastic drop of electrical resistance at the insulator–metal transition. Due to the presence of this structural transition, in many a quarters VO_2 and NbO_2 are termed as correlated insulators rather than Mott-insulators [27]. The canonical Mott insulators are considered to be those materials like $NiS_{2-x}Se_x$ and $(V_{1-x}Cr_x)_2O_3$, which do not show any structural transition. In this class of canonical Mott insulators, $(V_{0.99}Cr_{0.1})_2O_3$ and $NiS_{1.45}Se_{0.55}$ show thermally assisted metal–insulator transition [27]. In the second group of Mott insulators, an electric-field-assisted migration of ions takes place in the interface region of Mott-insulators and gives rise to a local filling controlled Mott insulator–metal transition. Delocalization of the Mott-insulating state in such systems can give rise to bipolar resistive transition. In the third group, the destabilization of the Mott-insulating state takes place in a strong electric field due to Zener dielectric breakdown [28]. This is a purely electronic mechanism and the insulator to metal transition takes place without any change in the lattice symmetry. Here the electric field excites pairs of doubly occupied sites (termed as doublons) and holes (sites with no electrons) in the Mott-insulator material, and destabilization of the Mott-insulating state happens when their density becomes sufficiently high, thus leading to an insulator-to-metal transition. For dc electric fields, the production rate of doublon–hole pairs shows a threshold behavior [29]. Denoting this tunneling threshold field by E_{th}, for small electron repulsion U, it behaves as $E_{th} \propto \Delta_{Mott}^2$, where Δ_{Mott} is the Mott–Hubbard gap. In the case of ac electric fields, there is a crossover from a weakly excited state to a strongly excited state with increasing field strength [29]. There are, however, reports that in some narrow-gap Mott-insulators like $GaTa_4Se_{8-x}Te_x$, the behaviour of the threshold electric field E_{th} is not in accord with a Zener breakdown mechanism. The observed behaviour rather points to an avalanche breakdown, where E_{th} increases as a power law of the Mott–Hubbard gap, in surprising agreement with the universal law $E_{th} \propto \Delta^{2.5}$ reported for avalanche breakdown in semiconductors [30]. However, in contrast with the conventional semiconductors, the delay time for the avalanche in such Mott insulators is over three orders of magnitude higher.

8.3.1 Resistor network model

The thermal and electrical mechanisms for resistive switching in the potential Mott devices have been the subject of intense debate during the last decade. In the classic Mott-insulator VO_2 the insulator–metal transition is in a large part thought to be due to heating by the applied electric field in the micrometer-scale [31]. The fluorescence spectra of rare-earth doped micron sized particles were used as a local temperature sensing mechanism at the surface of VO_2 thin layers to reach this conclusion. As such this work would suggest that materials like VO_2 will not be suitable for fast switches, since the heating effects are usually quite slow. This point, however, can be counterargued by the experimental facts that the Mott insulator–metal transition in VO_2 is quite fast and completes in less than a few nanoseconds, and hence it should not hinder applications [32]. On the other hand, in the narrow gap Mott-insulating compounds AM_4Q_8 (where A = Ga or Ge; M = V, Nb or Ta; and Q = S or Se) there exist experimental evidence that resistive switching cannot arise from Joule heating effects [26].

A theoretical model based on a realistic two-dimensional resistor network provides an interesting platform to compare with the key experimental results [26]. The thermal mechanism of resistive switching may cause the lowering of the resistance by the formation of a highly conductive hot filament in a Mott insulator. The hypothesis of the formation of such hot conductive filaments is tested within this resistor network model through the activated dependence of the resistivity to simulate the resistance and local thermal variations of the sample. In this resistor network there are 25 lines of cells between the electrodes, and each cell consists of four resistors connected to a single node in cross shape. In addition, the individual cell has four thermal conductances κ and a heat capacity c. The four resistors of a given cell have the same value, which depend on the temperature of the cell, $T(x, y)$, following the activation law [26]

$$r(x, y) = r_0 \exp\left(\frac{E_A}{k_B T(x, y)}\right) \tag{8.11}$$

where E_A is the activation energy for the electrical transport, k_B the Boltzmann constant, and r_0 an arbitrary prefactor. Periodic boundary conditions are assumed in the x direction. The values of E_A and r_0 are chosen to fit $r(x, y)$ accurately with the temperature dependence of the experimentally obtained resistivity. The simulation time is discretized in time steps. The standard nodal technique is used to solve the 2D resistor network in an applied voltage at each time step. To start with, a temperature T_A is set for all the sites. In real situations there will be defects present in the bulk material and imperfections in the electrodes. Resistance of a few cells (<1%) are set to a fixed value r_{defect} 10^4 times lower than $r(x, y)$ at T_A to represent such defects and imperfections. The change in temperature at each site is estimated from the calculation of local currents. This is achieved through the heat flow equation by expanding $\nabla^2 T(x, y)$ by means of the discrete Laplace approximation [26]:

$$c\frac{\partial T(x, y)}{\partial t} = p(x, y) - \kappa\left(4T(x, y) - \sum_i^{FN} T_i\right) \tag{8.12}$$

where FN in the superscript of the summation stands for the first neighbours, c is the heat capacity of each cell, $p(x, y)$ the electric power generated in the cell, and κ the thermal conductance between neighboring sites, which is same for all the networks. The boundary condition is that the cells at the top and at the bottom exchange heat

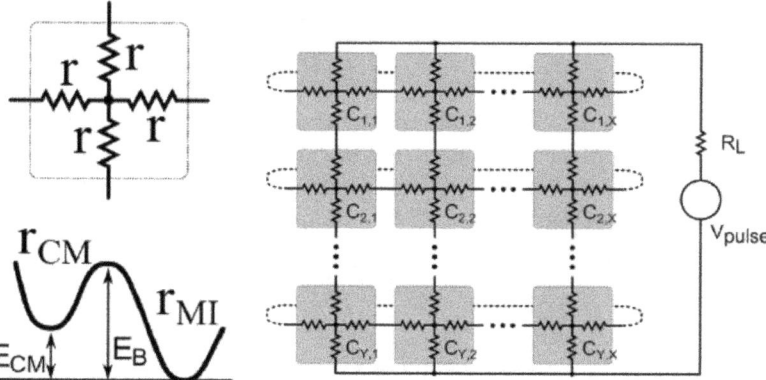

Figure 8.12. Schematic representation of a pure electronic model of the resistor network along with the energy barrier between the Mott insulator state and the correlated metal state. (Reprinted with permission from [26]. Copyright 2014 by the American Physical Society.)

with the three neighbours and a virtual cell with a fixed temperature T_A representing the electrodes. The matrix $r(x, y)$ is updated with this new temperature profile, and the simulation continues to the next time step. The results of the simulation show a gradual decrease of the resistance for all applied voltages due to Joule heating, which eventually leads to a stable low resistance state. The sequences of the evolution of the temperature profile at a different simulation time reveal the formation of a hot percolative path bridging the electrodes. The resistivity of this hot percolative path decreases due to the activated temperature dependence. This leads to an overall reduction of the sample resistance, and this reduction can be of several orders of magnitude for a large voltage pulse. However, this Joule heating model is not adequate to reproduce the experimental results on Mott insulators, especially the sharp drop of the resistance. Furthermore, the model fails to predict the existence of a threshold voltage and the voltage regulation effect observed after the transition.

To go beyond the pure thermal model, in the next step a pure electronic model of the resistor network is considered [26]. This model assumes that the active layer is composed of cells that may be in either of these two states: a low-resistivity correlated metal (CM) state and a high-resistivity Mott insulator (MI) state. The cells are modeled by four resistors with value R_{MI} when it is in the MI state or R_{CM} when it is in the CM state, with $R_{CM} < R_{MI}$. Figure 8.12 presents the array of such cells connected to the pulse source V through the limiting resistor R_L. An energy barrier E_B separates the MI and the CM state in each cell. In thermal equilibrium the device is initially assumed to be in a high resistance MI state. A small portion of the sample (of at least a few tenths of nanometers) is represented by each cell, such that the local resistivity may have a well-defined value [26].

A local electric field ϵ destabilizes the electronic configuration of a cell in the MI state by effectively reducing the E_B barrier, thus resulting in a probability [26]:

$$P_{\mathrm{CM}}(x, y) = \exp\left(-\frac{E_B - q|\epsilon(x, y)|}{k_B T(x, y)}\right) \qquad (8.13)$$

for the local conversion of the MI state to the CM state. Initially $T(x, y)$ is maintained at a fixed uniform value during the simulations. When a cell is converted into the highly conductive CM state, its internal electric field becomes negligible, and the cell will then have a probability [26]:

$$P_{\mathrm{MI}}(x, y) = \exp\left(-\frac{E_B - E_{\mathrm{CM}}}{k_B T(x, y)}\right) \qquad (8.14)$$

for switching back to the MI state. Here E_{CM} is the energy corresponding to the CM state (see figure 8.12). The time evolution of this system was simulated in the same manner as was done in the thermal model. However, it was evaluated from the computed voltage drops at each cell and by means of equations (8.13) and (8.14) whether a cell must change its state or not, and the values of the resistors of the network were updated accordingly. The results of this simulation agreed well with the experimental observations, namely the sudden drop of the resistance above a voltage threshold, the scaling of time delay with the voltage, and the voltage regulation effect after switching.

Depending on the applied voltage two different regimes were predicted from the simulations with this purely electronic model: a low voltage regime where it remains in the high resistance MI state and a high voltage regime where it undergoes a transition to a low resistance CM state. In the low voltage regime, the system consists of largely a homogeneous MI state and a few isolated CM cells. With the increase in the applied electric field a first order phase transition from the MI to CM state takes place. In this high voltage regime as a part of this first order transition process, at a threshold voltage a high local field region will be the nucleation point for the sudden formation of a percolative metallic cluster that connects the electrodes. This electronic model provides a reasonable explanation of the experimentally observed sudden drop of the resistance above a voltage threshold in various Mott-insulator materials. These points will be discussed further in section 8.4.1 in the context of experimental results on switching in the narrow gap Mott-insulating compounds AM_4Q_8.

Although the electronic model can explain well the characteristic features of the resistive switching in Mott insulators, it remains an experimental fact that the resistivity of the Mott insulators is quite sensitive to local changes in temperature. So, in any realistic system a combination of the electronic and the thermal models is necessary to capture the interplay between the Mott insulator–metal transition and the Joule heating. To this end the temperature at each cell is calculated as described earlier, and the switching probabilities P_{CM} and P_{MI} (see equations (8.13) and (8.14)) depend on this calculated local temperature. In addition, the resistances in the MI state are now considered to be temperature dependent, following equation (8.11). Within this combined model it is observed that the increase in temperature before switching is minimal. The switching behavior is dominated by the electric-field-driven Mott insulator–metal transition. Post switching, first there is a rapid increase in the temperature of the filament, and that is followed by the stabilization of a state with a higher conductance after a voltage dependent time-scale. Thus, both electronic and thermal mechanisms play their role but at different time scales. The voltage regulation after switching is observed at short time scale, while at longer times this stabilization may be lost due to filament Joule heating, which eventually may lead to the destruction of the sample.

8.3.2 Dissipative Hubbard model

The resistive network model discussed above is really a heuristic approach. To this end a full many-body out-of-equilibrium treatment of the Mott transition is required to understand the resistive switching mechanism from the microscopic level. This will also help to settle a question central to the ongoing intense debate, that is whether the thermal and electronic mechanisms are mutually exclusive or there is some scope of compatibility [33]. A microscopic theory for resistive switching is indeed a rather formidable research problem, but will be quite useful for engineering the future Mott devices. In recent times there have been effort to study electric field induced resistive switching using a dissipative Hubbard model (see [34] and references therein). The non-equilibrium steady states are accessed non-perturbatively in both the field and the electronic interactions using a nonequilibrium dynamical mean-field theory (DMFT).

A minimal quantum driven dissipative model has been recently proposed, which reproduces the main experimental features of resistive switching in VO_2 [33]. The model represents a situation where a slab of correlated electrons of length L, represented by the Hubbard model, is placed between two metallic leads connected in series with a DC voltage generator and a resistor R. The DC voltage generator delivers a voltage V_T, which creates a voltage bias V_S across the Mott-insulator sample and drives a current I through the circuit, so that $V_T = V_S + RI$. Energy relaxation inside the bulk sample is also considered in addition to the dissipation by the two metallic leads at the boundaries. The resistor R is crucial to reveal a nontrivial regime of negative dI/dVs, and dissipation is crucial to avoid overheating the sample [33]. At the starting point of the theoretical calculation, the Hamiltonian H was divided into three parts [33]:

1. The correlated electron sample is represented by a Hubbard model on a finite two dimensional square lattice: $H_{Hubbard}$.
2. The two metallic leads and the dissipative reservoirs of Fermions are represented by $H_{leads+bath}$.
3. The electric-field induced electrostatic potential originating from the choice to work on nonequilibrium DMFT in the Coulomb gauge, is represented by H_E.

The self-consistent mean-field Hartree–Fock theory was used to treat the Coulomb interaction in $H_{Hubbard}$, which produced a phase transition from a high-temperature low-U paramagnetic metal and a low-temperature large-U antiferromagnetic insulator. Here U represents the on-site Coulomb interaction. The problem is then solved self-consistently for any given DC generator voltage V_T without assuming a specific voltage profile in the sample. In this calculation the order parameter $\Delta = F(\Delta_{Hubbard}; T_{bath}; E; \Gamma)$ distinguishes the low temperature antiferromagnetic insulating state, and the high temperature paramagnetic metallic state arises as the self-consistent result from the balance of the electron–electron interactions, the nonequilibrium drive and the dissipative mechanisms. The parameter Γ is the hybridization parameter, which sets the rate at which particles and energy are exchanged with the environment. The readers are referred to the article by Li *et al* [33] for details of the theoretical calculations. The resistive switching mechanism is revealed by the local distribution function, which at finite electric fields deviate from the Fermi–Dirac distribution both in the metallic and insulating state. In the insulating phase, there are significant nonequilibrium excitations beyond the Mott–Hubbard gap Δ. The total number of nonequilibrium excitations above the bath chemical potential is well described by the Fermi surface averaged Zener tunneling rate [33]. This clearly suggests that the electronic mechanism is at the origin of resistive switching. The quasi-particles in the lower Hubbard band of the Mott-insulating material are accelerated by the electric field. They acquire a finite probability to tunnel across the gap Δ and populate the upper band, thus turning the system metallic.

In a purely thermal mechanism, the nonequilibrium drive effectively enters this problem only as an effective temperature $T_{eff}(E)$, caused by Joule heating. The experimentally observed very sharp and hysteretic resistive switching, however, indicates that the underlying mechanism must be beyond a simple reparametrization of the equilibrium theory. In equilibrium situation of zero electric field, there is only one stable solution of antiferromagnetic insulating state with a finite Δ for $T_{bath} < T_N$, and Δ goes to zero continuously as the paramagnetic state is reached for $T_{bath} > T_N$. In the nonequilibrium situation i.e. $E > 0$ there are two stable solutions for $T_{bath} < T_N$: (i) the antiferromagnetic insulating state at the equilibrium value $\Delta = \Delta_0$, and (ii) the paramagnetic state at $\Delta = 0$. This paramagnetic state is explained within the thermal mechanism with a high effective temperature $T_{eff} \propto E/\Gamma$ caused by Joule heating on the metallic side. This is prevented on the insulating side by the large Mott–Hubbard gap. The bistability of the order parameter Δ results in heterogeneous phases during the resistive switching.

8.4 Experimental situation on Mott devices

In recent times there had been many reports on Mott devices involving various kinds of Mott-insulators, including organic materials. The aim of the present book, however, is to introduce the basic concepts and working principles of various emerging classes of Mott devices by taking examples of some well studied Mott-insulating materials, rather than providing a complete review of this rapidly growing field. The subsequent discussions will be focused on the experimental works involving two-terminal Mott switches and memory devices, Mott-FET, photonic devices, thermal and chemical sensors, and Mott-insulators in energy storage and energy harvesting applications.

8.4.1 Mott switches and memory devices

The electrical switching phenomenon involves an electrical field induced rapid and reversible transition between a high-resistance state and a low-resistance state. There are generally two kinds of switching, namely threshold switching and memory switching, which are volatile and non-volatile, respectively, in nature. One of the early studies on the two terminal Mott switches involved sandwich metal/oxide/metal (MOM) structures based on VO_2 thin film [35]. This sandwich MOM device is shown in figure 8.13(a), which was fabricated by the anodic oxidation of the vanadium metal substrates synthesized from vacuum deposited vanadium layers. During the process of electro-forming, channels consisting of the vanadium dioxide phase were found to be formed in the initial anodic films [35]. Gold electrodes were then evaporated on the oxide film to form the MOM structure. The voltage–current characteristic of this MOM device is shown in figure 8.13(b), which displays the typical S-shaped characteristics. The threshold voltage (V_{th}) and current (I_{th}) parameters at room temperatures are of the order of 1–10 V and 10–100 μA, respectively, depending on the samples. The threshold voltage V_{th} is about 2 V for the sample shown in figure 8.13(a). Devices with such negative differential resistance have potential applications in electronics as switches, memory elements, and micro-sensors [36]. An electronic switch device based on Si–SiO$_2$–VO$_2$ structure on p-type Si-substrate with a silicon dioxide layer thickness of 70 nm (see figure 8.14) was also reported [35], where the injection of electrons from Si into the VO_2 was carried out. Experiments showed that the V_{th} in such a switching device increased with the decrease in temperature, which was attributed to the decreased VO_2 channel conductivity at lower temperatures [32]. In subsequent years more types of switching devices involving improved quality of VO_2 thin-films have been studied. One such example is shown in figure 8.15(a), where VO_2 thin films were deposited on sapphire substrates using the method of laser ablation, and Au/Cr electrodes were patterned on

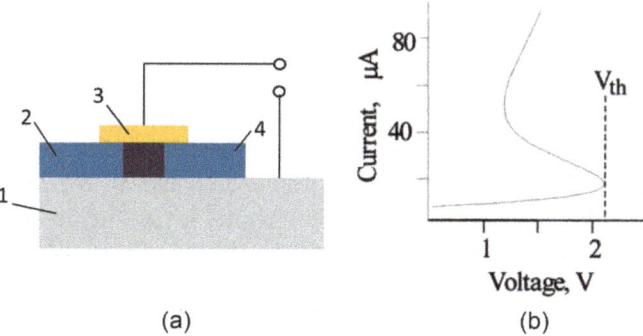

(a)   (b)

Figure 8.13. (a) Schematic diagram of the VO_2 thin film based sandwich metal–oxide–metal (MOM) structure: 1, vanadium; 2, vanadium oxide film; 3, gold electrical contact; and 4, VO_2 channel. (b) The voltage–current characteristic of the MOM structure. (Reproduced from [35]. (c) IOP Publishing. Reproduced with permission. All rights reserved.)

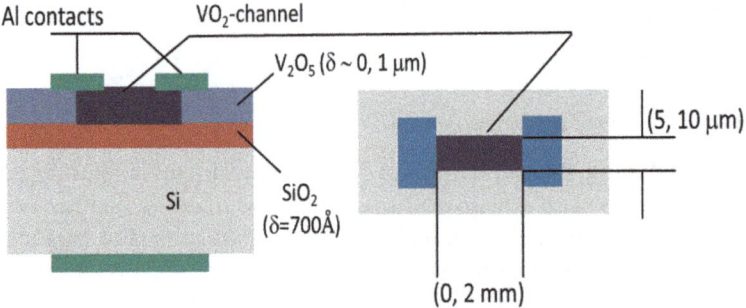

Figure 8.14. Schematic diagram of an planar electronic switch device based on Si–SiO$_2$–VO$_2$ structure. (Reproduced from [35]. (c) IOP Publishing. Reproduced with permission. All rights reserved.

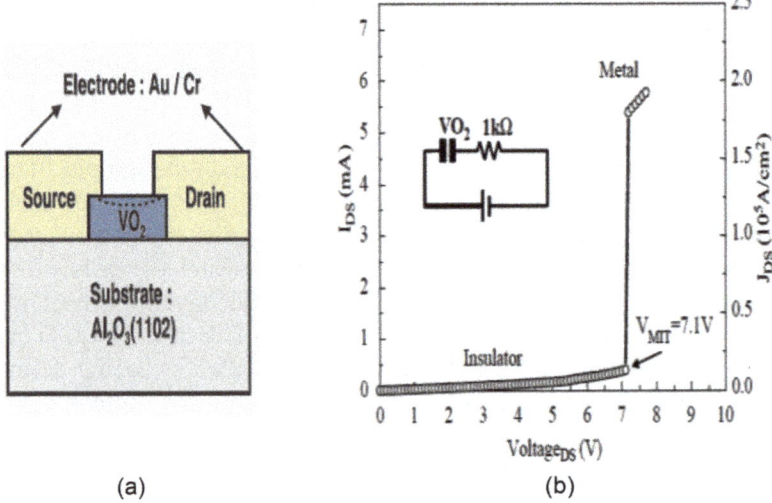

Figure 8.15. (a) Schematic diagram of a VO$_2$ thin-film based switching device grown on sapphire substrate. (Reproduced from [37]. (c) 2004 IOP Publishing Ltd. Deutsche Physikalische Gesellschaft. All rights reserved.) (b) Metal–insulator transition induced by a DC voltage in the VO$_2$ thin film of the device. The inset of the figure displays the circuit used for experiments. I_{DS} and J_{DS} are the drain–source current and the corresponding current density, respectively. (Reprinted from [38], copyright 2005, with permission from Elseveir.)

the VO$_2$ using a lift-off process [37, 38]. The Mott metal–insulator transition is clearly observed on application of a DC voltage to this device connected in series with 1 kΩ resistor (see figure 8.15(b)). The resulting current increases slowly with increase in the applied voltage, and then at the threshold voltage of 7.1 V jumps abruptly to a very high value. This device with suitable modification can be used as voltage-dependent varistor for protecting circuits from electric surges and electrostatic discharge [39]. In addition to such planar junction devices, electric field induced Mott metal–insulator transition has also been observed in out-of-plane VO$_2$ based Mott devices [32]. Since the early 2000 there has been a significant number of studies on resistive switching in thin film structures based on various transition metal oxides Mott insulators including NbO$_2$, MnO$_2$ and Fe$_3$O$_4$ [36].

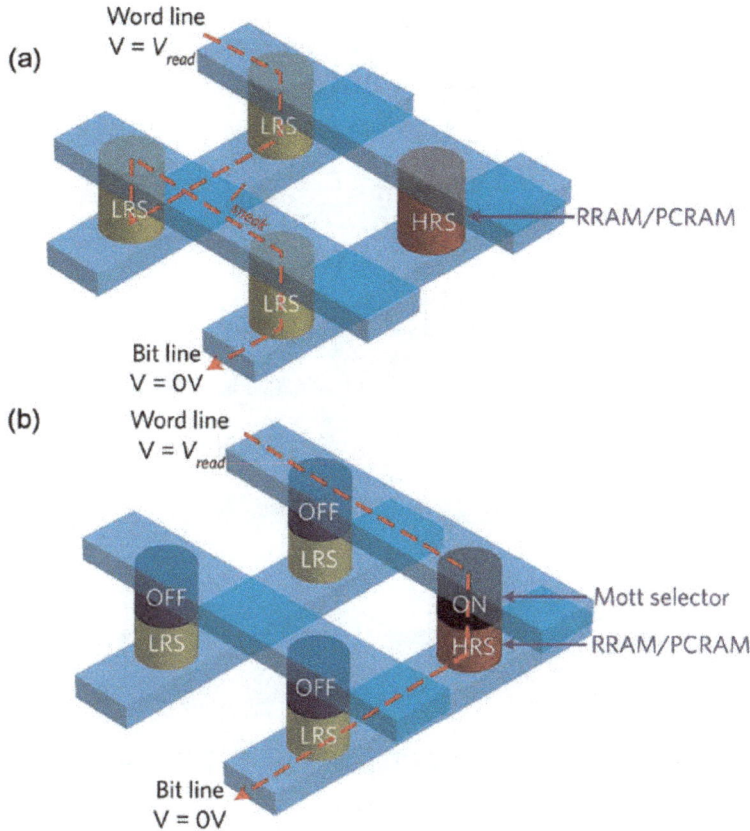

Figure 8.16. Schematic diagram of a crossbar array with high-density stackable packing of nonvolatile memories (ReRAMs or PCRAMs): (a) each unit only contains a memory and (b) each unit contains a memory and Mott resistive switch as selector. ((c) 2015 IEEE. Reprinted, with permission, from [3].)

The materials showing a volatile resistive switching behavior can be used as selectors in ReRAM crossbar arrays for reducing the undesired sneak currents. Figure 8.16(a) shows a crossbar nonvolatile memory arrays with a single ReRAM or PCRAM as the memory unit, where the current can flow through undesired links, leading to read/write errors [3]. This gives rise to the need of a selector device like a diode or transistor with asymmetric current–voltage characteristics to be integrated with each cell to form a 1-selector–1-ReRAM (1S1R) array. Figure 8.16(b) shows a 1S1R crossbar array with Mott-insulator resistive switches as the selectors. These Mott-insulator-based selectors do not have any junction; hence they are more stackable, compact, and easier to fabricate in comparison to diodes and transistors. In addition, Mott selectors are advantageous for bipolar switching ReRAMs, since unlike in diodes large current is allowed to pass in Mott selectors in both directions. During the last decade there have been several studies investigating selector properties of Mott insulators VO_2 and NbO_2 [3], and figure 8.17 shows a nanoscale 3D vertical ReRAM with NbO_2 selector. The basic 3D structure of the device

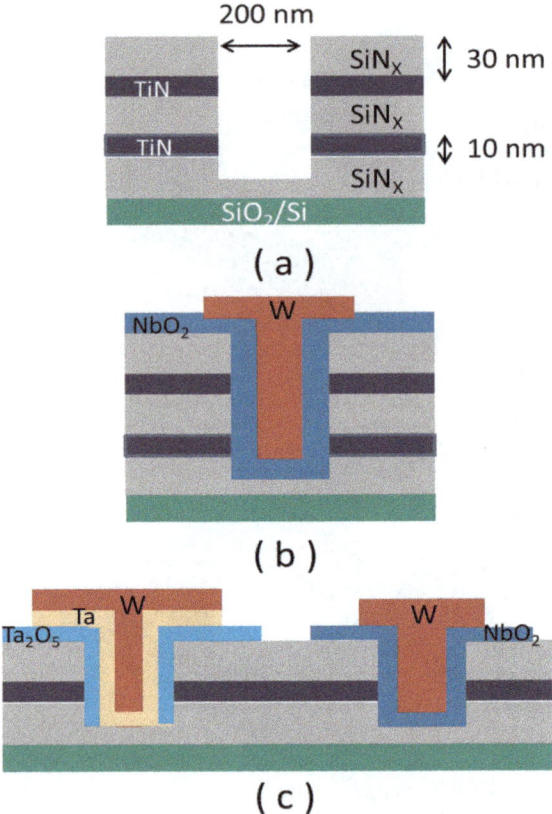

Figure 8.17. Schematic diagram of a 1S1R structure with 3D vertical ReRAM and Mott selector: (a) basic structure with stacked SiN and TiN layers, (b) NbO$_2$ Mott selector and (c) 1S1R device. ((c) 2013 IEEE. Reprinted, with permission, from [40].)

consists of 30 nm SiN and 10 nm TiN layers sequentially deposited on SiO$_2$/Si substrate by chemical vapour deposition process, and a 200 nm diameter hole created subsequently by conventional photolithography and reactive ion etching processes [40]. After that a NbO$_2$ layer with a tungsten top electrode was deposited on this basic structure. In the 1S1R structure, NbO$_2$ selectors were fabricated with shared TiN layer as a middle electrode (see figure 8.17).

From a technological points of view, the Mott-insulator material NiO has certain advantage among the transition meal oxides because NiSi serves as an interconnection material in nanoscale semiconductor devices, hence element contamination is eliminated. NiO exhibits stable unipolar switching properties, and ReRAM made of NiO has low operation current [41]. It may be noted here that the as-fabricated devices usually do not show threshold or memory switching effects. For the required device performance an initial process of electroforming is necessary, which involves an application of suitable voltage pulses higher in magnitude than the subsequent operating pulses [36]. The electroforming process can cause structural and electronic changes in the device material, and such changes can happen in a localized region or throughout the bulk of the material. As deposited NiO film is usually insulating in nature, and after electroforming this becomes a metal through the formation of conducting filaments created during electroforming process. The filamentary model of ReRAM explains the functioning of a NiO device in the following way. The conducting filaments in NiO will be ruptured by the Joule heating caused by strong current when the applied voltage reaches a certain V_{RESET}, and NiO material would become an insulator [41]. The conducting filaments, however, will reappear above another threshold voltage V_{SET}, which is higher than V_{RESET} but lower than the electroforming voltage V_{EFORM}, and NiO material becomes a metal again. Although the electroforming process is a non-destructive process (unlike dielectric breakdown), its characteristics of different materials are generally not well defined and exhibit large variations, even for devices fabricated under the same conditions [36].

By using a chemical solution deposition method, NiO thin films can be synthesized as a metal in its virgin state for ReRAM devices, where the subsequent electroforming step is not necessary [42]. This metallic state is achieved by deliberate carbonyl (CO) doping of NiO thin films. It may be recalled here that NiO is not a Mott–Hubbard insulator but a charge transfer insulator (see section 4.2), where the on-site electron–electron interaction energy or Hubbard U is greater than the charge-transfer energy Δ of electron transfer from the $3d^n$ to $3d^{n+1}$ energy level via intermediate oxygen ligand. The ReRAM switching mechanism in such NiO devices can be explained in terms of a critical electron population described within the Mott–Hubbard picture involving Hubbard U and charge transfer energy Δ.

Figure 8.18(a) shows the schematic I–V characteristics of the low resistance (ON) and high resistance (OFF) states of a NiO ReRAM model device. It may be noted that the virgin state of the device is a metallic ON state where the strong screening effect due to high electron concentration reduces the electrostatic Coulomb interaction. In figure 8.18(a) it is counter intuitive at first sight that an increased applied voltage causes a metal to insulator transition. In a non-equilibrium situation of the device there can be a variation in local electron concentration inside NiO [41]. During the process of establishing a stable current in the device, electrons near the anode are extracted out while electrons from the cathode enter the region adjacent to it. This causes variations in the net electron concentration in NiO component until the current is stable under a certain applied voltage (see figure 8.19(a)). The higher the applied voltage, the stronger would be the electron deficit near the anode. Eventually a threshold voltage V_{RESET} is reached, when the electron concentration n_c in this region fulfills the Mott criterion $n_c^{1/3}a_0 \sim 0.25$ for metal–insulator transition and the total current in the device drops sharply (see figure 8.18(a)). This conjecture that the metal to insulator transition occurring only in the anodic region gets experimental support from the results of transmission electron microscopy [43]. From a microscopic point of view, there is sufficient separation of Fermi levels in the leads of the ReRAM device; this along with the average quasi-Fermi level through the device is shown in figure 8.19(b). We have seen in section 4.2 that in NiO the p band of oxygen appears at energy between the lower and upper Hubbard bands. The p band strongly hybridizes with the d band, and there is an energy gap Δ

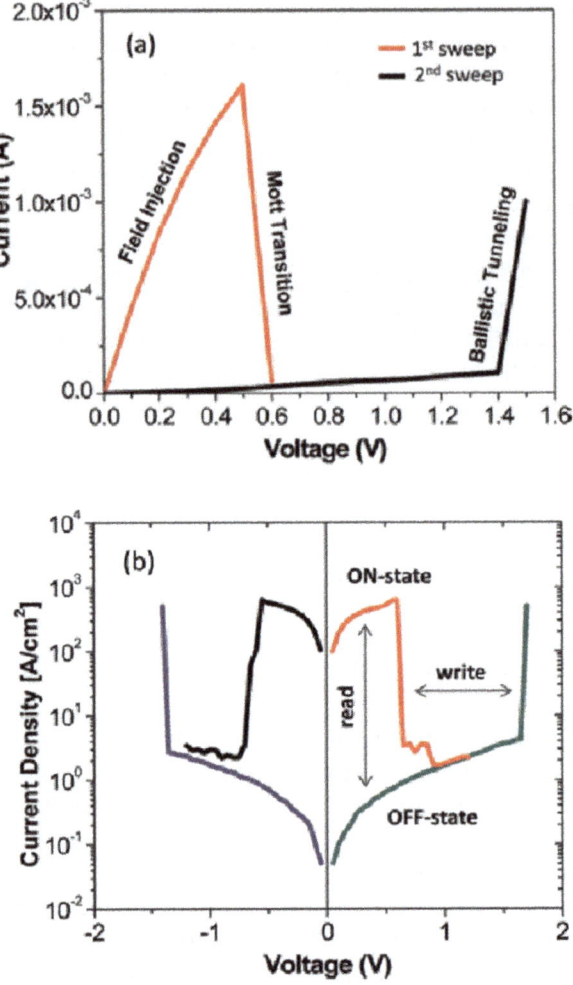

Figure 8.18. (a) Schematic diagram of a NiO ReRAM model device with proposed mechanisms for switching; (b) Unipolar resistive memory switching of NiO films showing wide writing and reading memory windows. (Reprinted from [42] with the permission of American Institute of Physics.)

between the p band and the upper Hubbard band. The threshold voltage V_{RESET} is expressed in terms of this charge transfer energy gap Δ as $V_{RESET} = \Delta q$, where q is the elementary charge [42]. In this charge transfer insulating state of NiO $U > \Delta$, and with further increase in the applied voltage an insulator to metal transition takes place at another threshold voltage $V_{SET} = Uq$. A high tunneling current sets in when the applied voltage reaches V_{SET}, and it compensates the electron deficit and recovers the electron concentration in the region near the anode. Figure 8.18(b) shows the unipolar resistive memory switching of a NiO film.

We will now present a nice example of how high density non-volatile memory can be achieved by combining two Mott insulators: (1) NiO, which acts as a memory element storing data through bistable resistance switching, and (2) VO$_2$, which acts

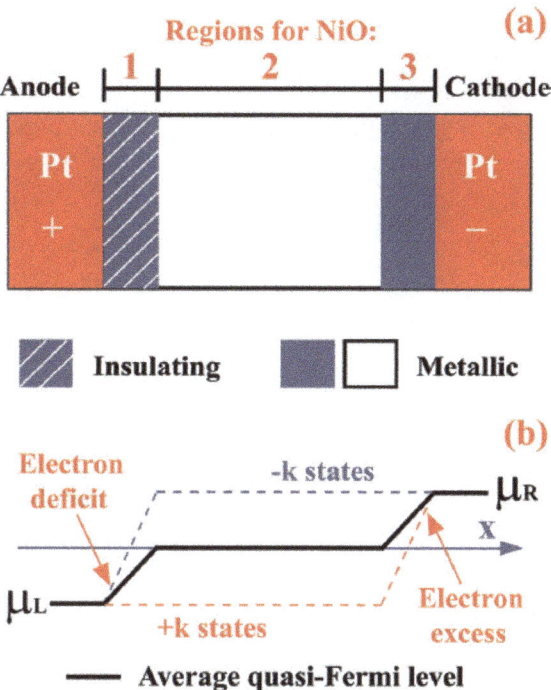

Figure 8.19. (a) Schematic diagram of NiO ReRAM device structure showing three regions in NiO. (b) Quasi-Fermi levels through the device at the threshold voltage V_{RESET}; bold line represents the average quasi-Fermi level. (Reprinted from [41] with the permission of American Institute of Physics.)

as a switch element. The conceptual architecture of this memory device with a three-dimensional stack cross bar structure is shown in figure 8.20(a), where one bit cell of the array consists of a memory element and a switch element between conductive lines on top (word line) and bottom (bit line). If a memory cell is to be accessed randomly, each memory needs connection with a switch element to prevent reading interference between neighboring cells. Figure 8.20(b) illustrates what happens in a simple 2×2 cross-point cell array without any such switching element. While reading the information of the cell in the high resistance state (HRS) surrounded by three cells in the low resistance state (LRS), the reading current can easily flow through the surrounding LRS cells, and hence gives rise to erroneous LRS information. Incorporation of a VO_2 switch element with rectifying properties, such alternate paths are eliminated by applying an appropriate voltage to all other unselected cells. The reading current now only flows through the cell selected to be accessed (see figure 8.20(c)). In summary, the interference between neighboring cells is prevented by connecting the NiO memory elements to the VO_2 switch elements. A schematic of the individual Pt–NiO–Pt–VO_2–Pt cell structure is shown in figure 8.20(d). Programming characteristics of combined Mott-insulator switch and memory elements are presented in figure 8.21, highlighting both the bi-stable resistive switching of the Pt/NiO/Pt memory element and the threshold switching of the Pt/VO_2/Pt switch element [44]. The cell remains inactive in region (a) since the

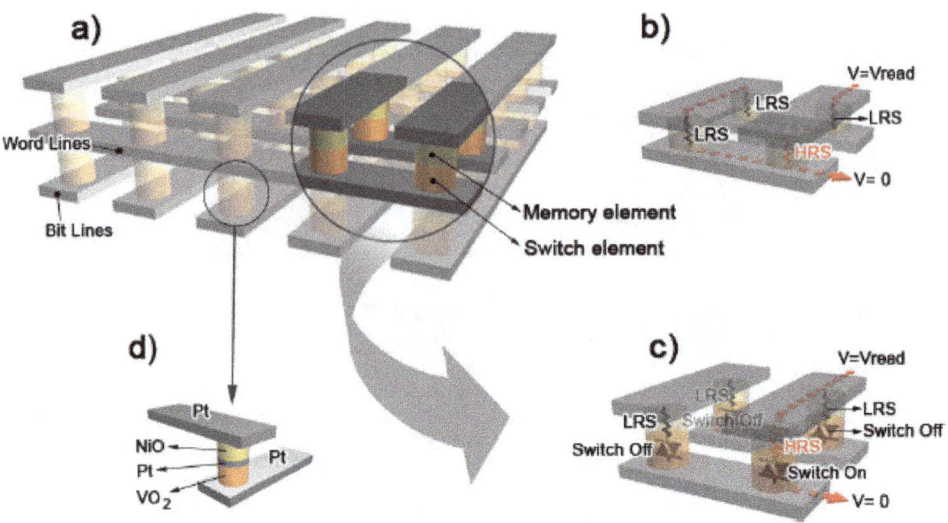

Figure 8.20. (a) Schematic presentation of a cross bar memory structure, where one bit cell of the array consists of a memory element and a switch element between conductive lines on top (word line) and bottom (bit line). (b) Reading interference in an array consisting of 2 × 2 cells without switch elements. (c) Reading operation in an array consisting of 2 × 2 cells with switch elements. (d) Detailed structure of a single cell consisting of a Pt/NiO/Pt memory element and a Pt/VO$_2$/Pt switch element. (Reproduced from [44] with the permission of John Wiley & Sons.)

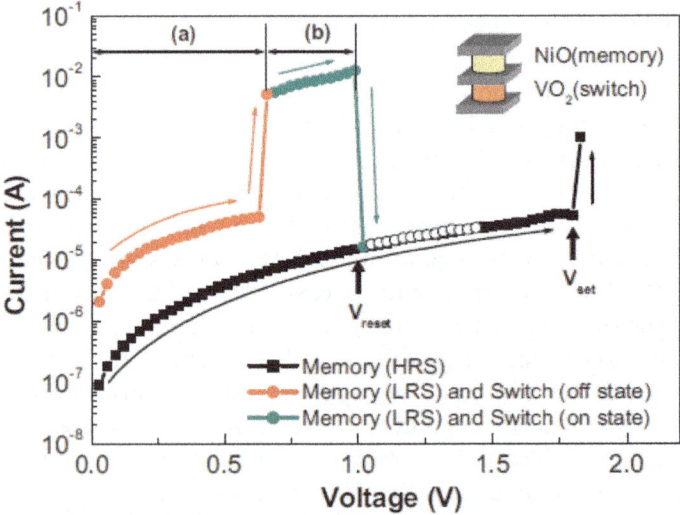

Figure 8.21. Programming characteristics of combined NiO switch and VO$_2$ memory elements. (Reproduced from [44] with permission from John Wiley & Sons.)

switch element is in the OFF state. The cell becomes active in region (b), because the switch element is in the ON state. The stored information is read by applying an appropriate reading voltage in region (b). The cell can be accessed and programmed by applying a writing voltage comparable to V_{SET} or V_{RESET}, since both are higher than threshold voltage. In this way one can exclusively access a single cell by applying read or write voltage to that cell, while maintaining a voltage belonging to region (a) in all the other cells. This kind of memory structure can have extremely fast programming speed of several tens of nano-seconds due to the fast resistance switching property of Mott-insulators.

The narrow gap Mott insulator compounds AM_4Q_8 (where A = Ga or Ge; M = V, Nb or Ta; and Q = S or Se) exhibit both a volatile and a non-volatile unipolar resistive switching [27], and this behaviour is rationalized within the model of resistor network with competing insulator and metal phases (see section 8.3.1). The volatile resistive switching originates from an avalanche breakdown mechanism. The switching is triggered above a threshold electric field of a few kV cm^{-1} through the creation of a metallic/conductive percolating path. The number of metallic sites continues to grow after the creation of the percolating metallic filament as long as the electric field is kept on, and this leads to an increase in the filament diameter. Since Mott metal–insulator transition is a first order phase transition, the framework of nucleation and growth processes is readily applicable here. The stabilization of an equilibrium phase becomes possible only above a critical size. Hence a stabilized metallic filament of a suitable size achieved after application of electric field pulses of strength considerably greater than the threshold field for volatile resistive switching gives rise to the non-volatile resistive switching. The application of a very long electric pulse can be used to destroy the nonvolatile metallic filaments and bring back the material to a value very close to the virgin state i.e. the state before application of any electric field [27]. A Au/GaV$_4$S$_8$/Au metal–insulator–metal structure involving GaV$_4$S$_8$ polycrystalline thin layer is shown in figure 8.22(a), where a non-volatile resistive switching could be induced using short electric pulses in the 500 ns–10 μs range [27]. The resistance versus temperature plot of the device (see figure 8.22(b)) shows a change from an insulating state in the virgin condition to a conductive state after application of the electric field pulse. A substantial difference between the two states still remain at room temperature. Using a voltage pulse protocol involving a series of seven identical short pulses of large amplitude to generate the conducting state and a single long and low amplitude pulse to get back the insulating state, a reversible switching between the insulating and conducting states was observed in this Au/GaV$_4$S$_8$/Au structure. The stability of high and low resistive states has been investigated at room temperature, and both the states exhibit good retention time (see figures 8.22(c) and (d)), which is essential for the data storage. Overall this kind of resistive switching, which is also possible in other canonical Mott-insulators like $(V_{1-x}Cr_x)_2O_3$ and $NiS_{2-x}Se_x$, has promising features such as a resistive switching ratio exceeding 10^3, cycling endurance reaching more than 65,000 resistive switching cycles, data retention time till 10 years and writing speed below 100 ns [27].

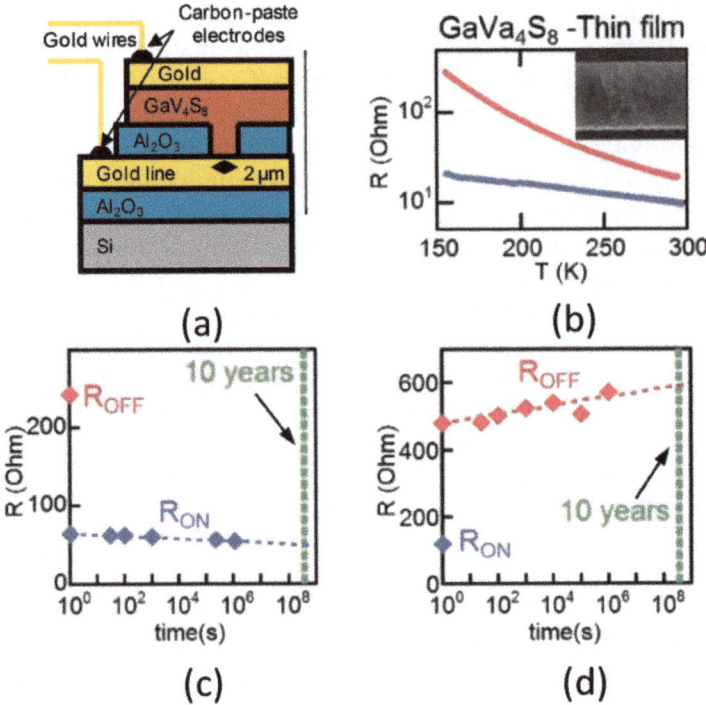

Figure 8.22. (a) Schematic diagram of a Au/GaV$_4$S$_8$/Au metal–insulator–metal structure; (b) comparison of the temperature dependence of high (magenta line) and low (blue line) resistive states; (c) and (d) resistance versus time plot for the low and high resistance states, respectively. (Reproduced from [27] with the permission of John Wiley & Sons.)

8.4.2 Mott memristor and neuristor

Standard electrical circuits consist of three fundamental two-terminal passive elements namely resistors, capacitors and inductors. The function of such electrical circuits is defined in terms of four basic variables namely current (i), voltage (v), electrical charge (q) and magnetic flux (ϕ). The passive two-terminal elements can dissipate or store energy but cannot generate energy. The response of each of these circuit elements is defined by a simple linear relationship between two of the four basic variables (see figure 8.23). There are six possible mathematical relations between pairs of the four fundamental circuit variables i, v, q and ϕ [45]. Two of these arise from the definitions of charge and magnetic flux in terms of current and voltage, $q = i \int dt$ and $\phi = v \int dt$, respectively. Then there are three relations, which define the properties of three passive circuit elements: (1) resistance (R) is the rate of change of voltage with current, (2) capacitance (C) is the rate of change of charge with voltage, and (3) inductance (L) is the rate of change of flux with current. In 1971 Leon Chua pointed out that the remaining sixth possible relation linking magnetic flux and charge would lead to a fourth passive element, which he named the 'memristor' [46] (see figure 8.23). This new element memristor has now been realized

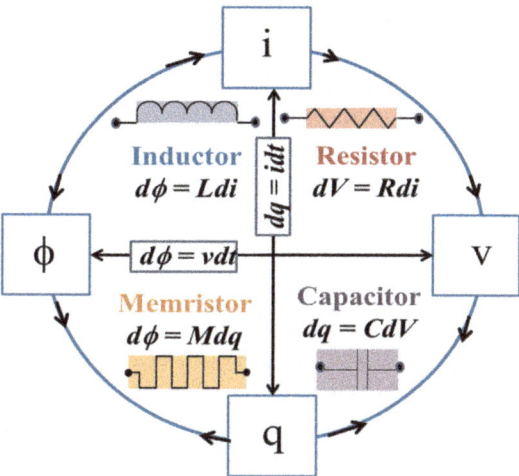

Figure 8.23. Schematic representation of the relations between four fundamental circuit variables: current, voltage, electrical charge, magnetic flux and four passive circuit elements resistor, capacitor, inductor and memristor.

physically in 2008 by a group of researchers from Hewlett-Packard [47, 48], and since then significant interest has been generated in this field by the predictions that such devices may play key roles in developing various kinds of novel devices including ultra dense information storage and neuromorphic circuits [49].

The memristor relates the current to the voltage, but the resistance of this two-terminal device depends on the entire dynamical history of the charge flowing in the system. This is in contrast with the ohmic resistance, which is associated with the instantaneous value of the applied voltage. Thus, the memristor is a resistor with memory, where the resistance depends on the state and history of the resistive element. The first memristive system reported [47, 48] was a layered Pt–TiO$_2$–Pt junction device exhibiting fast bipolar nonvolatile switching. The change in resistance in the device structure takes place due to the ionic motion of oxygen vacancies activated by current flow. The positively charged oxygen vacancies in the TiO$_2$ layer drifts under an applied electric field, which causes changes to the electronic barrier at the Pt/TiO$_2$ interface. Conducting channels are created when vacancies drift towards the interface, which shunt the electronic barrier to switch the device ON. On the other hand, the drift of vacancies away from the interface destroys such channels and recovers the electronic barrier to switch the device OFF. The oxygen vacancies in the device structure do not easily go back to their original position when the power source is turned off, and the device retains its new resistance state. In 2008 Chua and his collaborators generalized the concept of memory devices to capacitors and inductors by introducing two more memory devices namely 'memcapacitor' and 'meminductor', where the capacitance and inductance, respectively, depend on the state and history of the device [49]. However, in contrast with the memristor, a memcapacitor or a meminductor can store energy. In the context of the present book we shall now concentrate only on the properties of memristors.

If x denotes a set of n state variables describing the internal state of a system, an nth-order current controlled memristive system is described by the following equations [49]:

$$V_M(t) = R(x, I, t)I(t) \tag{8.15}$$

$$\dot{x} = f(x, I, t). \tag{8.16}$$

Here $V_M(t)$ and $I(t)$ represent the voltage and current across the device, and R is called the memristance with physical units of ohm. The memristive devices with $R(x, I, t) > 0$ are passive and they cannot store energy, which implies that $V_M = 0$ whenever $I = 0$. Furthermore, if $\dot{x} = f(x, I, t) = 0$ has a steady-state solution, a memristor would behave as a linear resistor in the limit of infinite frequency and as a non-linear resistor in the limit of zero frequency. At very low frequencies, irrespective of the underlying physical mechanisms defining the state of the memristor, it will have enough time to adjust its value of resistance to an instantaneous value of the control parameter namely current or voltage and the memristor behaves as a non-linear resistor with non-ohmic response. There would not be enough time, however, at very high frequencies for any kind of resistance change during a period of variation of the control parameter, and the memristor behaves linearly as an usual resistor with ohmic response.

In thin film samples of suitable Mott insulators the memristive behaviour can be driven by the Mott insulator-to-metal transition, and we highlight this behaviour with the example of VO_2 based memristive device (see figure 8.24). The operating temperature of the device needs to be in the temperature regime near the Mott insulator–metal transition, where due to the first order nature of the transition the temperature dependence of the resistance shows distinct hysteresis (see figure 8.24). Electric pulses of 50 V are applied to take the device to a new resistive state, and the corresponding currents are monitored. The current (I) versus voltage (V) curve of this VO_2 device shows a nonlinear response for voltages above 20 V (see figure 8.24). As discussed above such a non-ohmic response is a typical characteristic of a memristor. The hysteresis associated with the first order Mott insulator to metal phase transition in VO_2 drives the memory aspect of the memristor. The memory in this VO_2 device remains between the pulses, and also when the applied voltage has been set to zero for some time. This behaviour is illustrated in figure 8.25, which shows that the slope of the I–V curve of each subsequent pulse picks up where the last pulse left off. This memory is expected to remain indefinitely in an ideal memristor, but all the Mott memristors studied so far exhibited finite storage time [50]. The proximity of the near room temperature Mott metal–insulator transition in VO_2 with large magnitude of resistivity, and the fact that both of these parameters can be suitably taylored through the thin-film growth and nano-patterning, makes the VO_2 memristive devices particularly attractive for practical

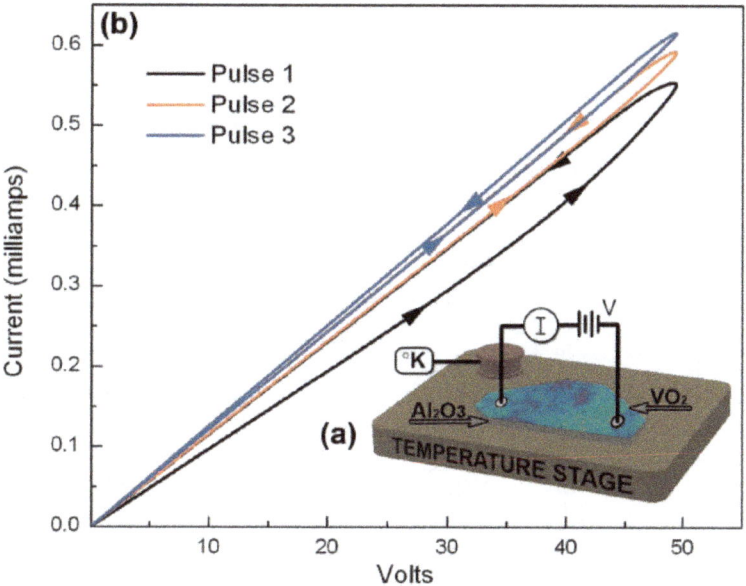

Figure 8.24. (a) Schematic of the VO_2 Mott-memristive device using a 25 mm^2 VO_2 thin-film. (b) Three current–voltage *I–V* curves for this VO_2 device showing the characteristic nonlinear hysteretic behavior of a memristive system. (Reprinted from [50] with the permission of American Institute of Physics.)

applications. In addition, Mott metal–insulator transition in VO_2 can also be triggered by photo-excitations and static electric field, which opens up the possibilities of memristive optoelectronics. However, this is a rather evolving field and various other potential Mott insulators are also currently under investigation. One such example is NbO_2, which will be discussed below in the context of some newer and novel applications, namely neuristor.

The artificial neural circuits are drawing considerable attention in recent times with a long-term interest of building computers mimicking the human brain. Information is generated in neurons and transmitted through electrical and chemical signals. Each neuron consists of: (i) a body cell, which generates electrical spikes, (ii) an axon, which sends signals to other neurons, and (iii) dendrites, which receive signals. On the other hand, synapse is a structure that permits the signal to transfer from the axon terminal of one neuron cell and the dendrites of another. The connection strength of synapse can be modulated, and this is believed to lay the foundation of memories. The electrical spiking signals in neurons are known as action potentials, which are caused by the change in the sodium (Na) and potassium (K) ion concentrations within and outside the neuron cell body. The Hodgkin–Huxley model [51] for action potential generation in biological axons is central for understanding the computational capability of the nervous system and emulating its functionality [52]. Na^+ and K^+-ion channels dynamically permit or prevent polarizing currents to charge or discharge through the cell membrane. These ion channels change conductance markedly if a cell body is sufficiently polarized through its dendritic inputs. A voltage spike or action potential is then triggered,

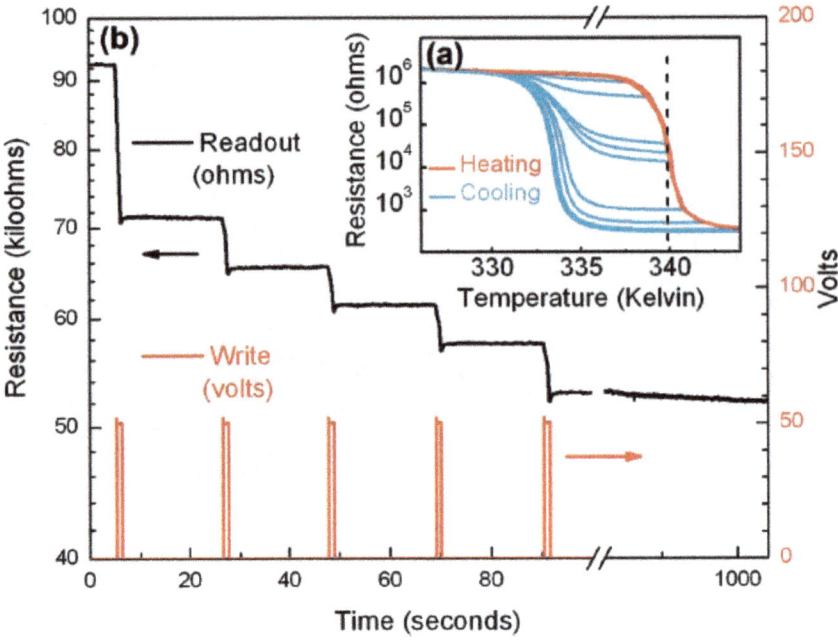

Figure 8.25. (a) Resistivity–temperature plots for the VO_2 Mott-memristor shown in figure 8.25(a) illustrating the hysteretic nature of the Mott insulator–metal phase transition. The vertical dotted line shows the temperature of the device operation. (b) Demonstration of information storage in the VO_2 Mott-memristor. Each 50 V pulse triggers the transition to a new resistivity. (Reprinted from [50] with the permission of American Institute of Physics.)

which travels along the axon. This process of spiking is considered to be the fundamental process of computation in biology [52].

From the viewpoint of circuit theory, the Na^+ and K^+-ion channels of the Hodgkin–Huxley model are mathematically similar to two distinct memristors. However, the neuromorphic computing based on conventional CMOS technology would require a large number of transistors to simulate single neuron or synapse. There is also inefficiency inherent in the CMOS technology in simulating brains, and that is due to the digital versus analog way, respectively, that CMOS circuits and the brain work. Mott insulators provide some promise here, and it is possible to mimic the behavior of neurons initiating action potentials by a neuristor built from two Mott threshold switches along with resistors and capacitors. This device can represent the features essential for action-potential-based computing: threshold-driven spiking, loss less spike propagation at a constant velocity with uniform spike shape, and a refractory period [52].

Figure 8.26 shows the circuit diagram of a neuristor, developed using two nominally identical and oppositely dc-biased NbO_2 Mott memristors, M_1 and M_2 each with two parallel capacitances C_1 and C_2. The two channels are energized with opposite polarity voltages, to mimic the Na^+ and K^+-ion channels of the Hodgkin–Huxley model, and they are coupled through a load resistor R_{L2}. In addition, the

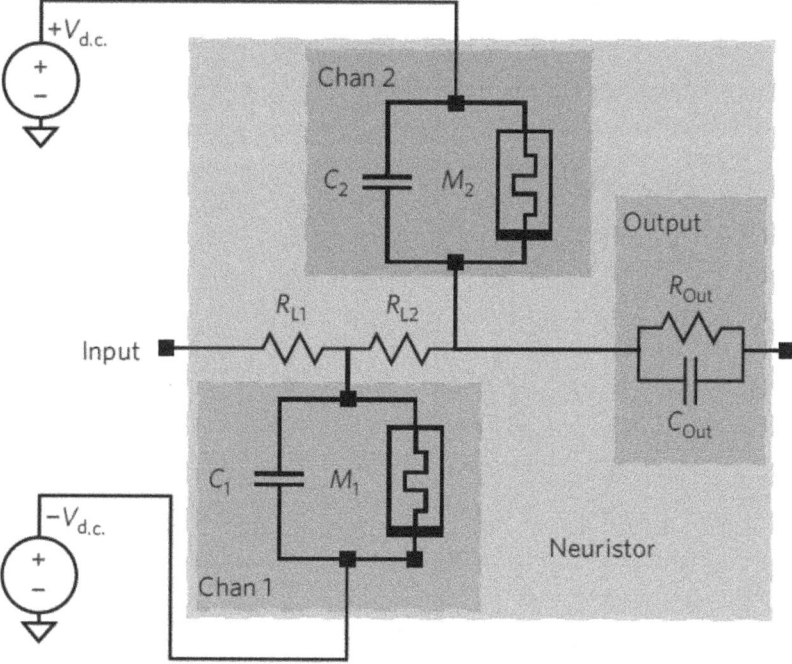

Figure 8.26. Schematic of a Mott neuristor circuit made out of two NbO$_2$ based Mott memristors and two normal capacitors. (Reproduced from [52] with the permission of Springer Nature.)

circuit consists of an input resistance R_{L1} and output impedance (R_{Out} and C_{Out}). The capacitors are charged by the input current until the voltage across M$_1$ and M$_2$ reaches the value of their threshold voltage for Mott insulator–metal transition. Beyond this threshold voltage the NbO$_2$ Mott insulator goes into the metallic phase, and the capacitors C_1 and C_2 get discharged. The opening and closing of the ion channels can be correlated here to the turning ON and OFF, respectively, of the Mott switches. In analogy to the biological systems, the offset between the charging/discharging time of the capacitors results in action potential voltage spikes. The response of the neuristor device under two different stimulating input pulses is illustrated in figure 8.27. The temporal profile of voltage output looks like that of biological action potentials when the input voltage to the device is larger than the threshold voltage i.e. super-threshold, and there is a maximum gain in the output voltage (see figures 8.27(a) and (b)). On the other hand, the output voltage is low when the input voltage is below the threshold voltage i.e. sub-threshold (see figures 8.27(c) and (d)). This response of the Mott neuristor mimics the all-or-nothing rule for neurons [52]. Figure 8.28 shows how the control of the inter-spike timing intervals and the spike width can be achieved with a different combination of the value of capacitances C_1 and C_2. Such Mott neuristors are particularly interesting from a technological point of view, since they switch rapidly in less than 1 ns time with low transition energy (less than 100 fJ), scale at least to tens of nanometers, and are compatible with conventional CMOS materials and processes [52].

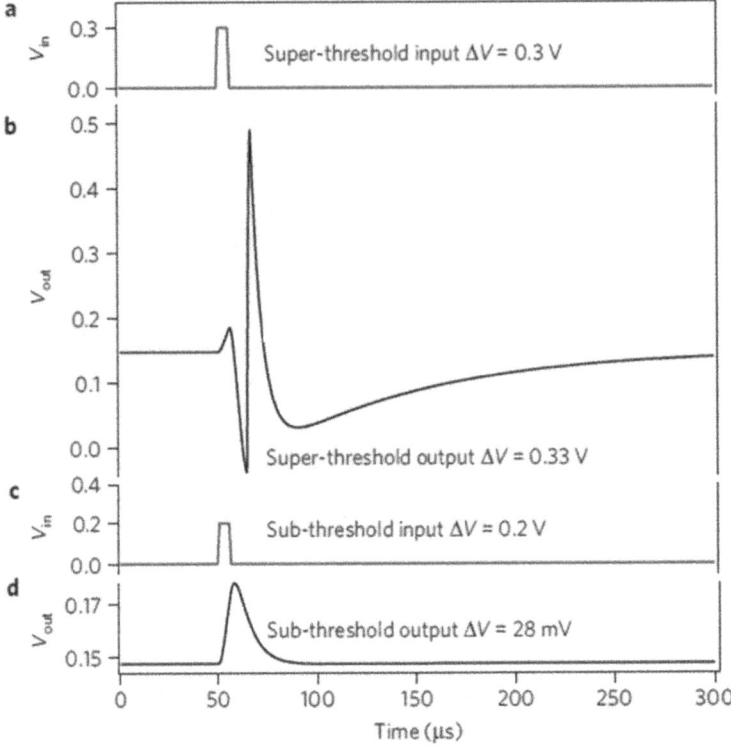

Figure 8.27. Schematic diagram showing simulated all-or-nothing response of the Mott neuristor: (a) and (b) superthreshold input and the corresponding spike output; (c) and (d) subthreshold input and the corresponding attenuated output. (Reproduced from [52] with the permission of Springer Nature.)

8.4.3 Mott field effect transistors

We describe here a VO_2 based Mott-FET reported in one of the early works on Mott devices [37]. This device structure consists of a VO_2 channel, a gate dielectric layer on top of this VO_2 channel, a gate electrode on top of the gate dielectric layer, and source and drain electrodes at both ends of the channel. The operating temperature of the device is set above the Mott metal–insulator transition temperature of VO_2. Without application of the gate voltage the VO_2 channel remains in the metallic state, but goes to the dielectric state once a gate voltage is applied. The situation is similar to a MOSFET working in the depletion mode. Figure 8.29 shows the schematic of this VO_2 based Mott-FET, where VO_2 thin films were deposited on SiO_2/Si substrate using the method of laser ablation [37]. In this Mott-FET, the SiO_2 layer underneath the VO_2 thin film serves as the gate oxide, and Au/Cr and WSi were used for the source–drain and gate contacts, respectively. Figure 8.30(a) shows the source–drain current density versus electric field applied in the VO_2 channel at room temperature in the open gate condition. The hysteresis loops are characteristic of the first-order nature of Mott metal–insulator transition, and it indicates that the device is turned ON and OFF at effective electric fields of ~4.8 and ~2.8 MV m^{-1}, respectively. The effect of the gate voltage on the device can be seen in figure 8.30(b),

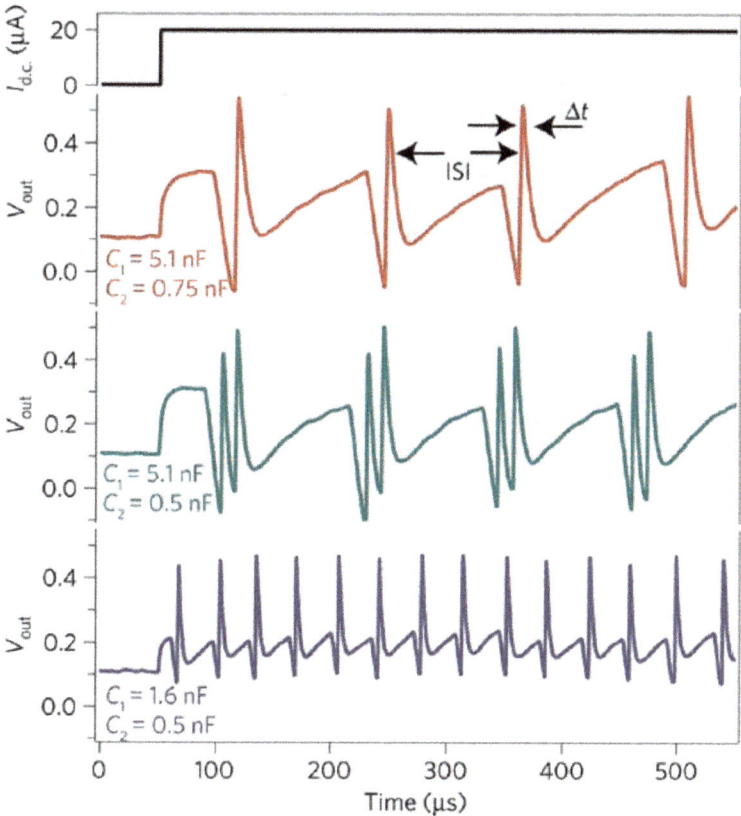

Figure 8.28. Various types of spike patterns that can be achieved in single Mott neuristor circuit by adjusting the capacitor values in the circuit. (Reproduced from [52] with the permission of Springer Nature.)

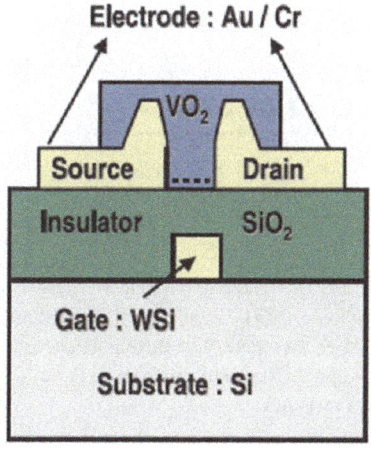

Figure 8.29. Schematic diagram of a VO_2-based Mott-FET structure with SiO_2 as gate insulator. (Reproduced from [37]. (c) 2004 IOP Publishing Ltd. Deutsche Physikalische Gesellschaft. All rights reserved.)

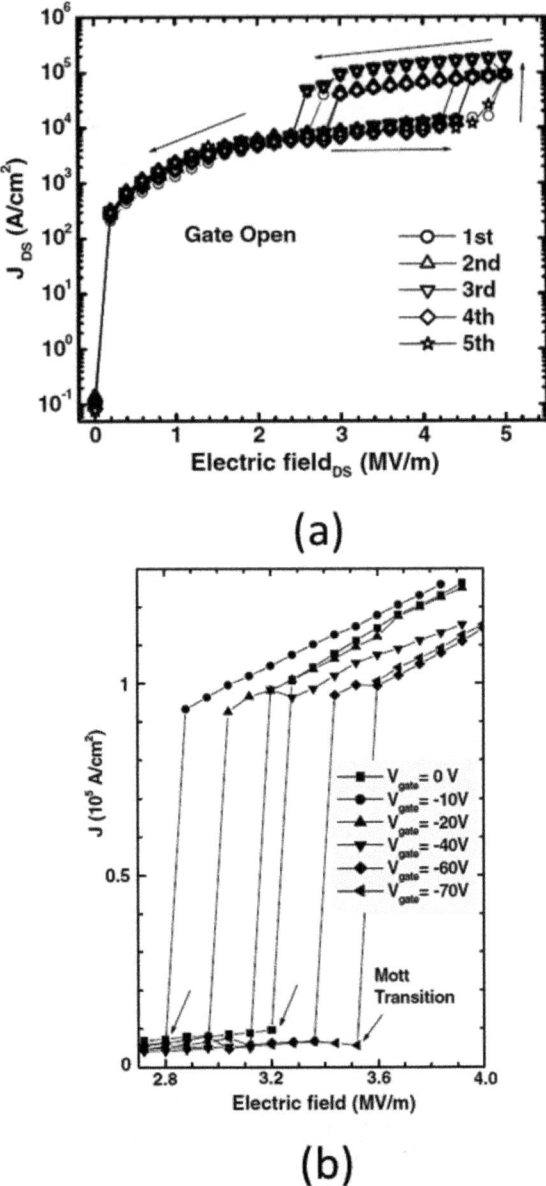

(a)

(b)

Figure 8.30. (a) Drain to source current (J_{DS}) versus applied electric field (E_{DS}) plot for the VO$_2$-based Mott-FET structure in the absence of any gate voltage. (b) J_{DS} versus E_{DS} (highlighting the region near the insulator to metal transition) in the presence of applied gate voltage. The threshold electric field value increases with increasing negative gate voltage from $V_g = -10$ V. (Reproduced from [37]. (c) 2004 IOP Publishing Ltd. Deutsche Physikalische Gesellschaft. All rights reserved.

where the source–drain current density is plotted as a function of applied electric field, under different applied negative gate voltages. The threshold voltage for the Mott metal–insulator transition i.e. the turn-ON voltage of the Mott-FET increases with the magnitude of the negative gate bias voltage. This is attributed to the decrease in conductivity due to increased hole carrier density and the disappearance of electrons in the VO_2 channel [37].

There have been numerous studies of pure electrostatic effect on the VO_2 based Mott-FETs since early 2000. The influence of the fabrication of multiple components of the device, including the gate oxide deposition on the VO_2 film characteristics has been investigated in details (see [53] and [54]). However, only small changes in the VO_2 channel resistance and/or the temperature of Mott metal–insulator transition could be achieved in such devices, possibly due to the insufficient electric field available with conventional solid dielectrics. The metallic nature of the VO_2 channel material introduces limitations on the external field penetration, and limits the conducting layer to a surface layer of thickness to the order of the Debye screening length. The estimation of the Debye screening length in VO_2 suggests that 1–3 nm thick high quality VO_2 thin-films will be required for the desired device performance. Hence, the main difficulty in implementing the VO_2-based Mott FET lies in the small thickness of the metal phase layer, and the resulting low magnitudes of the electric field induced impedance change at the Mott metal–insulator transition.

In the VO_2 Mott-FET structures the magnitude of the electric fields provided by conventional gate dielectrics was limited by their dielectric properties. To go beyond this limit of conventional electrostatic method, electric-double-layer technique involving an organic ionic liquid has been employed to develop VO_2 Mott-FET structure [55]. Ionic liquid is a salt with organic cations and inorganic anions, which at the temperature of Mott-FET operation is in a molten or in glass-like state. Unlike an ionic solution, an ionic liquid does not contain solvents but only cations and anions, and has comparatively large ionic conductivity but small electronic conductivity. The ionic liquid covers the channel surface, source and drain of the Mott-FET. It is also in contact with a gate electrode, which is isolated from the channel. Cations (anions) accumulate at the liquid/solid surface when a positive (negative) bias is applied to the gate electrode with respect to source/drain. This in turn will result in electrostatic doping of electrons (holes) into the FET channel, provided there are no electrochemical reactions. The accumulated ions in ionic liquid form a layered structure at the electrolyte–oxide interface. This is known as the electric double layer (EDL), and the electric field of this charge layer creates an electron-enriched region in the VO_2 channel, which leads to the Mott metal–insulator transition [55]. It has been suggested that electrostatic surface charge accumulation is accompanied by a collective lattice deformation in the VO_2 channel along the c-axis direction, which results into delocalization of previously localized electrons in the bulk VO_2 film, leading to a three-dimensional metallic ground state with high carrier density throughout the film. This event is shown schematically in figure 8.31. The detailed investigation involving VO_2 channels with thickness varying between 10 to 70 nm showed that the electrostatic surface charge accumulation could trigger carrier delocalization in the bulk VO_2 film irrespective

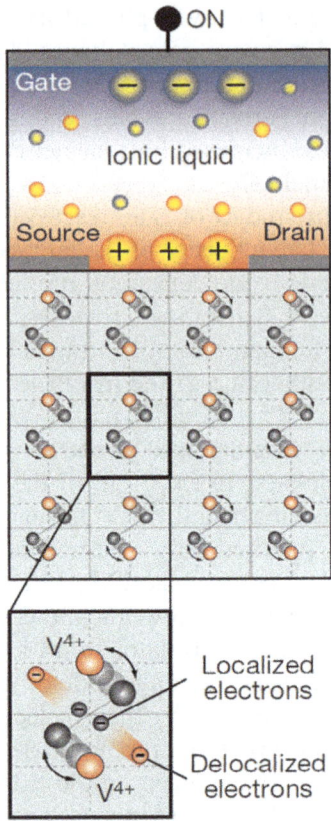

Figure 8.31. Schematic diagram of an ion liquid gated VO_2 based Mott-FET. (Reproduced from [55] with the permission of Springer Nature.)

of the film thickness, and beyond the fundamental limit of the screening effect [55]. The resultant change in channel resistance was up to 50% at a gate voltage of +2 V with suppression of the MIT to temperatures below ~10 K. However, in a subsequent work it was claimed that the metallic state was rather induced by the electric field-induced migration of oxygen from the VO_2 channel into the ionic liquid [56]. An important question remains here, whether the ion liquid gating results in a structural phase transition [55] or the initially insulating VO_2 film continues to be in the monoclinic phase and the metallic behaviour rather originates from the formation of the oxygen vacancies. This question has now been addressed subsequently with *in situ* synchrotron x-ray diffraction and absorption experiments, which reveal that the whole VO_2 film on gating undergoes giant structural changes with lattice expansion up to ~3% near room temperature [57]. These structural changes are fully reversible on reverse gating, and is in striking contrast to the 10 times smaller contraction VO_2 undergoes in the thermally induced insulator–metal transition. The reversible structural changes, however, take place only in VO_2 structures in which channels evolved from chains of edge-sharing VO_6 octahedra not lying in the plane of the films. This observation suggests that these channels are the

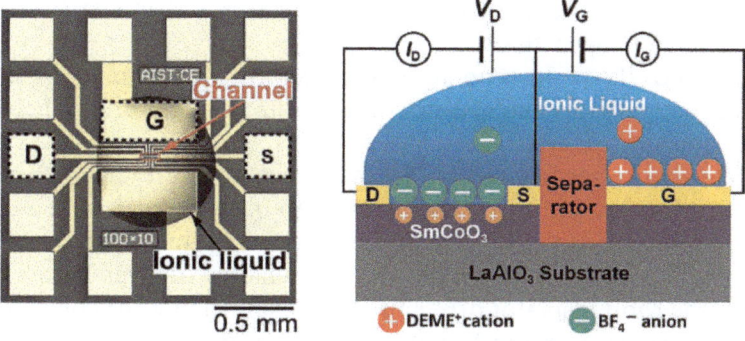

Figure 8.32. Optical micrograph of a SMCoO$_3$-based electric-double-layer Mott-FET showing the config-uration of the channel, ionic liquid, and electrodes (left), and a schematic cross-sectional view of this Mott-FET (right). The source, drain, and gate electrodes are denoted as S, D, and G, respectively. (Reproduced from [58] with the permission of John Wiley & Sons.)

paths along which the gate-induced oxygen migration to ionic liquid takes place [57]. In summary the metal–insulator transition in the VO$_2$ channel may be understood in the following way: the surface layer of the VO$_2$-channel within the screening length first becomes metallic along with a structural transition, and the subsequent propagation of this structural deformation from the surface to the bulk region of VO$_2$ material then causes the delocalization of electrons in the entire bulk region. However, it needs to be assessed whether the large amount of stress produced during this process is actually detrimental for device applications or not.

The electric-double-layer (EDL) technique involving an organic ionic liquid has also been employed to develop Mott-FET structure with SmCoO$_3$, which is a Mott-insulator with a non-magnetic ground state and shows a temperature-driven Mott metal–insulator transition without any accompanying structural transition [58]. Figure 8.32 shows the optical micrograph and a schematic cross-sectional view of such SmCoO$_3$-based EDL Mott-FET. This EDL Mott-FET showed a switching ratio of as large as two decades at room temperature. The experimental results suggested that the resistive switching in this Mott-FEL originated purely from the electrostatic control of the crossover from an insulating channel to a barely-insulating channel caused by hole doping (akin to band filling control) of this Mott insulator SmCoO$_3$ [58]. Unlike in the case of VO$_2$, the electromigration of oxygen ions or vacancies is not involved in the resistance change. From the thickness dependence of the drain current I_D in the ON state, the screening length was estimated to be about 5 nm, which suggested that this was indeed a case of a two-dimensional Mott-transistor.

The are several technological challenges on the feasibility of EDL-FETs as regular commercial devices. The liquid form of the gate is a basic disadvantage to start with, which gives rise to concerns on stability, scalability and integration [17]. Relaxation time in EDL-FETs is also longer in comparison with conventional oxide dielectric based FETS due to the slow motion of ions, which is not very good for ultrafast switching applications. The EDL-FETs may not also be very suitable for high-temperature applications because of the degraded chemical stability of ionic liquids at elevated temperatures.

Going beyond the conventional oxide dielectric and ionic liquids as gate materials, Mott transistors can also be conceptualized in the form of hetero-junction modulation-doped Mott transistor. This is the Mott counterpart of a conventional junction field-effect transistor (JFET) or hetero-structure field-effect transistor (HFET). The Mott-insulator in such a device can be modulated with a degenerately doped conventional band insulator. In a modulated Mott-FET or MMFET, an introduction of modest space charge through charge transfer across the hetero-junction into the Mott insulator drives the Mott channel into the conducting state. A voltage applied across the interface controls the electron transfer from the doped band insulator to the Mott-insulator and initiates transistor action by inducing Mott insulator-to-metal transition. A MMFET is sometimes called a 'charge gain' device, where gating only 1%–10% of the channel electrons releases an entire conduction band of electrons in the Mott material [59]. There has been some experimental works involving $SrTiO_3$ and rare-earth nickelates like $NdNiO_3$ to assess the potential of the MMFET, but significant challenges still remain on the way to develop a robust MMFET structure.

8.4.4 Thermal and chemical sensors

The thermally induced sharp change in the electrical resistivity of VO_2 across the Mott metal–insulator transition can be used in thermal switching and detection. We will discuss this here with the example of a programmable critical temperature sensor where the applied voltage is controlled by a program [60]. The two-terminal device consists of ~100 nm VO_2 thin film channel (dimension 50 μ × 20 μ), grown on Al_2O_3 substrate by the sol–gel method, and nickel metal electrodes. Figure 8.33 shows the temperature dependence of the electrical conductivity of this VO_2 thermal sensor device. With the change in applied voltage from 1 to 22 V, the jump in the temperature dependence of conductivity becomes abrupt when the applied voltage is above 5 V. The Mott metal–insulator transition temperature gradually shifts from 68 °C at bias voltage of 1 V to room temperature at a bias voltage of 21 V. The distinct linear region of the conductivity versus temperature plot after the abrupt jump (see figure 8.33) was attributed to the intermediate phase of VO_2 between the insulator–metal transition and the higher temperature structural phase transition. A prototype thermal sensor was demonstrated using this characteristic property of the variation of the metal–insulator transition temperature of VO_2 with the applied bias voltages [60]. The possibility also exists for a thermal capacitor device with a VO_2 thin film as an active layer, which can have more than one order of magnitude capacitance change from room temperature to 100 °C due to the dielectric constant change of VO_2 across the metal–insulator transition [32].

The sensitivity of the metal–insulator transition and the associated change in conductance of VO_2 single crystal nanowire to the small changes in molecular composition, pressure, and temperature of the ambient gas environment have been utilized to demonstrate a gas sensor thermistor device [61]. This device

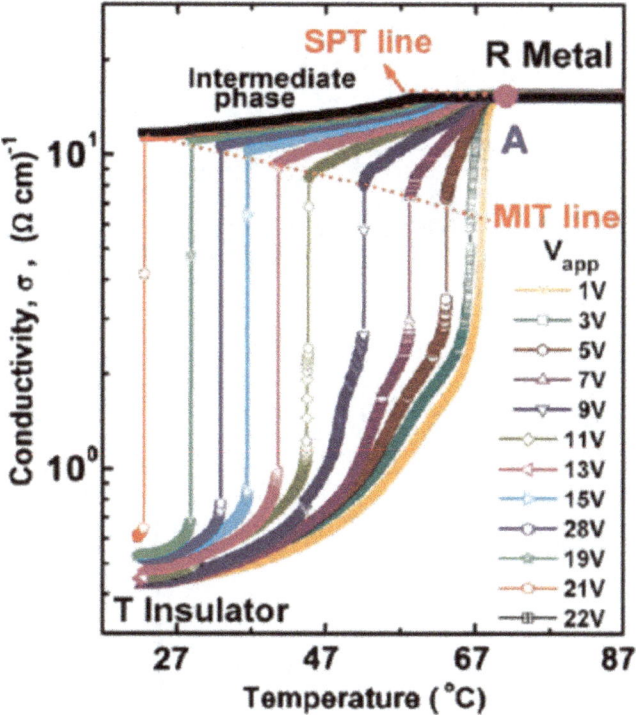

Figure 8.33. Temperature dependences of the conductivity of a programmable VO$_2$-based temperature sensing device at constant applied voltages. (Reprinted from [60] with the permission of American Institute of Physics.)

(see figure 8.34(a)) is made up of VO$_2$ nanowires grown on a SiO$_2$/Si substrate, and droplets of liquid metal alloys (Ga–In–Sn) act as soft contacts on both ends of the VO$_2$ nanowire. An externally applied periodic voltage bias sustains the Joule self heating of the device. Figure 8.34(b) presents the *I–V* characteristics of this gas sensor device under different argon gas pressures. Thermal loss of the gas sensor to the environment increases with argon gas pressure, and causes a shift of the threshold voltage for Mott insulator–metal transition to larger values. This leads to a significantly different *I–V* curve for different gas pressures.

8.4.5 Photovoltaic applications

A conventional band semiconductor based solar cell operates on the same principle as that of a photodiode, which is a *p–n* junction (see section 7.4) operated under the condition of reverse bias. When an EM wave penetrates to the depletion region of the photodiode, the electric field in that region separates the electron–hole pair generated by the EM wave. As a result, an electric current known as photo-current flows in the external circuit. The photo generated holes drift in the depletion region, diffuse into the neutral *p* region, and then combine with electrons entered from the negative electrode [14]. In a similar manner photo-generated electrons drift in the

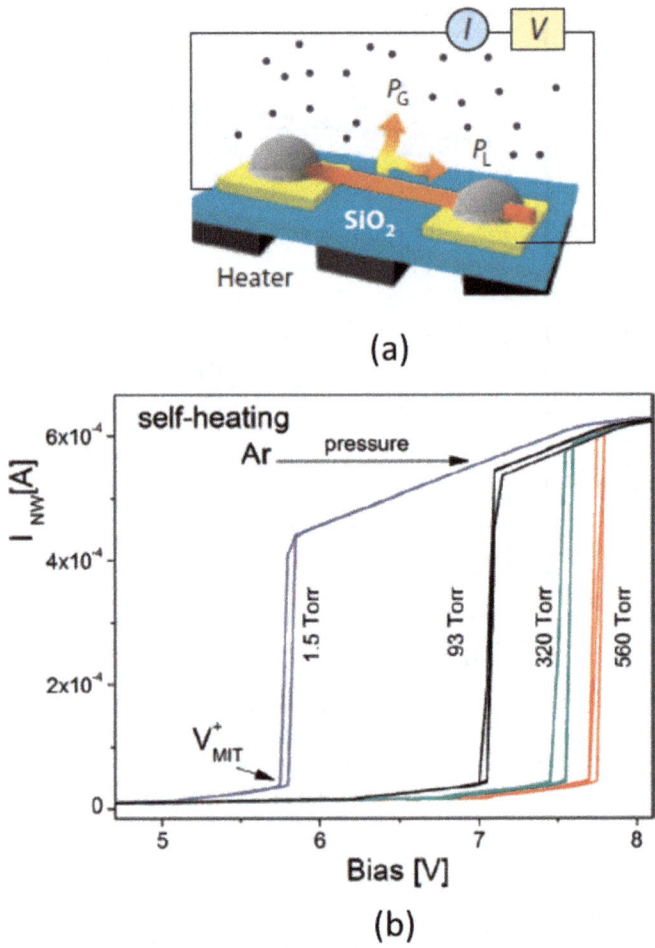

Figure 8.34. (a) Schematic of a VO$_2$ nanowire gas sensor thermistor device. (b) *I–V* characteristics of the gas sensor device at different Ar pressures. (Reprinted with permission from [61]. Copyright 2009 American Chemical Society.

opposite direction. On the other hand, the EM wave penetrating within a diffusion length outside the depletion region will also generate carriers, which will diffuse into the depletion region and drift across the depletion region to the other side. These neutral regions can be regarded as resistive extensions of electrodes to the depletion region. The photocurrent will depend on the number of electron–hole pairs generated and the drift velocities of the carriers. The depletion layer in a *p–n* diode needs to be sufficiently thick to allow a large fraction of the incident light to be absorbed so as to increase the quantum efficiency, but also thin enough to reduce the transit time, especially for high-frequency operation. Thus, there is a trade-off between the response speed and quantum efficiency [14].

A solar cell is, however, a large-area device and covers a wide range of the optical spectrum of solar radiation. Nowadays, a solar cell is a major candidate for

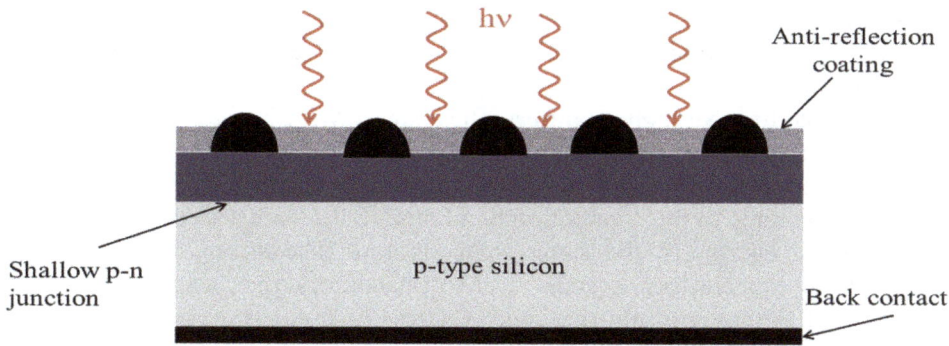

Figure 8.35. Schematic diagram of a silicon p–n junction solar cell.

sustainable energy source, because it converts sunlight directly to electricity with good efficiency and without any environmental hazard. A p–n junction solar cell (see figure 8.35) consists of (i) a shallow p–n junction formed on the surface, (ii) a front ohmic contact stripe and fingers, (iii) a back ohmic contact that covers the entire back surface, and (iv) an anti-reflection coating on the front surface. In a solar cell only the photons that have energy greater than E_g contribute energy E_g to the cell output, and the energy greater than E_g is wasted as heat. The electron–hole pairs created in the depletion layer become separated by the built-in electric field. The energy gap of the semiconductor determines the built-in-voltage, which limits the potential difference. Since the solar spectrum is limited, the current generated by solar radiation decreases with the increase in E_g in the semiconductor material.

There are two categories of solar cells: wafer-based and thin-film solar cells [14]. Silicon is the most important semiconductor for solar cells, and the silicon passivated emitter and rear locally-diffused (PERL) cell with reported conversion efficiency of 24% is a widely used solar cell design. On the other hand III–V compound semiconductors and their alloy systems provide wide choices of bandgaps with closely matched lattice constants, and they are ideal for producing tandem solar cells with reported efficiency reaching up to 30% in GaInP/GaAs tandem cell. The biggest problem with the conventional Si solar cell involving a thick layer of single crystalline silicon is cost, because such silicon material is an expensive commodity. In this regard the solar cell made with amorphous silicon thin films deposited directly on low-cost large-area substrates with reported efficiency of ~15% is a viable lower-cost alternative. A wide band gap semiconductor copper indium diselenide (CIS)-based solar cell with a conversion efficiency of 6% was reported in the early 1970s, and the efficiency was increased to 10% in the CdS/CIS solar cell in the early 1980s. Subsequently with the partial substitution of Ga in CIS and thus using Copper indium gallium diselenide (CIGS) with larger optical bandgap, the conversion efficiency in CdS/CIGS solar cells has increased steadily through the 1990s from 15% to presently at more than 20%. The conversion efficiency of CIGS depends on the substrate on which CIGS polycrystalline thin films are deposited, and the best efficiency seems to have been achieved in glass substrates. Extensive research is currently focused on the third generation photovoltaic cells involving dye-sensitized solar cells (DSSC), organic solar cells and quantum-dot solar cells, with the aim to provide higher efficiency and lower cost per watt of electricity generated [14].

It is possible to have high quantum efficiency in Mott-insulators [62, 63], which makes them suitable for photovoltaic applications. In Mott insulators the photo-excited electron utilize its strong Coulomb interaction with another valence electron to promote the same to the conduction band, and thus create a second electron–hole pair using the energy of the same solar photon. The strong electron–electron interaction in the electron system of Mott-insulators then can lead to a fast conversion of the initially photoexcited electron–hole pair into multiple electron–hole pairs on a time scale (femtoseconds) much faster than the typical time scale (picoseconds) of electron–phonon interaction [63]. The energy of a single incident photon can thus be used to create multiple electron–hole pairs rather than thermal-izing the excess energy absorbed by photo-excited carriers. Furthermore, in Mott-insulator oxide based hetero-structures, the polar discontinuity at interfaces can be utilized to create an electrical field across the photo-absorbing Mott insulator layer to separate the photo-excited electron–hole pairs [64].

The Mott insulator $LaVO_3$ has a bandgap of 1.1–1.2 eV, which lies within the optimal range for solar cell applications. The photovoltaic properties of $LaVO_3$ thin films have been recently evaluated for solar cell performance [65]. Such $LaVO_3$ films show a band gap of 1.08 eV, and strong light absorption over a wide wavelength range in the solar spectrum. However, a high concentration of defects is commonly observed in Mott insulating films. These defects can act as recombination centers, which in turn will degrade the photovoltaic conversion efficiency. Recently high quality $LaVO_3$ thin films have been grown by hybrid molecular beam epitaxy, which have defect densities of in-gap states up to two orders of magnitude lower compared to the $LaVO_3$ films reported earlier, and a factor of three lower than $LaVO_3$ bulk single crystals [64]. Figure 8.36 shows photo-responsivity of a two-terminal test structure (see inset of figure 8.36) based on such high quality La-rich, stoichiometric and V-rich $LaVO_3$ films. There is an abrupt increase in the photo-responsivity at 1.12 ± 0.01 eV irrespective of the stoichiometry of $LaVO_3$ films, which indicates the band gap energy of $LaVO_3$. The observed photo-responsivity is significantly higher in the device with stoichiometric $LaVO_3$ film. This may be correlated to longer carrier life time due to a lower recombination rate associated with lower defect concentration in the stoichiometric $LaVO_3$ films.

8.4.6 Energy storage applications

Since their introduction in 1990 [66] rechargeable Li-ion batteries have undergone rapid development process, and they are now ubiquitous in the field of electronics. Currently with their high energy density and ultrahigh discharge rates, Li-ion batteries seem to be the technology of choice for powering electric vehicles, and also for grid applications in providing back-up for wind and solar energy. In Li-ion batteries electrical energy is generated by the conversion of chemical energy through reduction–oxidation (redox) reactions at the anode and cathode with the movement

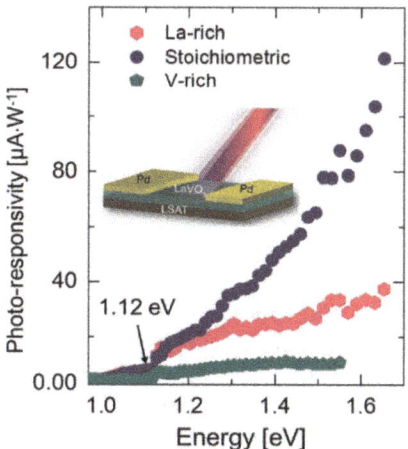

Figure 8.36. Photo-responsivity of a two-terminal device (see inset) based on La-rich, stoichiometric and V-rich $LaVO_3$ films. (Reprinted with permission from [64]. Copyright 2017 American Chemical Society.)

of Li ions between the cathode and anode, and the electrical charge is stored in the bulk of a material. The electrolyte allows the movement of Li ions, but restricts the movement of electrons. When the battery is charging (discharging), lithium ions are extracted from the cathode (anode) and inserted into the anode (cathode). The process of lithium moving into (out of) the cathode is referred to as intercalation (deintercalation).

The cell potential provides the driving force causing the ion movements, and this is directly related to the thermodynamic changes in energy due to the corresponding chemical reactions. The cell potential in Li-ion batteries is determined by the difference between the chemical potential of the Li in the anode and cathode. The functioning of the battery will depend on charge-transfer, charge-carrier and mass transport within the bulk and across interfaces, and also on the phase transition in the concerned materials due to changes in the concentration of Li ions. $LiCoO_2$, $LiMn_2O_4$, $LiFePO_4$ and $Li_2Ru_{1-y}Mn_yO_3$ have been identified as suitable positive electrode material for large scale applications [67, 68]. In some of these materials, especially in $LiFePO_4$ and $Li_2Ru_{1-y}Mn_yO_3$, strong electron–electron interaction and Mott physics are definitely playing important roles. There are three main parameters which define the performance in batteries: the voltage, the current capacity per unit-weight measured in milliamp-hours per gram i.e. the specific capacity, and the number of charge–discharge cycles in which the performance remains within the 50% of the initial values.

The room temperature cycle life of $Li/Li_2Ru_{0.6}Mn_{0.4}O_3$ cells is shown in figure 8.37(a). The cells were cycled in a galvano-static mode between 2 and 4.6 V at a rate of C/5 (1 Li in 5 h) [68]. The reversible cell capacity initially drops from 250 mA h g^{-1} to 220 mA h g^{-1} during the first cycle and then stabilizes after 40 cycles to the value of 210 mA h g^{-1} (see figure 8.37(b)). Thus the material has an attractive specific capacity at an average voltage of 3.6 V together with good cycling

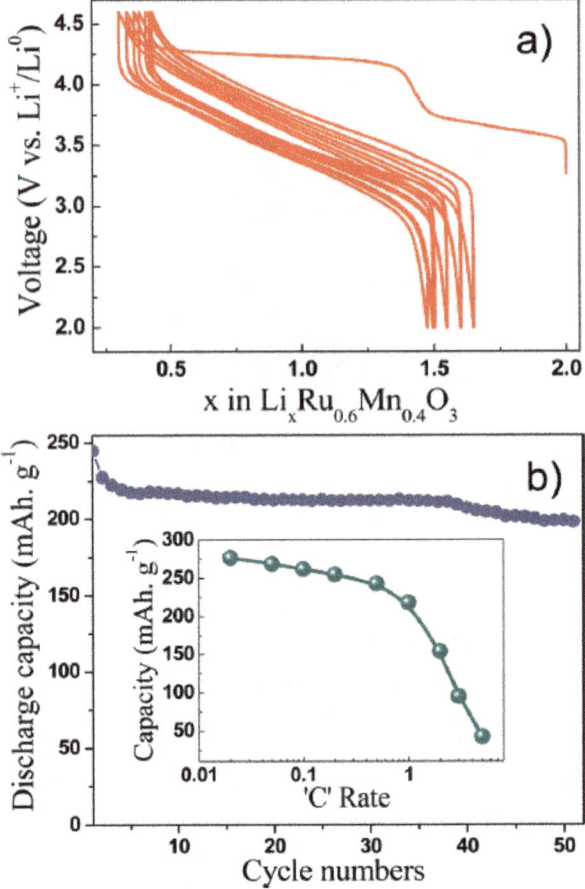

Figure 8.37. (a) Composition versus voltage of $Li_2Ru_{0.6}Mn_{0.4}O_3$ vs Li^+/Li_0 cells cycled at C/5 rate. (b) Capacity retention and the power rate capability of the same cell. (Reprinted with permission from [68]. Copyright 2013 American Chemical Society.)

performance. Moreover, this material can still deliver more than 200 mA h g^{-1} i.e. more than 90% of its initial capacity at 1 C (see inset of figure 8.37(b)), hence it has a good rate capability.

We shall now discuss the explicit role of strong electron–electron interaction and Mott physics in the functioning of $Li/Li_2Ru_{0.6}Mn_{0.4}O_3$ cells. The structure of the cell material $Li/Li_2Ru_{0.6}Mn_{0.4}O_3$ has Li by itself both in the layers and also in the transition metal-oxide layers. X-ray diffraction study showed that structural phase transition took place during the process of cycling [68]. Furthermore, detailed x-ray photoemission spectroscopy studies carried out in successive charging and discharging processes revealed that more than a Li per formula unit left the sample on charging with simultaneous changes in valence of Ru and Mn from 4+ to 5+ and 4+ to 3+ respectively, and evidence of the reversibility of the O^{2-} to O^- anionic process. It was argued that while Mn participated in the redox mechanism, it was the oxygen

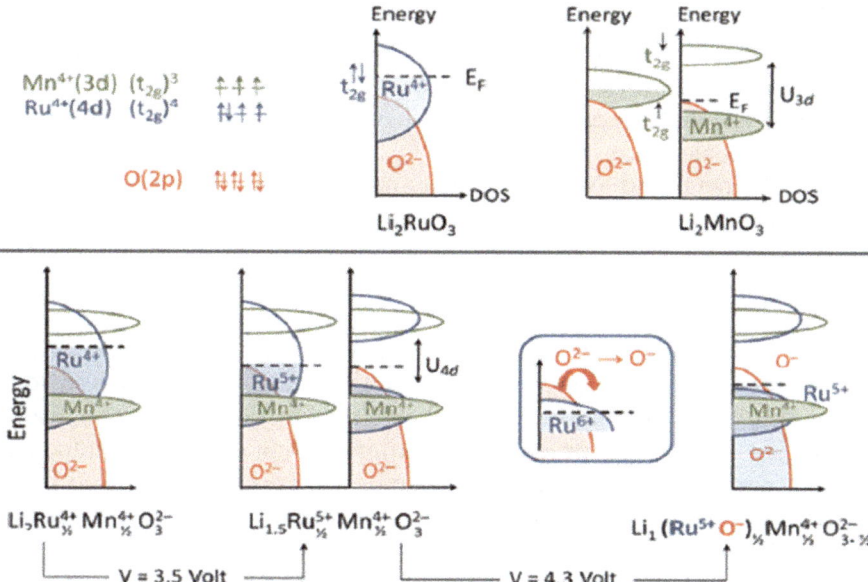

Figure 8.38. Schematic representations of the electronic density of states of Li_2RuO_3 and Li_2MnO_3 and $Li_{2-x}Ru_{1/2}Mn_{1/2}O_3$. The Fermi level E_F is represented by a horizontal dotted line. (Reprinted with permission from [68]. Copyright 2013 American Chemical Society.)

redox process which formed an integral part of the mechanism accounting for the high capacity shown by the $Li_2Ru_{0.6}Mn_{0.4}O_3$ cell materials and also other potential battery materials like Li_2MnO_3–$LiMO_2$ (M = Ni, Co) composites [68]. The participation of oxygen ligand in the redox activity of the system may be explained with the help of schematic representations of the density of states of Li_2RuO_3 and Li_2MnO_3 and $Li_{2-x}Ru_{1/2}Mn_{1/2}O_3$ (see figure 8.38). This presents a qualitative picture of the t_{2g}-like band in the Li_2RuO_3 and Li_2MnO_3 systems, and the evolution of the $Li_xRu_{1-y}Mn_yO_3$ electronic structure upon oxidation. Li_2RuO_3 shows a large 2/3-filled Ru^{4+} (t_{2g})-band where the four electrons are fully delocalized over the Ru–O and Ru–Ru bonds. On the other hand, Li_2MnO_3 shows a narrower Mn^{4+} (t_{2g})-band where the electrons are localized on the Mn atom. This band is half-filled, and it splits through the on-site Coulomb repulsion i.e. Hubbard U, into one low-lying band filled with three spin-up ($t_{2g\uparrow}$) and one high-lying fully empty band of spin-down ($t_{2g\downarrow}$). The electronic structure of $Li_2Ru_{0.5}Mn_{0.5}O_3$ is then characterized by a Fermi level (E_F) lying in the Ru^{4+} (t_{2g})-band [68]. When removing 0.5 Li from $Li_2Ru_{0.5}Mn_{0.5}O_3$, the Ru^{4+} is first oxidized into Ru^{5+} with an average potential of 3.5 V, and the t_{2g}-band of Ru^{5+} merges with the top portion of the O(2p)-band [68]. Further oxidation of Ru^{5+} from the $Li_{1.5}Ru_xMn_{1-x}O_3$ structure results in the creation of a virtual Ru^{6+} state, which can be described as Ru^{5+} plus a hole in the O^{2-} band (see figure 8.38). This electron transfer is favoured by the splitting of the half-filled t_{2g}-band of the Ru^{5+} by on site Coulomb repulsion of Ru 4d electrons U_{4d}. The destabilization of the oxidized O^--2p levels triggers the formation of oxygen vacancies. Thus in $Li_2Ru_{1-y}Mn_yO_3$ materials, there are two oxidation processes: the irreversible removal of oxygen and the oxidation of O^{2-}. The experimental studies showed that the cycling-driven structural phase transition, however, cannot be linked to the creation of the oxygen vacancies, which solely appear during the first charge [68]. It was suggested that the phase transition might be due to the anionic redox process, which by a hither to unknown mechanism could trigger both $3d$ metal and oxygen vacancies migration and hence favour the growth of a new phase.

8.4.7 Photonic devices

There is a growing demand of high-performance and cost-effective transparent conductors for display technologies, photovoltaics, smart windows and solid-state lighting industries. The fundamental challenge for designing such transparent conductors lies in finding a proper combination of high electrical conductivity and optical transparency. Such competing demands are presently addressed by increasing carrier concentration in a wide-bandgap semiconductor with low effective carrier mass through heavy doping.

The development of thin films exhibiting high electrical conductivity and high optical transparency in the visible light spectrum requires minimization of photon absorption and reflection, and at the same time maintaining a high carrier concentration and low carrier scattering. The free career reflection edge is represented by the screened plasma energy:

$$\hbar\omega_p = \hbar\left(\frac{e}{\sqrt{\epsilon_0\epsilon_r}}\right)\sqrt{\frac{n}{m^*}} \tag{8.17}$$

where ϵ_0 and ϵ_r are vacuum and relative permittivity, respectively and e is the elemental charge. The reflection edge can be minimized with an appropriate ratio of the free carrier concentration n and the effective carrier mass m^*. For an ideal transparent conductor, the free carrier reflection edge should be below 1.75 eV, whereas absorption due to the strong interband optical transition is above 3.25 eV. As discussed earlier in section 1.1, the electrical conductivity of free carriers is expressed as:

$$\sigma = e^2\tau\left(\frac{n}{m^*}\right). \tag{8.18}$$

This electrical conductivity needs to be maximized for a transparent conductor. From equations (8.17) and (8.18) it is clear that the ratio n/m^* is the key factor for the performance of transparent conducting materials. This may be maximized to enhance the electrical conductivity, but at the same time keeping the free carrier reflection edge below the visible light spectrum. Presently the design strategy for state-of-the-art transparent conductors is to choose a wide-bandgap oxide semiconductor as a host material to ensure that interband transitions occur above the visible range (>3.25 eV), and degenerately dope the same to increase the electrical conductivity [69]. However, the solubility limit of the dopants and pronounced self-compensation do not allow the maximum carrier concentrations to go beyond

3×10^{21} cm^{-3}. A second limitation arises from the reduction in total scattering time τ due to an enhanced ionized and neutral impurity scattering from the high dopant concentration. The screened plasma energy $\hbar\omega_p$ is much below the visible spectral range for these carrier concentrations; as a result oxide semiconductors with a low m^* are preferred for further enhancement of the ratio n/m^*. Reduction in the film thickness can diminish the free carrier reflection to some extent. However, the scaling down to less than 10 nm range causes a reduction in τ due to enhanced surface scattering for thicknesses below the electron mean free path, thus affecting the conductivity negatively. Thus most of the present transparent conductors are oxides and they are commonly termed as transparent conducting oxides (TCO). Currently tin-doped indium oxide (ITO) has the best combination of high optical transparency and high electrical conductivity among all the TCOs.

Strategies to optimize transparent conducting oxide materials seem to be becoming exhausted over the past decade, and the continued rise in costs of ITO due to indium scarcity necessitates alternative solutions. In this direction, control of the carrier effective mass in Mott-insulators by tailoring the electron correlation strength can provide an interesting alternative approach for the design of transparent conductors [69]. We have earlier discussed in section 5.2.2, how Mott metal–insulator transition is understood in terms of the divergence of the quasi-particle effective mass within the Brinkman–Rice framework. This deviation from the non-interacting free electron behaviour with an effective mass $m^* = m^*_{band}$ is parametrized by incorporating the strong electron–electron interaction through the carrier mass renormalization $m^* = m^*_{band}/Z_k$, where Z_k is the renormalization factor. In the case of non-interacting electrons $Z_k = 1$, and it gets reduced with the increase in electron correlation, either by reducing the conduction bandwidth W or increasing on site Coulomb interaction i.e. Hubbard U. Thus the carrier effective mass m^* increases with the electron–electron interaction, and diverges ultimately in the Mott insulating state where $Z_k = 0$. The mass renormalization factor or the correlation strength Z_k for selected correlated metals along with their room-temperature conductivities are shown in figure 8.39. The conductivity decreases with the increase in Z_k, but the values are comparable with the degenerately doped wide bandgap semiconductors ZnO and tin-doped InO even in the vicinity of the Mott metal–insulator transition. The candidate materials shown in figure 8.39 represent only a small fraction of the potential materials that could be considered as transparent conductors [69]. Figure 8.40 shows the room-temperature conductivity and reduced plasma energies reported for degenerately doped wide bandgap semiconductors ZnO and InO, conventional metals, and the two correlated metals SrVO$_3$ and CaVO$_3$. Both of these correlated metals SrVO$_3$ and CaVO$_3$ have conductivity exceeding that of the conventional TCOs and a reduced plasma energy smaller than 1.75 eV, which bring them closer to the level of ideal transparent conductors. Figure 8.41(a) shows the normal incidence transmission spectrum of SrVO$_3$ and CaVO$_3$ films with varying thicknesses grown on 0.5 mm-thick 001 (LaAlO$_3$)$_{0.3}$(Sr$_2$AlTaO$_6$)$_{0.7}$ (LSAT) and 001 SrLaAlO$_4$ (SLAO) substrates, respectively, from the infrared (0.5 eV) to the ultraviolet (6.2 eV) region. Photographs of the samples on a coloured background are shown in figure 8.41(b). With increasing film thickness, transmission was

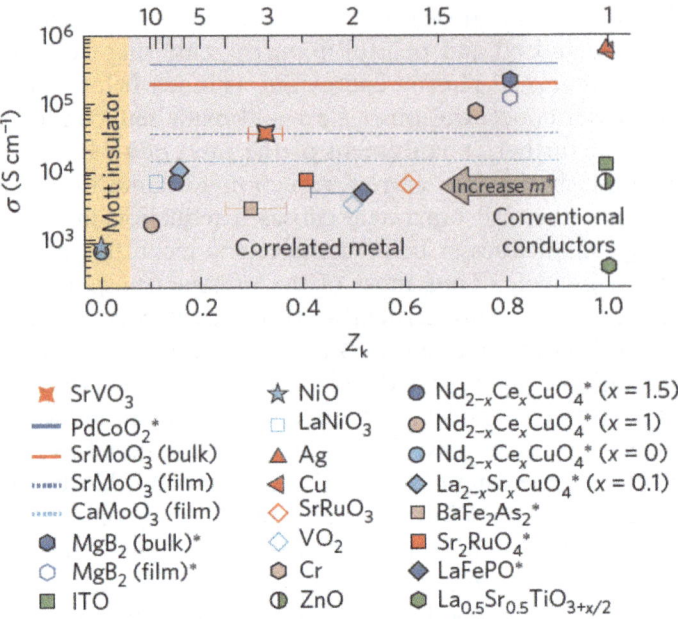

Figure 8.39. Electrical conductivity as a function of correlation strength Z_k for various correlated metals and wide band semiconductors. (Reproduced from [69] with the permission of Springer Nature.)

suppressed for energies above ~3.3 eV, marking the onset of strong interband optical transitions [69]. Discovery of new transparent conductors in the class of correlated metals beyond $SrVO_3$ and $CaVO_3$ will need design effort involving first the identification of a potential correlated metal close to the Mott metal–insulator transition. This new material must sustain pronounced correlation effects with sufficient itinerant carrier concentration, and at the same time have proper combination of band position and band dispersion, and thus Fermi surface size and shape, to suppress interband transitions.

8.4.8 Interesting applications of nuclear fuel material UO_2

Uranium dioxide (UO_2) is the primary material used worldwide for nuclear fuel, which releases enormous amount of useful heat upon nuclear fission. In a nuclear reactor, the nuclear reaction heats up a pellet made of either UO_2 or its mixture with PuO_2 and ThO_2, and the heat is converted to electrical energy. The thermal conductivity of UO_2, however, is very low. This leads to the inefficient transfer of the heat from the core of the nuclear fuel pellet to its outer area, and in turn brings a set of complex problems such as cracking and premature degradation of the fuel pellets. Finding ways to improve the thermal conductivity in UO_2 has always been a subject of considerable research interest, and the recent identification of UO_2 as a Mott-insulator (see section 4.4) has provided some pathway for possible solutions [70]. It has been found that the only efficient heat carriers in UO_2 are the longitudinal acoustic phonons. The heat transfer process does not involve

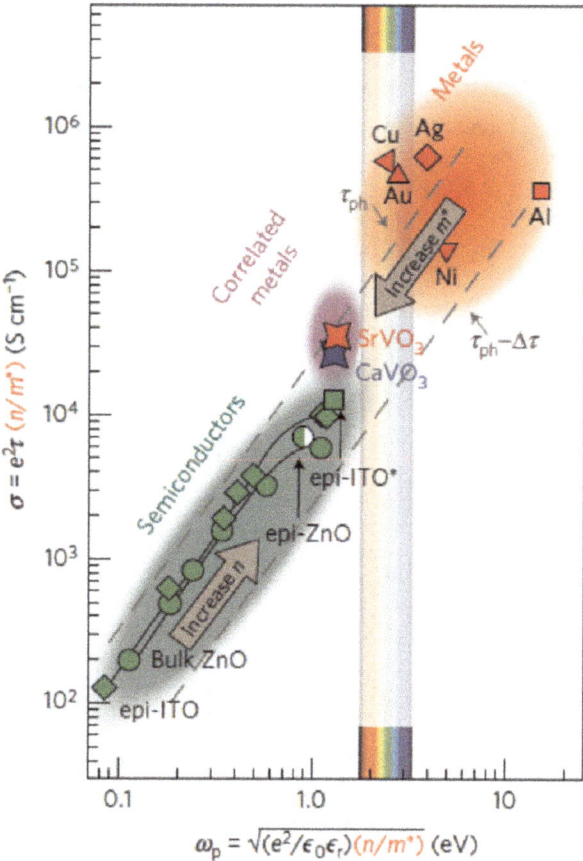

Figure 8.40. Room-temperature conductivity versus reduced plasma energies plot for degenerately doped wide bandgap semiconductors ZnO and InO, conventional metals, and the two correlated metals SrVO$_3$ and CaVO$_3$. (Reproduced from [69] with the permission of Springer Nature.)

the dispersive longitudinal optical modes because of the large anharmonic effect in UO$_2$. This gives a very important input to the nuclear material design, that the anharmonicity in UO$_2$ needs to be suppressed for more efficient nuclear fuels. One possible way is to mix UO$_2$ with elements filling in the cubic interstitials of the lattice, thus preventing large ionic excursions.

UO$_2$ has electrical and electronic properties equivalent to or even better than the properties of conventional band semiconductors like Si, Ge, and GaAs [71]. The energy band gap value of UO$_2$ lies between those of Si and GaAs; intrinsic electrical conductivity of UO$_2$ is approximately the same as GaAs, and the dielectric constant nearly double that of Si and GaAs. The larger dielectric constant implies that in principle the UO$_2$ integrated circuits can be made much denser than conventional Si-based integrated circuits. In addition, UO$_2$ can tolerate much higher operating temperatures up to 2500 K than Si or GaAs (<473 K). In light of the identification of UO$_2$ as a Mott-insulator and the discussions in the sections above on various devices

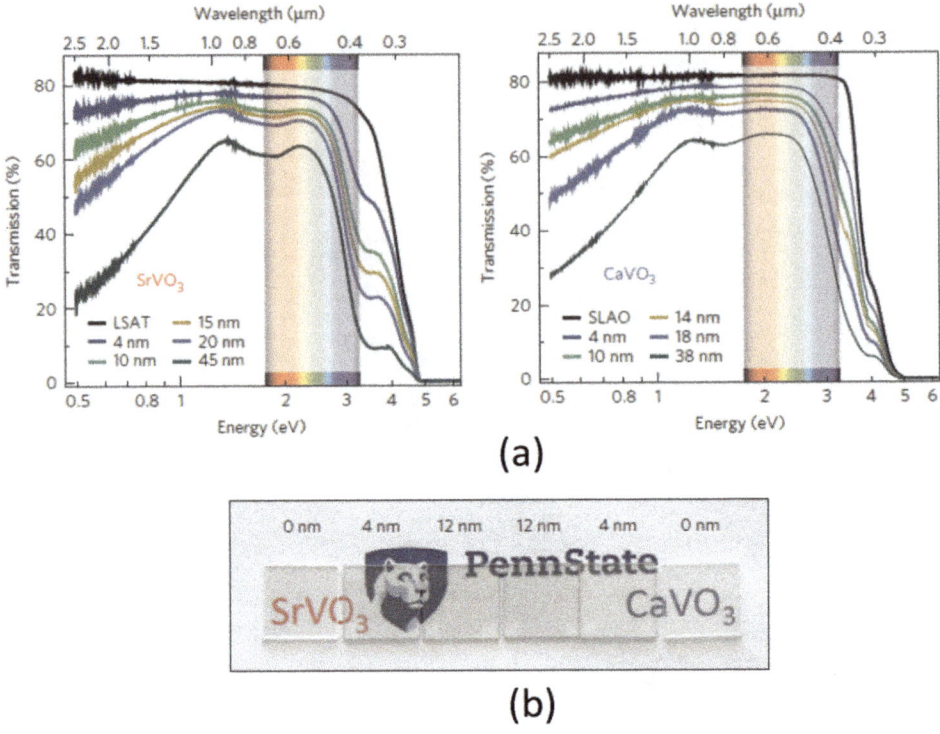

Figure 8.41. (a) Transmission spectrum of SrVO$_3$ films (left) and CaVO$_3$ (right) with varying thicknesses on LSAT and SLAO substrates, respectively. (b) Photographs of 4 nm and 12 nm SrVO$_3$ and CaVO$_3$ films on LSAT and SLAO substrates, respectively. (Reproduced from [69] with the permission of Springer Nature.)

based on Mott physics, it appears that UO$_2$ has the potential as materials for high-performance devices including solar cells. With so much of depleted uranium based materials available at the nuclear reactor sites all over the world, it is a bit surprising that there is not much effort in this direction, which may be due to the nuclear tag attached with the material. There is some report of p–n junction and bipolar p–n–p transistor being fabricated from UO$_2$, which has potential for solar cell applications [72]. Such UO$_2$-based devices may also find applications as solid state neutron detector [73].

It has been reported recently that single crystals of UO$_2$ when subjected to strong magnetic fields in the magnetic state below 30 K show an abrupt appearance of positive linear magneto-striction [74]. This linear term also reverses sign upon reversal of the magnetic field. This is a hallmark of piezomagnetism, which is the magnetic counterpart to piezoelectricity. A magnetic moment can be induced in piezomagnetic crystals by application of a physical stress. A switching phenomenon is observed in UO$_2$ at ±18 T, and that persists during subsequent field reversals. This phenomenon demonstrates the existence of a robust magneto-elastic memory in UO$_2$, which makes this material the hardest piezomagnet known [74].

8.4.9 Solid state refrigerant for magnetic cooling technology

The magnetic field induced isothermal entropy change or adiabatic temperature change in a magnetic solid is known as the 'magneto-caloric effect' (MCE). In a MCE material randomly oriented magnetic moments are aligned by an applied magnetic field in a magnetic-cooling cycle, resulting in the reduction of magnetic entropy. In turn the material is heated via the increase of its lattice entropy. This heat is then removed from the material to its surroundings by a heat-transfer medium. On removing the magnetic field, the magnetic moments become randomized causing an increase in the magnetic entropy, which leads to cooling of the MCE material below the ambient temperature. MCE offers the prospect of energy efficient magnetic cooling technology, which has a potential to reduce global energy consumption and avoid the need of ozone depleting and greenhouse gas chemicals [75].

The MCE of a material is completely characterized when the behaviour of the total entropy of the material as a function of both temperature and magnetic field is known. The infinitesimal isobaric–isothermal magnetic entropy change is related to the change in bulk magnetization M of a material as a function of temperature T and applied magnetic field H through the well known Maxwell relation [76]:

$$\left(\frac{\partial S_M(T, H)}{\partial H}\right)_T = \left(\frac{\partial M(T, H)}{\partial T}\right)_H.$$
(8.19)

This equation (8.19) after integration gives:

$$\Delta S_M(T) = \int_{H_1}^{H_2} \left(\frac{\partial M(T, H)}{\partial T}\right)_H dH.$$
(8.20)

The infinitesimal adiabatic temperature rise for the reversible adiabatic isobaric process is expressed as [76]:

$$dT(T, H) = -\left(\frac{T}{C(T, H)}\right)_H \left(\frac{\partial M(T, H)}{\partial T}\right)_H dH$$
(8.21)

where $C(T, H)$ is the temperature and magnetic field-dependent heat capacity at constant magnetic field. The adiabatic temperature change $\Delta T_{adiabatic}(T)_{\Delta H}$ is obtained by integrating equation (8.21):

$$\Delta T_{adiabatic}(T)_{\Delta H} = \int_{H1}^{H2} dT(T, H) = -\int_{H1}^{H2} \left(\frac{T}{C(T, H)}\right)_H \left(\frac{\partial M(T, H)}{\partial T}\right)_H dH.$$
(8.22)

It is quite apparent from the above equations that the potential MCE materials are expected to show a rapid variation of magnetization with temperature. The first order insulator to metal transition accompanied by a change of magnetic structure in various strongly correlated manganese oxide systems has turned out to be an interesting source of large magneto-caloric effect (see [77] and references within).

A large MCE has been observed in the vicinity of the first order transition from the antiferromagnetic insulator phase to ferromagnetic metallic phase in $Nd_{0.5}Sr_{0.5}MnO_3$ compound, which is actually three times larger than the conventional MCE observed near the higher temperature paramagnetic to ferromagnetic transition in the same compound [78]. In these magnetic materials, the coupling between spin and lattice degrees of freedom plays an important role in enhancing the MCE. A similar large MCE associated with a first order antiferromagnetic to ferromagnetic transition has also been reported in $Pr_{1-x}Ca_xMnO_3$, $Eu_{0.55}Sr_{0.45}MnO_3$, $La_{0.2}Nd_{0.47}Ca_{0.33}MnO_3$ and $La_{1-x}Sr_xMnO_3$. The magnetic field induced change in the configurational entropy associated with the structural transition and the magnetic inhomogeneity arising from the phase-coexistence of antiferromagnetic and ferromagnetic state across the disorder-influenced first order insulator–metal phase transition, are responsible for the large observed MCE. The various features associated with the magnetic field and temperature induced first order magneto-structural transition in manganese oxide systems and other known materials with giant magneto-caloric effect are quite similar in nature [5]. Among these MCE materials manganese oxide based strongly correlated electron systems are thought to be suitable for a variety of large- and small-scale magnetic refrigeration applications in the temperature range of 100–375 K [77]. These materials have actually been tested in magnetic refrigeration test devices [79, 80]. The relatively low cost of materials and easy processing route makes such materials quite promising for future magneto-caloric applications.

8.4.10 Interesting applications of TaS_2

The discussions on various Mott devices so far are almost entirely based on the different oxide compounds. The only exception was the narrow gap Mott insulator compounds AM_4Q_8 (where A = Ga or Ge; M = V, Nb or Ta; and Q = S or Se) exhibiting both a volatile and a non-volatile unipolar resistive switching. In chapter 6 we discussed that in the layered dichalcogenide $1T\text{-}TaS_2$ with decrease in temperature, competition arising from repulsive Coulomb interaction between electrons, lattice strain and a Fermi surface instability, gives rise to a metallic state with an incommensurate (IC), a domain textured nearly-commensurate (NC) or an insulating commensurate (C) charge density wave (CDW) state. Both in-plane (IP) and out-of-plane (OP) strong electron correlation in $1T\text{-}TaS_2$ lead to an unusual low-temperature Mott-insulator state, where the electronic charge density within each layer is modulated by a three-directional in-plane charge density wave (CDW).

$1T\text{-}TaS_2$ develops an IC-CDW state with an associated lattice distortion below 550 K. This charge modulation sharpens on reducing the temperature to form star-shaped polaron clusters causing a transition to the NC state in the temperature region below 350 K. At room temperature, the $1T\text{-}TaS_2$ is in this NC state, which on further cooling undergoes a first order phase transition to the insulating C state near 180 K with an abrupt change in resistance. Ultrafast resistance switching from this low temperature C state in TaS_2 to a hidden (H) CDW state has been reported, induced by a 35 fs laser pulse [81], and also by 40 ps electrical pulse-injection [82]. To investigate the roles of Mott-insulating state of the individual TaS_2 layers and the

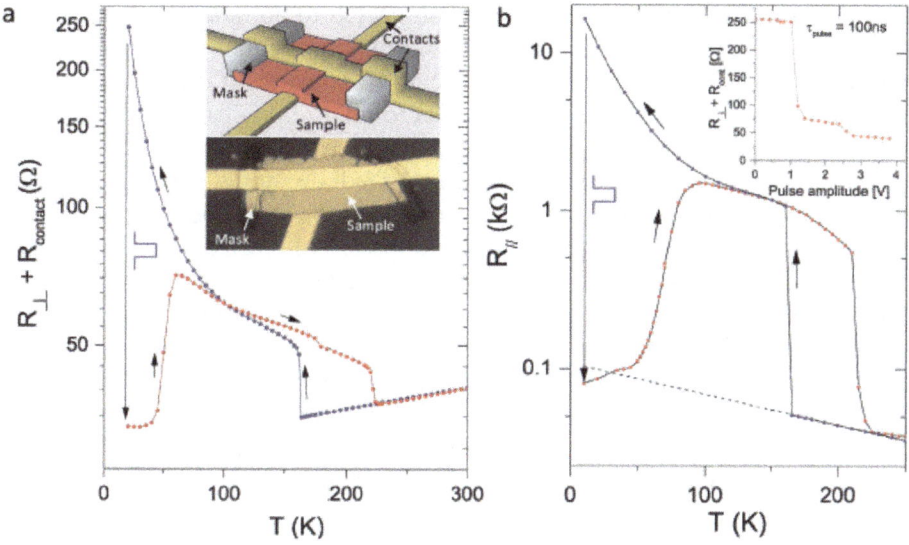

Figure 8.42. (a) The temperature dependence of the out of plane c-axis resistance of 1T-TaS$_2$ before (blue) and after (red) the application of a 10–50 µs pulse. The inset shows the two terminal experimental configuration and an image of the sample. (b) The temperature dependence of the in-plane resistance before (blue) and after (red) switching. The inset shows the threshold for switching along the c-axis with 100 ns pulses. (Reproduced from [83] with the permission of Springer Nature.)

three-dimensional stacking in the switching phenomena, three-dimensional resistivity switching experiments were performed with two contacts on either side of a 90 nm thick flake of 1T-TaS$_2$ at 20 K (see inset of figure 8.42(a)). A single electrical pulse of 10 V, 50 µ applied at 4 K causes an instant drop in resistance by more than two orders of magnitude (see main frame of figure 8.42(a)) [83]. With subsequent heating the resistance merges with the cooling curve at temperatures above 80 K, and then at the higher temperature the hysteretic nature of the first order phase transition between the C and NC state becomes clearly visible. The in-plane resistance switching is shown in figure 8.42(b) for comparison. The measured critical threshold current densities for switching are quite similar, $i_\perp^T \sim 10^{-3}$ A µm^{-2} and $i_\parallel^T \sim 1.7 \times 10^{-3}$ A µm^{-2}, respectively, while the applied threshold electric fields are vastly different for c-axis and IP switching, $E_\perp^T \sim 100$ V µm^{-1} and $E_\parallel^T \sim 1$ V µm^{-1}, respectively. The value of the resistance in the hidden H-CDW state after switching is different from that extrapolated from the high temperature NC-state. This indicates that the H-CDW state is not the same state, which can be reached by supercooling the NC state across the first order NC–C phase transition. The inset of figure 8.42(b) shows the final state resistance as a function of applied electric pulse amplitude. The final state resistance tends to saturate in two plateaus indicating the possibility of more than one metastable state.

The switched 1T-Ta$_2$ sample, after the 'write (W)' electrical pulse, remains in the hidden H-CDW state if the pulse length is less than 200 ms. This conducting H-CDW state is very stable within a large temperature interval, and is retained unless the sample is heated above 80 K or subjected to an 'erase (E)' electrical pulse of longer (2 s) duration. This longer E-pulse raises the temperature of the device by Joule heating, which causes a sharp switching of the resistance back to the high resistivity state. Switching has been observed in the 1T-TaS$_2$ samples for pulse lengths less than 1 s down to ~30 ps, that is by the rise time of the pulse source [82]. Bi-stable switching is also possible with intermediate states. The extremely sharp switching thresholds indicate that a collective many-body ordering process may be responsible for both 'write' and 'erase' processes. All-electronic switching mechanism means that the switching energy in such devices can be quite small with a promise of achieving sub-atto-joule/bit values in small size devices [83]. Moreover the adjacent cells in such devices can be switched independently without cross-talk, which implies that useful thin-film devices with unoriented films may be fabricated in either cross-bar or lateral stripline geometry [83]. This will offer flexible design options for new ultrafast low-temperature memristive memory devices based on CDW state switching.

Self-assessment questions and exercises

1. What are the characteristic features of a first order phase transition in general?
2. What is meant by the 'field effect' in a MOSFET?
3. In real cases practically no change is observed in the channel conductance until the gate voltage in a MOSFET reaches a threshold voltage V_T. Which properties of MOSFET structure/material dictate the magnitude of V_T?
4. What are the limitations of a practical MOSFET, which prevents it from performing as a perfect switch?
5. What is the subthreshold slope of a conventional MOSFET and what prevents it to go below a minimum value? What is the minimum change in the surface potential required in a conventional MOSFET to modulate the drain current I_D by one order of magnitude at room temperature?
6. What is the essential difference in the response of electronic band gap in MOSFET and Mott-FET due to the 'field effect'?
7. What is the subthreshold slope in a Mott-FET? How does it compare with respect to the subthreshold slope in a conventional MOSFET?
8. What are the possible advantages of a Mott-insulator based Landau switch over a conventional band semiconductor based Boltzmann switch?
9. Assuming the charge compressibility $\partial n/\partial E_F$ in a Mott-insulator at a finite temperature to be 0.5/(eV unit cell), what is the expected Thomas–Fermi screening length in this material?
10. What are the practical difficulties in implementing the Mott-FET for real life commercial applications? What are the current possibilities to circumvent such difficulties?
11. What are various mechanisms, which can cause an electric field induced MIT in a two-terminal Mott device?

12. What is the indication of Joule heating by the current passing through a Mott device, and how can this dissipation of heat energy be minimized in the device?
13. What physical properties of a Mott-insulator enable a two terminal Mott-device to function both as a volatile and non-volatile memory device?
14. What is the difference between the switching mechanism in two-terminal Mott devices and ReRAM (described in chapter 7)?
15. Show that a high density non-volatile memory structure can be achieved by combining two Mott insulators, one functioning as a memory element storing data and the other as a switch element.
16. What is a memristor and how its functional properties differ from a conventional resistor?
17. What is the reason of higher quantum efficiency in Mott-insulator based solar cells in comparison to conventional band semiconductor based solar cells?
18. What is the current design strategy for state-of-the-art transparent conductors involving band semiconductor materials, and what are the limitations in this direction? In which way can Mott-insulators be useful here?
19. How can the hidden-CDW state in TaS_2 be used in an ultrafast memory application?

References

[1] Newns D M, Misewich J A and Zhou C 1996 Nanoscale Mott-transition Molecular Field Effect Transistor *US Patent* YO996-06
[2] Zhou C, Newns D M, Misewich J A and Pattnaik P C 1997 *Appl. Phys. Lett.* **70** 598
[3] Zhou Y and Ramanathan S 2015 *Proc. IEEE* **103** 1289
[4] Chaikin P M and Lubensky T C 1995 Principle of Condensed Matter Physics (Cambridge: Cambridge University Press)
[5] Roy S B 2013 *J. Phys.: Condens. Matter* **25** 183201
[6] Imry Y and Wortis M 1979 *Phys. Rev.* B **19** 3580
[7] Soibel A, Zeldov E, Rappaport M, Myasoedov Y, Tamegai T, Ooi S, Konczykowski M and Geshkenbein V B 2000 *Nature* **406** 282
[8] Roy S B, Perkins G K, Chattopadhyay M K, Nigam A K, Sokhey K J S, Chaddah P, Caplin A D and Cohen L F 2004 *Phys. Rev. Lett.* **92** 147203
[9] Manekar M A and Roy S B 2008 *J. Phys.: Condens. Matter* **20** 325208
[10] Moore R G, Zhang J, Nascimento V B, Jin R, Guo J, Wang G T, Fang Z, Mandrus D and Plummer E W 2007 *Science* **318** 615
[11] Oka T and Nagaosa N 2005 *Phys. Rev. Lett.* **95** 266403
[12] Mott N F 1990 *Metal-Insulator Transition* 2nd edn (London: Taylor and Francis)
[13] Tokura Y, Kawasaki M and Nagaosa N 2017 *Nat. Phys.* **13** 1156
[14] Sze S M and Lee M 2012 *Semiconductor Devices: Physics and Technology* 3rd edn (New York: Wiley)
[15] Ahn C H *et al* 2006 *Rev. Mod. Phys.* **78** 1185
[16] Neamen D A 2012 *Semiconductor Devices and Physics: Basic Principles* 4th edn (New York: McGraw-Hill)

[17] Zhou Y and Ramanathan S 2013 *Crit. Rev. Solid State Mater. Sci.* **38** 286

[18] Ferain I, Colinge C A and Colinge J-P 2011 *Nature* **479** 310

[19] Ionescu A M and Riel H 2011 *Nature* **479** 329

[20] Salahuddin S and Datta S 2008 *Nano Lett.* **8** 405

[21] Jo J and Shin C 2016 *IEEE Electron Device Lett.* **37** 245

[22] Wang X *et al* 2017 *npj 2D Mater. Appl.* **1** 384

[23] Fisher D S, Kotliar G and Moeller G 1995 *Phys. Rev.* B **52** 17112

[24] Imada M, Fujimori A and Tokura Y 1998 *Rev. Mod. Phys.* **70** 1040

[25] Roy S B and Chaddah P 2014 *Phys. Status Solidi* **252** 2010

[26] Stoliar P, Rozenberg M, Janod E, Corraze B, Tranchant J and Cario L 2014 *Phys. Rev.* B **90** 045146

[27] Janod E *et al* 2015 *Adv. Funct. Mater.* **25** 6287

[28] Oka T and Aoki H 2010 *Phys. Rev.* B **81** 033103

[29] Oka T 2012 *Phys. Rev.* B **86** 075148

[30] Guiot V, Cario L, Janod E, Corraze B, Phuoc V T, Rozenberg M, Stoliar P, Cren T and Roditchev D 2013 *Nat. Commun.* **4** 1722

[31] Zimmers A, Aigouy L, Mortier M, Sharoni A, Wang S, West K G, Ramirez J G and Schuller I K 2013 *Phys. Rev. Lett.* **110** 056601

[32] Yang Z, Ko C and Ramanathan S 2011 *Annu. Rev. Mater. Res.* **41** 337

[33] Li J, Aron C, Kotliar G and Han J E 2017 *Nano. Lett.* **17** 2994

[34] Li J, Aron C, Kotliar G and Han J E 2015 *Phys. Rev. Lett.* **114** 226403

[35] Stefanovich G, Pergament A and Stefanovich D 2000 *J. Phys.: Condens. Matter* **12** 8837

[36] Pergament A, Stefanovich G, Malinenko V and Velichko A 2015 *Adv. Condens. Mater. Phys.* **2015** 654840

[37] Kim H-T, Chae B-G, Youn D-H, Maeng S-L, Kim G, Kang K-Y and Lim Y-S 2004 *New J. Phys.* **6** 52

[38] Chae B-G, Kim H-T, Youn D-H and Kang K-Y 2005 *Physica* B **369** 76

[39] Kim B-J, Lee Y W, Choi S, Yun S J and Kim H-T 2010 *IEEE Electron Device Lett.* **31** 14

[40] Cha E *et al* 2013 *Proc. IEEE Int. Electron Devices Meeting* 10.5.1

[41] Xue K-H, Paz de Araujo C A, Celinska J and McWilliams C 2011 *J. Appl. Phys.* **109** 091602

[42] Celinska J, McWilliams C, Paz de Araujo C and Xue K-H 2011 *J. Appl. Phys.* **109** 091603

[43] Park G S, Li X-S, Kim D-C, Jung R-J, Lee M-J and Seo S 2007 *Appl. Phys. Lett.* **91** 222103

[44] Lee M J *et al* 2007 *Adv. Mater.* **19** 3919

[45] Tour J M and He T 2008 *Nature* **453** 42

[46] Chua L 1971 IEEE Trans *Circuit Theory* **18** 507

[47] Strukov D B, Snider G S, Stewart D R and Williams R S 2008 *Nature* **453** 80

[48] Yang J J 2008 *Nat. Nanotech.* **3** 429

[49] Di Ventra M, Pershin Y V and Chua L O 2009 *Proc. IEEE* **97** 1371

[50] Driscoll T, Kim H-T, Chae B-G, Di Ventra M and Basov D N 2009 *Appl. Phys. Lett.* **95** 043503

[51] Hodgkin A L and Huxley A F 1952 *J. Physiol.* **117** 500

[52] Pickett M D, Medeiros-Ribeiro G and Williams R S 2013 *Nat. Mater.* **12** 114

[53] Hormoz S and Ramanathan S 2010 *Solid-State Electron* **54** 654

[54] Pergament A L, Stefanovich G B and Velichko A A 2013 *J. Sel. Topics Nano Electron. Comput.* **1** 14

[55] Nakano M, Shibuya K, Okuyama D, Hatano T, Ono S, Kawasaki M, Iwasa Y and Tokura Y 2012 *Nature* **487** 459

[56] Jeong J, Aetukuri N, Graf T, Schladt T D, Samant M G and Parkin S S P 2013 *Science* **339** 1402

[57] Jeong J, Aetukuri N B, Passarello D, Conradson S D, Samant M G and Parkin S S P 2015 *Proc. Natl. Acad. Sci.* **112** 1013

[58] Xiang P-H, Asanuma S, Yamada H, Sato H, Inoue I H, Akoh H, Sawa A, Kawasaki M and Iwasa Y 2013 *Adv. Mater.* **25** 2158

[59] Son J, Rajan S, Stemmer S and Allen S J 2011 *J. Appl. Phys.* **110** 084503

[60] Kim B-J, Lee Y W, Chae B-G, Yun S J, Oh S-Y and Kim H-T 2007 *Appl. Phys. Lett.* **90** 023515

[61] Strelcov E, Lilach Y and Kolmakov A 2009 *Nanoletters* **9** 2322

[62] Manousakis E 2010 *Phys. Rev.* B **82** 125109

[63] Coulter J E, Manousakis E and Gali A 2014 *Phys. Rev.* B **90** 165142

[64] Zhang H T, Brahlek M, Ji X, Lei S, Lapano J, Freeland J, Gopalan V and Engel-Herbert R 2017 *ACS Appl. Mater. Interfaces* **9** 12556

[65] Wang L *et al* 2015 *Phys. Rev. Appl.* **3** 064015

[66] Nagaura T and Tozawa K 1990 *Prog. Batt. Sol. Cells* **9** 209

[67] Kang B and Ceder G 2009 *Nature* **458** 190

[68] Sathiya M, Ramesha K, Rousse G, Foix D, Gonbeau D, Prakash A S, Doublet M L, Hemalatha K and Tarascon J-M 2013 *Chem. Mater.* **25** 1121

[69] Zhang L *et al* 2016 *Nat. Mater.* **15** 204

[70] Yin Q and Savrasov Y 2008 *Phys. Rev. Lett.* **100** 225504

[71] Meek T, Hu M and Haire M J 2001 *Waste Management 2001 Symp. (Tucson, Arizona) February 25–March 1, 2001)*

[72] Meek T and von Roedern B 2009 *Vacuum* **83** 226

[73] Caruso A N 2010 *J. Phys.: Condens. Matter* **22** 443201

[74] Jaime M *et al* 2017 *Nat. Commun.* **8** 99

[75] Tishin A M and Spichkin Y I 2003 *The Magnetocaloric Effect and its Applications* (Bristol: Institute of Physics)

[76] Pecharsky V K and Gschneidner K A Jr 1999 *J. Appl. Phys.* **86** 565

[77] Phan M-H and Yu S-C 2007 *J. Magn. Magn. Mater.* **308** 325

[78] Sande P, Hueso L E, Miguéns D R, Rivas J, Rivadulla F and López-Quintela M A 2001 *Appl. Phys. Lett.* **79** 2040

[79] Theil Kuhn L, Pryds N, Bahl C R H and Smith A 2011 *J. Phys.: Conf. Ser.* **303** 012082

[80] Bahl C R H, Velázquez D, Nielsen K K, Engelbrecht K, Andersen K B, Bulatova R and Pryds N 2012 *Appl. Phys. Lett.* **100** 21905

[81] Stojchevska L, Vaskivskyi I, Mertelj T, Kusar P, Svetin D, Brazovskii S and Mihailovic D 2014 *Sci.* **344** 177

[82] Vaskivskyi I, Mihailovic I A, Brazovskii S, Gospodaric J, Mertelj T, Svetin D, Sutar P and Mihailovic D 2016 *Nat. Commun.* **7** 11442

[83] Svetin D, Vaskivskyi I, Brazovskii S and Mihailovic D 2017 *Sci. Rep.* **7** 46048

IOP Publishing

Mott Insulators
Physics and applications
Sindhunil Barman Roy

Appendix A

Some relevant experimental techniques

We discuss here some measurement techniques relevant to the subject matter of the present book namely Mott-insulators and related materials with strong electron–electron interaction and metal–insulator transition.

A.1 Two-probe and four-probe electrical resistance measurements

Two-probe techniques are used for general purpose resistance measurements and generation of the current (I)–voltage (V) curve. The heart of such experiments is often a commercially available ohmmeter. To determine the resistance of an unknown sample, Ohm's law is used: $R = V/I$. In an ohmmeter a known current I is passed through the sample with unknown resistance, and the voltage drop V across the sample is measured to determine the resistance of the sample under measurement (see figure A1). However, in two-probe measurements, apart from voltage drop across the sample the voltage measured also includes the voltage due to the lead resistance and contact resistance (between the current lead and the sample under measurement). The results of measurements in this method are fairly accurate when the resistance of the unknown sample is above a few ohms, because the lead resistance and contact resistance is usually of the order of a few milli-ohms. The situation however changes when measuring the resistance in small metallic samples where the resistance can be below the milli-ohm range. In such cases the measurement with two-probe techniques will be greatly influenced by the lead resistance and contact resistance, and the results can be erroneous. In such a situation one needs to use the four-probe measurement technique.

The four-probe technique [1] is widely used in the resistivity measurements of a small metallic sample especially in a variable temperature environment. In metallic samples depending on the temperature the measured resistance can be in the range of micro-ohm and there, unless eliminated, the lead and contact resistance will greatly

doi:10.1088/2053-2563/ab16c9ch9

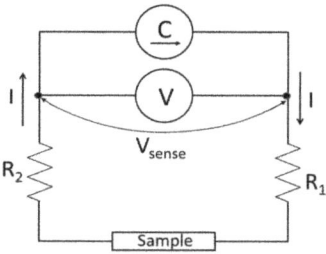

Figure A1. Schematic representation of the two-probe resistance measurement technique.

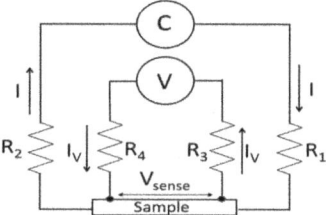

Figure A2. Schematic representation of the four-probe resistance measurement technique.

influence the results of the measurements. Even if the resistance of the metallic sample is relatively high, the change in resistance due to the change in sample temperature can be easily masked by the extraneous resistance. In the four-probe method a known current I generated by a current source is passed through two contact points in the unknown sample and the voltage drop is measured by a voltmeter across two separate contact points on the sample (see figure A2). Good current sources have very low impedance and good voltmeters have very high impedance. Hence, a very small current is drawn by the voltmeter during the measurement. Hence, the measured voltage is essentially due to the source current passing through the sample. Nowadays, a voltmeter with the sensitivity of even few nano-volt is easily available, and hence temperature dependent metallic resistivity of various kinds of metals can be measured with a large degree of accuracy. The same four-probe technique is regularly used in Hall-voltage measurements as well.

In practice a four-probe resistivity measurements is more involved in spite of using the high-resolution accurate current source and voltmeter. There can be influence of electrical noise voltage in the measuring of electrical circuits, which unless care is taken can easily even be of the order of micro-volts. There are certain procedures to reduce this electrical noise in the measurement electrical circuits [1]. While total elimination of electrical noise is almost impossible, with proper care noise level can be regularly reduced to a few tens of nano-volts. Low frequency ac-technique can be used for ac electrical resistivity (conductivity) measurement. The ac-technique often involves resistance bridges [1], which has the advantage of using a lock-in amplifier to measure the voltage drop across the sample. With a careful design of the circuit for the resistivity measurement, a lock-in amplifier has the advantage of detecting the electrical signal buried in the environment of background electrical noises.

A.2 Thermal conductivity measurements

Thermal conductivity k describes the transport of thermal energy or heat Q through a length or thickness L of a material under a steady state temperature difference of $(T_1 - T_2)$ $(T_1 > T_2)$ in the direction normal to a surface area A. Thermal conductivity in materials is due to heat transport by quanta of lattice vibrations (phonons) and by available free electrons. In insulators the heat is transported only by phonons, while both types of carriers are present in the metals. However, in metals heat transport is predominantly by electrons and the phonon contribution is relatively small. When measuring thermal conductivity of a metallic sample, it is usually made in the form of a bar. One end of the sample is heated electrically, while the other end is cooled with the help of a suitable cooling medium. The surface of the sample is thermally insulated. The heat loss through the sample surface can be estimated by subtracting the rate at which the heat enters the cooling medium from the rate of heat generation by the electrical source at the other end of the bar [2]. The heat lost from the surface for most metals is negligible in comparison with the heat flow through the bar. The temperatures T_1 and T_2 at two places of distance L apart on the bar shaped sample is measured with suitable temperature sensors. The average thermal conductivity within this assigned temperature range $(T_1 - T_2)$ is given by the following equation:

$$k = \frac{L}{A(T_1 - T_2)} \frac{dQ}{dT} \tag{A.1}$$

where A is the area of cross-section of the metallic bar. In the case $T_1 - T_2$ is small, thermal conductivity k can be considered as the thermal conductivity of the concerned metal at that mean temperature [2]. Thermal conductivity has the units of watts per meter-kelvin (W m^{-1} K^{-1}).

During an experiment the sample is put between a heat source with a known power output and a heat sink (see figure A3). This results in a temperature gradient across the given length L. This drop in temperature across the length of the sample is measured with the help of two thermometers placed at the two ends of this length L in the sample. The major challenges in this experiment involve: (i) determination of the heat flow rate through the sample in the presence of parasitic heat losses, and (ii) accurate measurement of temperature difference. Parasitic heat losses can originate

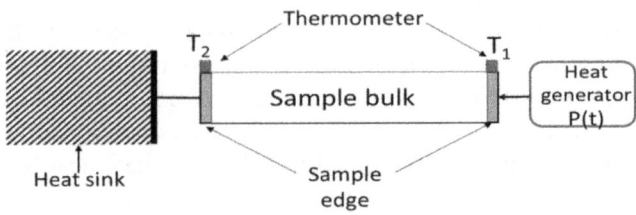

Figure A3. Schematic representation of the thermal conductivity measurement technique.

from convection and radiation to the surrounding and conduction through the wires of temperature sensors. It is desired that such parasitic heat losses should be less than 2% of the total heat flow through the sample. The measurements are often carried out under vacuum with radiation shields to reduce convection and radiation heat losses. There is also a concern of heat loss via conduction through the wires of temperature sensors. When a thermocouple is used as temperature sensor, thermocouples with small wire diameter and low thermal conductivity wires e.g. chromel–constantan are preferred.

A.3 Heat capacity measurements

If a material experiences a change in temperature from T_1 to T_2 during the transfer of Q units of heat, the average heat capacity of the materials is defined as [2]:

$$C_{Average} = \frac{Q}{(T_2 - T_1)}. \tag{A.2}$$

With the reduction of both Q and $T_2 - T_1$, the average heat capacity approaches a limiting value. This is defined as heat capacity C and expressed as:

$$C = Lt_{T_2 \to T_1} \frac{Q}{T_2 - T_1} \tag{A.3}$$

or

$$C = \frac{dQ}{dT} \tag{A.4}$$

at temperature T_1 and is measured in joules/Kelvin ($J\ K^{-1}$). It may be noted that heat capacity is an extensive quantity. One can also have an intensive quantity namely specific heat by dividing heat capacity with mass of the sample. The specific heat is an intensive quantity measured in joules per kilogram-kelvin ($J\ kg^{-1}\ K^{-1}$).

In the classical way of heat capacity measurements, a small heat input dQ causes a small temperature rise dT in the sample, and the heat capacity is determined by using the standard definition of heat capacity given in equation (A.4). This is the traditional adiabatic method of heat capacity measurement. However, in this technique the sample has to be thermally isolated, which requires a fairly large size sample to minimize the effects of stray heat leaks.

A paradigm shift in the heat capacity measurement technique took place in the late 1960s, when Sullivan and Seidel [3] reported how to measure heat capacity of small samples using an ac method employing a lock-in amplifier for accurate and sensitive detection of ac-electrical signals. In this method if an ac heating current of frequency $w/2$ is passed through a resistance heater attached to the sample, there will be an ac temperature response at frequency w [4]:

$$T_{ac} = Power/2wC\sqrt{1 + \frac{w^2}{\tau_1^2} + w^2\tau_2^2} + constant. \qquad (A.5)$$

Here τ_2 is the response time of the sample, heater and thermometer to the ac heat input, τ_1 is the sample to bath relaxation time, and C is the total heat capacity of the sample, its supporting wire, the heater and the thermometer. When τ_2 can be kept much less than $1/w$ and τ_1 much greater than $1/w$, then using equation (A.5) heat capacity can be expressed as:

$$C = Power/2wT_{ac}\sqrt{1 + constant}. \qquad (A.6)$$

Here the constant term is $2K_S/3K_{SB}$, with K_S being the thermal conductivity of the sample and K_{SB} being the sample-to-bath thermal conductivity. If the sample thermal conductivity K_S is kept large in comparison to K_{SB}, then equation (A.7) takes a simple form:

$$C = Power/2wT_{ac}. \qquad (A.7)$$

The major strength of the ac heat-capacity technique is its capability of measuring very small changes in heat capacity, hence it is a good method to study the temperature dependent change in the heat capacity of a given sample. However, there can be significant inaccuracy in this method caused by the uncertainty of the heat-capacity addenda contribution from the thermometer, heater, and connecting wires [4]. In addition, while the ac method can be used to obtain relative changes in the case of poor sample thermal conductivity, this method fails to measure an absolute value of heat capacity in the cases where the condition $\tau_1^{-1} < w < \tau_1^{-2}$ is not satisfied [4].

Further development in small sample calorimetry took place in the early 1970s with the introduction of the time-relaxation technique of heat capacity measurements [5]. In this method the sample mounted on a suitable platform is heated above a constant reference temperature T_0 by a small amount ΔT ($\Delta T/T_0 \sim 1\%$) and the heat is then turned off. The temperature then decays exponentially. It was shown by solving the one-dimensional heat-flow equation with proper boundary conditions that the heat capacity can be expressed as [4]:

$$C = K\tau_1. \qquad (A.8)$$

The time-relaxation method has the advantage that at a given reference temperature it is easy to signal average numerous decays to improve the signal-to-noise ratio. This method is also useful when the sample thermal conductivity is quite poor, even to the extent when $K_{sample}/K_{wires} \sim 1$ [4]. A disadvantage of the time constant method is that, for samples with small heat capacity, τ_1 can be relatively short (<50 ms). This problem, however, can be remedied with the use of fast recording electronics.

In actual experiments employing the time relaxation method, the small sample is mounted on a sapphire platform with low specific heat grease. The sapphire platform is connected to a heat sink with the help of suitable metallic (usually Au–Cu alloy) wires (see figure A4(a)). The heat sink provides the reference

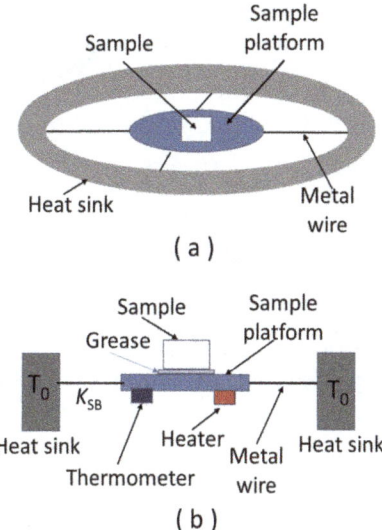

Figure A4. Schematic diagram showing time-relaxation method of heat capacity measurement: (a) sample platform connected by four metallic-wires to ring shaped heat sink; (b) detailed configuration of the sample platform with sample, heater and temperature sensor.

temperature. The sample heater and the temperature sensor are attached to the bottom of the sapphire platform (see figure A4(b)). In practice the wire leads of the temperature sensor and heater often provide the thermal link between the sample platform and heat sink. Experiment is usually done with keeping the sample-platform and heat-sink assembly in vacuum environment to avoid the extraneous effect of heat convection.

A.4 Optical conductivity

Optical spectroscopic measurements performed in the energy/frequency range from 1 meV to 10 eV have played a very important role in establishing the existing physical pictures of semiconductors and metals [6]. This technique has also contributed extensively in exploring various metals to highlight their interesting properties beyond the Drude model and driven by strong electron correlation (see [6] and references within).

We have studied in chapter 1 that ac-conductivity $\sigma(w)$ is the linear response function relating the current j is related to the frequency dependent applied electric field E: $j(w) = \sigma(w)E(w)$. There are real and imaginary contributions to this ac or optical-conductivity: $\sigma = \sigma_1(w) + i\sigma_2(w)$. Another important quantity is the dielectric function $\epsilon(w) = \epsilon_1(w) + i\epsilon_2(w)$. These two optical constants are related:

$$\sigma_1(w) = \frac{w}{4\pi}\epsilon_2(w) \tag{A.9}$$

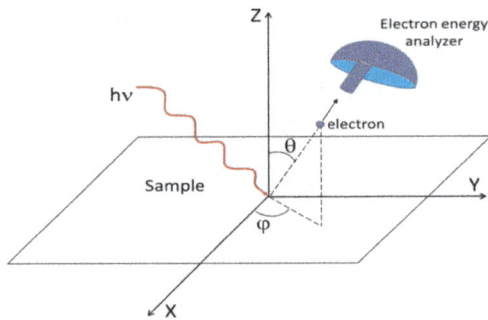

Figure A5. Schematic diagram showing the technique of angle resolved photo-emission spectroscopy (ARPES). (Reprinted figure with permission from [8]. Copyright (2003) by the American Physical Society.)

and

$$\sigma_2(w) = -\frac{w}{4\pi}[\epsilon_1(w) - 1]. \tag{A.10}$$

Spectra of the real part σ_1 of optical conductivity contains additive contributions from absorption mechanisms associated with various excitations and collective modes in solids, hence this can be directly accessed through optical experiments [6].

The complex optical constants can be obtained from one or several complementary procedures [6, 7].

1. Kramers–Kronig analysis of reflectance spectra $R(w)$ for opaque systems or transmittance spectra $T(w)$ for transparent systems.
2. Ellipsometric coefficients ψ or Δ for either transparent or opaque materials.
3. Various interferometric approaches, in particular, Mach–Zehnder interferometry.
4. THz time-domain spectroscopy.

Optical spectroscopy provide information averaged over the whole reciprocal-space, but their usual probing length is of the order of micro-meter. Due to this later property this technique has advantage over the surface-sensitive spectroscopic techniques in the study of the bulk properties of solid materials.

A.5 Photo-electron spectroscopy

Photo-electron spectroscopy is a general term which refers to all techniques based on the photoelectric effect [8]. The principle behind these techniques is the Einstein's photoelectric effect. In this class of spectroscopy, angle-resolved photo-emission spectroscopy (ARPES) is the most direct method for investigation of the electronic structure of solids. When monochromatic light or photons are incident on a material, they can induce the emission of photo-electrons, which contain the information on both energy and crystal momentum of the material.

The geometry of an ARPES experiment are shown schematically in figure A5. A beam of monochromatized photons obtained either from a gas-discharge lamp or a synchrotron radiation source is incident on a sample. In order to carry out momentum-resolved experiments, one needs to have a suitably aligned single crystal sample. The electrons are emitted by the photoelectric effect and travel into the vacuum in all directions. The kinetic energy E_{kin} of the emitted photo-electrons are measured for a given emission angle with the help of an electron energy analyzer characterized by a finite acceptance angle [8]. In this way, the momentum $\mathbf{p}$ of photo-electron is also completely determined. The modulus of the momentum is given by the relation $\mathbf{p} = \sqrt{2mE_{kin}}$ with its components parallel and perpendicular to the sample surface being obtained from the polar (θ) and azimuthal (ϕ) emission angles.

The kinetic energy E_{kin} of the emitted photo-electron can be written according to the energy conservation law as:

$$E_{kinetic} = h\nu - \Phi - |E_{binding}|. \tag{A.11}$$

Here, Φ is the work function of the material, $h\nu$ is the energy of the incident photon and E_b is the binding energy of the initial state. The momentum conservation case is more complicated, but it is simplified with the assumption that the momentum of the incident photon is negligible. This later assumption holds good in the case of the usual low photon energy ARPES with $h\nu < 1000$ eV. With the momentum conservation law, the component of the crystal momentum $\mathbf{k}_\parallel$ parallel to the crystal surface in the initial state can be expressed as:

$$\mathbf{k}_\parallel = \frac{1}{\hbar}\sqrt{2mE_{kin}} \sin\theta. \tag{A.12}$$

Due to the breaking of the transitional symmetry across the crystal surface the perpendicular component of the crystal momentum $\mathbf{k}_\perp$ is not conserved. However, this can be estimated by considering the inner potential of the system with the help of a photon energy scan. Thus, it is possible to obtain the electronic band structure of the occupied electronic states with the experimentally acquired knowledge of the energy and emission angle of photo-electrons.

A complementary technique is inverse photo-emission spectroscopy (IPES), which involves an incident electron on the solid sample filling an empty state and causing emission of a photon to conserve energy. IPES experiment is most often carried out with a detector that is limited to a fixed photon energy range. This is known as Bremsstrahlung isochromat spectroscopy (BIS).

ARPES is an important technique for electronic structure studies of strongly correlated electron systems [8]. Strong electron correlation may not affect the carrier density of the material, but it changes the effective mass of the carrier in the band. In the independent-electron framework, electrons occupy the valence bands and the conduction bands are empty in the band insulators. In this situation with fixed crystal momentum, energy and momentum conservation leads to sharp δ-function peaks in the measured kinetic energies spectrum of the photo-electrons. However, in the presence of electron–electron correlation, the remaining electrons in the band react to the hole that is left behind, and there are an infinite number of possible excitations. This gives rise to a continuous energy spectrum with broadening of the peaks and even extra structures. All these features of PES spectrum can be correlated to the results of theoretical analysis namely one-electron Green's function and the electron self-energy.

References

[1] Richardson R C and Smith E N 1998 *Experimental Techniques in Condensed Matter Physics at Low Temperatures* (Reading, MA: Addison Wesley)

[2] Dittman R H and Zimansky M W 1997 *Heat and Thermodynamics* (New Delhi: Tata McGraw-Hill)

[3] Sullivan P F and Seidel G 1968 *Phys. Rev.* **173** 679

[4] Stewart G R 1983 *Rev. Sci. Instrum.* **54** 1

[5] Bachmann R *et al* 1972 *Rev. Sci. Instrum.* **43** 205

[6] Basov D N, Averitt R D, van der Marel D, Dressel M and Haule K 2011 *Rev. Mod. Phys.* **83** 471

[7] Dressel M and Grüner G 2000 *Electrodynamics of Solids: Optical Properties of Electrons in Matter* (Cambridge: Cambridge University Press)

[8] Damascelli A, Hussain Z and Shen Z-X 2003 *Rev. Mod. Phys.* **75** 473

IOP Publishing

Mott Insulators
Physics and applications
Sindhunil Barman Roy

Appendix B

Fermi–Dirac distribution function

We shall derive here the distribution function or occupation probability $f(E, T)$ for electron gas at finite temperatures. The distribution we are talking about arises where various quantum states are in equilibrium with each other. In deriving the distribution function we shall closely follow the treatments given in the book by Ibach and Luth [1].

We start with an atomic system with one-electron energy level E_j. In a solid these energy levels will lie quite close to one another, and there we will consider new energy levels E_i. Each of this new energy levels will be degenerate and have many E_j. The degeneracy and the occupation number of these new levels E_i are denoted by g_i and n_i, respectively, both of which can be large numbers. Now electrons must obey the Pauli exclusion principle, which ensures that $n_i \leqslant g_i$. When all the energy states in a system are in equilibrium with each other, the free energy (F) of the system must be minimum with respect to a variation in the relative occupation numbers of the levels. This means:

$$\delta F = \sum_i \frac{\partial F}{\partial n_i} \delta n_i = 0 \tag{B.1}$$

along with the condition for conservation of particle numbers:

$$\sum_i \delta n_i = 0. \tag{B.2}$$

In the event of exchange of electrons between two arbitrary levels p and q, the equilibrium condition is given by:

$$\frac{\partial F}{\partial n_p} \delta n_p + \frac{\partial F}{\partial n_q} \delta n_q = 0 \tag{B.3}$$

and

$$\delta n_p + \delta n_q = 0. \tag{B.4}$$

It follows from the above relations that the derivatives of the free energy with respect to the occupation numbers needs to be equal:

$$\frac{\partial F}{\partial n_p} = \frac{\partial F}{\partial n_q}. \tag{B.5}$$

The two levels p and q were chosen randomly, hence all $\partial F / \partial n_i$ must be the same. This variation of free energy F with respect to the occupation number n_i is called the *chemical potential* μ.

The free energy of the electron gas is expressed as:

$$F = U - TS. \tag{B.6}$$

Here U is the internal energy:

$$U = \sum_i n_i E_i \tag{B.7}$$

and S is the entropy:

$$S = k_B \ln P \tag{B.8}$$

where P is the number of possible ways the electrons can be distributed among the available energy states and k_B is the Boltzmann constant.

The number of ways an electron can be placed in an energy level E_i is g_i. If a second electron is now to be accommodated in the same energy level, the number of ways this can be done is $g_i - 1$, and so on. We can now see that the possible ways n_i electrons can be placed in definite positions within the energy level E_i can be expressed as:

$$g_i(g_i - 1)(g_i - 2)(g_i - 3) \cdots (g_i - n_i + 1) = \frac{g_i!}{(g_i - n_i)!}. \tag{B.9}$$

Arrangements with just an exchange of electrons within the energy level, however, are not distinguishable, and there can be $n!$ such possibilities. Hence, the total number of distinguishable possibilities of putting n_i electrons in the level E_i is expressed as:

$$\frac{g_i!}{n!(g_i - n_i)!}. \tag{B.10}$$

The number of possibilities P for the complete system is then given by the product of all possibilities of occupying each

$$P = \prod_i \frac{g_i!}{n!(g_i - n_i)!}. \tag{B.11}$$

Using equation (B.8) we can write:

$$S = k_B \sum_i [\ln g_i! - \ln n_i! - \ln(g_i - n_i)!].$$

(B.12)

Now, the factorials in the above equation can be replaced by using Stirling's approximation for large value of n:

$$\ln n! \approx n \ln n - n.$$

(B.13)

We can now get the expression for the chemical potential μ as:

$$\mu = \frac{\partial F}{\partial n_i} = E_i + k_B T \ln \frac{n_i}{g_i - n_i}.$$

(B.14)

The above equation (B.14) is rearranged to get an expression for the occupation number n_i:

$$n_i = g_i(e^{(E_i - \mu)/k_B T} + 1)^{-1}.$$

(B.15)

Using equation (B.15) the probability that a quantum mechanical state is occupied by an electron can be expressed in the form of a distribution function $f(E, T)$ known as the Fermi–Dirac distribution function, which is given by:

$$f(E, T) = \frac{1}{e^{(E-\mu)/k_B T}} + 1.$$

(B.16)

The Fermi–Dirac distribution function is the probability distribution function for all the fermions i.e. the particles with half-integer spin.

At $T = 0$ K, the Fermi–Dirac distribution function has a value of 1 for $E < \mu$ and 0 for $E > \mu$. Thus at, $T = 0$ K the chemical potential of electron gas is equal to the Fermi energy:

$$\mu(T = 0 \text{ K}) = E_F.$$

(B.17)

Thus, the Fermi–Dirac distribution function is often expressed as:

$$f(E, T) = \frac{1}{e^{(E-E_F)/k_B T}} + 1.$$

(B.18)

Reference

[1] Ibach H and Lüth H 2009 *Solid State Physics: An Introduction to Principles of Materials Science* (Berlin: Springer)

Appendix C

Idea of second quantization

In this appendix we will provide a short introduction to the idea of second quantization. This technique has been used extensively in the treatments of many body physics, and has been introduced formally in many books dealing with the subjects of many body physics and quantum field theory [1–4]. However, for a beginner I found the introduction to the second quantization to be very illustrative as well as illuminating in the book by Lancaster and Blundell entitled *Quantum Field Theory for the Gifted Amateur* [5]. Here I closely follow the treatment given in that book.

Before going to second quantization, it is a natural question to ask: what is first quantization? Well, one of the concepts which came to the forefront with the arrival of quantum mechanics is that the things that had been considered so far as particles could also be conceived as waves. For example electrons, protons and neutrons are not simply particles, they can also be described in terms of a wave equation, that is Schrödinger equation. This idea is known as first quantization. Within the realm of first quantization: particles behave like waves.

However, it was also realized that waves could possess particle like properties too. Electromagnetic waves and lattice vibrations in a solid are not simply periodic changes in some medium, but they are perceived to behave like particles. That gave rise to the idea of quanta of electromagnetic wave and lattice vibrations, namely photons and phonons, respectively. This is the idea of second quantization, and within its realm: waves behave like particles.

In many body calculations it becomes relatively easy to solve a problem if it is cast in the second quantization form rather than solving the complicated differential wave equations in a tedious manner. That is why the wide usage of the second quantization formalism in the correlated many electron systems. We shall now

doi:10.1088/2053-2563/ab16c9ch11

elaborate this formalism of second quantization and its usefulness with the help of the well known problem of a harmonic oscillator.

C.1 Harmonic oscillator problem

The Schrödinger equation of harmonic oscillator problem is written as:

$$\left(-\frac{\hbar^2}{2m}\frac{\partial^2}{\partial x^2} + \frac{1}{2}Kx^2\right)\psi = E\psi. \tag{C.1}$$

The first quantization or standard way to solve to this equation, involves the series-solution method (see any standard quantum mechanics book), and the solutions are given by:

$$\psi_n(\xi) = \frac{1}{\sqrt{2^n n!}}\left(\frac{mw}{\pi\hbar}\right)^{1/4} H_n(\xi)e^{-\frac{\xi^2}{2}}. \tag{C.2}$$

Here $H_n(\xi)$ is Hermite polynomial, $\xi = \sqrt{mw/\hbar}x$ and $w = \sqrt{K/m}$. The eigenfunctions or solutions have wave-like nature but they have particle like qualities as well, which is of course evident from the eigen-values of the harmonic oscillator problem given by:

$$E_n = \left(n + \frac{1}{2}\right)\hbar w. \tag{C.3}$$

The equation (C.3) actually presents a ladder of energy levels (see figure C1). It may, however, be noted that the energy is not zero when $n = 0$ but $E_0 = \hbar w/2$, which is known as the zero-point energy. Since the energy levels form a ladder, one needs to add a quantum of energy $\hbar w$ move up the ladder by one step. It is as if one is adding a particle [5]. We will now make this picture more concrete. To achieve this we first

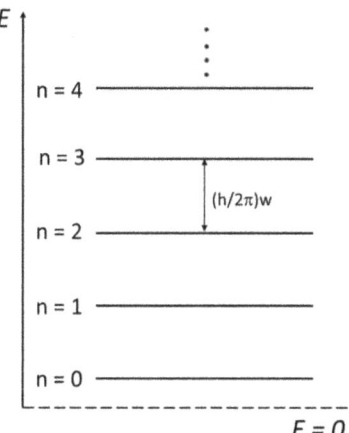

Figure C1. Schematic representation of the energy ladder of the simple harmonic oscillator. Note that the lowest energy level is not $E = 0$ but $E = \frac{1}{2}\hbar w$.

write down the Hamiltonian of this harmonic oscillator problem in terms of the quantum mechanical position operator $\hat{x}$ and momentum operator $\hat{p}$ the following way:

$$H = \frac{\hat{p}^2}{2m} + \frac{1}{2}mw^2\hat{x}^2 \tag{C.4}$$

where the spring constant $K = mw^2$. We now try to factorize this Hamiltonian by expressing it as:

$$\frac{1}{2}mw^2\left(\hat{x} - \frac{i}{mw}\hat{p}\right)\left(\hat{x} + \frac{i}{mw}\hat{p}\right). \tag{C.5}$$

However, in quantum mechanics the operators $\hat{x}$ and $\hat{p}$ do not commute. Hence, equation (C.5) does not really represent the Hamiltonian H, rather it gives the following equation:

$$\frac{1}{2}mw^2\left(\hat{x} - \frac{i}{mw}\hat{p}\right)\left(\hat{x} + \frac{i}{mw}\hat{p}\right) = \frac{\hat{p}^2}{2m} + \frac{1}{2}mw^2\hat{x}^2 + \frac{iw}{2}[\hat{x}, \hat{p}]. \tag{C.6}$$

Here $[\hat{x}, \hat{p}] \equiv \hat{x}\hat{p} - \hat{p}\hat{x} = i\hbar$ is known as the commutator of $\hat{x}$ and $\hat{p}$. With this equation (C.6) can be rewritten as:

$$\frac{1}{2}mw^2\left(\hat{x} - \frac{i}{mw}\hat{p}\right)\left(\hat{x} + \frac{i}{mw}\hat{p}\right) = \hat{H} - \frac{\hbar w}{2}. \tag{C.7}$$

This is nearly the Hamiltonian of the harmonic oscillator, with a minus term $\frac{\hbar w}{2}$ which is the zero point energy. The factorization therefore works and one realizes that the operators $\hat{x} - \frac{i}{mw}\hat{p}$ and $\hat{x} + \frac{i}{mw}\hat{p}$ are going to be useful [5]. Since $\hat{x}$ and $\hat{p}$ are Hermitian, the operators $\hat{x} - \frac{i}{mw}\hat{p}$ and $\hat{x} + \frac{i}{mw}\hat{p}$ are adjoints of each other [5]. Therefore these operators themselves are not Hermitian, and cannot correspond to any observable. But we will see that these operators have useful properties. They are given special names $\hat{a}^+$ and $\hat{a}$ after including a multiplicative constant $\sqrt{mw/2\hbar}$:

$$\hat{a} = \sqrt{mw/2\hbar}\left(\hat{x} + \frac{i}{mw}\hat{p}\right) \tag{C.8}$$

and

$$\hat{a}^+ = \sqrt{mw/2\hbar}\left(\hat{x} - \frac{i}{mw}\hat{p}\right). \tag{C.9}$$

In terms of $\hat{a}^+$ and $\hat{a}$ the Hamiltonian can be written as:

$$\hat{H} = \hbar w\left(\hat{a}^+\hat{a} + \frac{1}{2}\right). \tag{C.10}$$

If $\hat{a}^+\hat{a}$ has an eigenstate $|n\rangle$ with eigenvalue n, then $\hat{H}$ will also have an eigenstate $|n\rangle$ with eigenvalue $(n + 1/2)\hbar w$. This is indeed the eigenvalue of harmonic oscillator as given in equation (C.3). However, it needs to be established that n takes values 0, 1, 2, 3, ... To this end the first step is to show that $n \geqslant 0$. This can be done by noting that:

$$\langle n|\hat{a}^+\hat{a}|n\rangle = |\hat{a}|n\rangle|^2 \geqslant 0. \tag{C.11}$$

The next step is to show that n takes only integer values. Before that we introduce a number operator $\hat{n}$ defined as:

$$\hat{n} = \hat{a}^+\hat{a} \tag{C.12}$$

and hence write,

$$\hat{n}|n\rangle = n|n\rangle. \tag{C.13}$$

The quantity n marks the energy level on the ladder (see figure C1) that the system has reached, or the number of quanta that must have been added to the system when it was in its ground state [5]. The Hamiltonian can now be rewritten as:

$$\hat{H} = \hbar w\left(\hat{n} + \frac{1}{2}\right) \tag{C.14}$$

and hence,

$$\hat{H}|n\rangle = \hbar w\left(n + \frac{1}{2}\right)|n\rangle \tag{C.15}$$

so that $|n\rangle$ is also an eigenstate of the Hamiltonian. So $|n\rangle$ can be a suitable substitute for the complicated eigenfunction given by equation (C.2).

We will now see that the eigenvalue n takes the integer numbers. For this let us look at the property of the state defined by $\hat{a}^+|n\rangle$. Operating on this with the number operator we get:

$$\hat{n}\hat{a}^+|n\rangle = \hat{a}^+\hat{a}\hat{a}^+|n\rangle. \tag{C.16}$$

Now, the commutator $[\hat{a}, \hat{a}^+]$ can be shown to be equal to 1. Hence, one can write:

$$\hat{a}\hat{a}^+ = 1 + \hat{a}^+\hat{a}. \tag{C.17}$$

This follows that:

$$\hat{n}\hat{a}^+|n\rangle = (n + 1)\hat{a}^+|n\rangle. \tag{C.18}$$

This example indicates that the state $\hat{a}^+|n\rangle$ is an eigenstate of $\hat{H}$, but with an eigenvalue one higher than the state $|n\rangle$ [5]. The operator $\hat{a}^+$ has the effect of adding one quantum of energy, and thus this operator is known as a raising operator.

On the other hand operating on the state $\hat{a}|n\rangle$ with the number operator we get:

$$\hat{n}\hat{a}|n\rangle = \hat{a}^+\hat{a}\hat{a}|n\rangle. \tag{C.19}$$

Again using the the commutator relation $[\hat{a}, \hat{a}^+] = 1$, it can be shown that:

$$\hat{a}^+\hat{a} = \hat{a}\hat{a}^+ - 1. \tag{C.20}$$

This leads to:

$$\hat{n}\hat{a}|n\rangle = (n - 1)\hat{a}|n\rangle. \tag{C.21}$$

This example shows that the state $\hat{a}|n\rangle$ is an eigenstate of $\hat{H}$, but with an eigenvalue one lower than the state $|n\rangle$ [5]. The operator $\hat{a}$ has the effect of reducing one quantum of energy, and thus this operator is known as a lowering operator.

If one keeps on operating the state $|n\rangle$ with lowering operator $\hat{a}$ one eventually goes to the ground state $|0\rangle$ of the harmonic oscillator. Further operation with the operator $\hat{a}$ will annihilate the state completely since $\hat{a}|0\rangle = 0$. For this reason the operator $\hat{a}$ is also known as the annihilation operator.

On the other hand if one continues to operate with the raising operator $\hat{a}^+$ starting from the ground $|0\rangle$ of the harmonic oscillator, that leads to:

$$\hat{a}^+|0\rangle = |1\rangle \tag{C.22}$$

$$\hat{a}^+|1\rangle = \sqrt{2}|2\rangle \implies |2\rangle = \frac{(\hat{a}^+)^2}{\sqrt{2}}|0\rangle \tag{C.23}$$

$$\hat{a}^+|2\rangle = \sqrt{3}|3\rangle \implies |3\rangle = \frac{(\hat{a}^+)^3}{\sqrt{3 \times 2}}|0\rangle \tag{C.24}$$

and so on. Thus, we can write in general:

$$|n\rangle = \frac{(\hat{a}^+)^n}{\sqrt{n!}}|0\rangle. \tag{C.25}$$

We can see that the state $|n\rangle$ can be obtained by repeated application of the operator $\hat{a}^+$, hence it is also known as the creation operator. It can be visualized that the operator $\hat{a}^+$ acts to generate a quantum of energy and move the oscillator up the ladder by one step (see figure C2). In the same vein the operator $\hat{a}$ annihilates a quantum of energy and move the oscillator down by one step in the ladder (see figure C2). These quanta of energy behave like particles, and the particles are added or subtracted by the application of creation and annihilation operators, respectively. In other words, within the framework of second quantization the harmonic oscillator has particle-like eigenfunctions. Thus, the harmonic oscillator problem, which is essentially a wave problem, has turned into a problem of particles. At the same time the methodology of solving the problem has become relatively simple and transparent.

In the next step the problem can be generalized to N number of uncoupled harmonic oscillators with the general state of the system written as $|n_1, n_2, n_3, \ldots, n_n\rangle$. This is known as occupation number representation. This

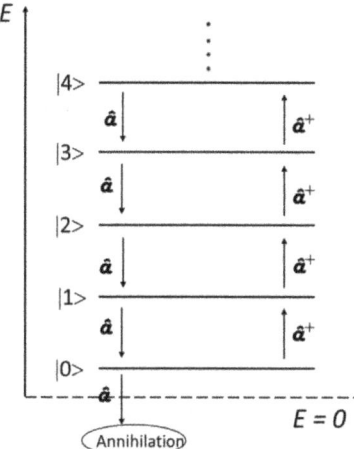

Figure C2. Schematic representation of the effect of creation ($\hat{a}^+$) and annihilation ($\hat{a}$) operators. While $\hat{a}^+$ moves the oscillator up one step in the energy ladder, $\hat{a}$ moves it down by one step. Dropping down the lowest step in the energy ladder $|0\rangle$ is the operation $\hat{a}|0\rangle$. This results in zero i.e. annihilation.

general state can be expressed by using the same technique for the single harmonic oscillator as:

$$|n_1, n_2, n_3, \dots, n_n\rangle = \frac{1}{\sqrt{n_1! n_2! \cdots n_N!}} (\hat{a}_1^+)^{n_1}, (\hat{a}_2^+)^{n_2} \cdots (\hat{a}_N^+)^{n_N} |0, 0\dots\rangle. \qquad (C.26)$$

The operation here is on the ground state (or vacuum state as commonly termed in literature) with a product of creation operators necessary to put n_1 quanta of energy into oscillator number 1, n_2 quanta of energy into oscillator number 2, etc [5]. The general state can be expressed in more compact form:

$$|n_k\rangle = \prod_k \frac{1}{\sqrt{n_k!}} (\hat{a}^+)^{n_k} |0\rangle. \qquad (C.27)$$

Here $|0\rangle$ is the vacuum state.

At a finite temperature the atoms (or structural units) of a solid vibrate around their mean position at the lattice sites. Thus they can be thought of as a collection of harmonic oscillators. But each structural unit in a solid is coupled to its neighbours, hence we cannot take the problem as a collection of harmonic oscillators. In fact a solid with N number atoms will generate N number vibration patterns or excitation modes involving all the N atoms. However, it can be shown that such modes in a solid can be represented as a set of totally independent oscillators. This is because it is possible to Fourier transform the problem, so that although the atoms are coupled in real space, the excitation modes are uncoupled in reciprocal space [5]. These modes are called phonons. Each phonon mode labeled by k behaves like a harmonic oscillator, and can be given integer multiples of the quantum of energy $\hbar w_k$. Thus, it provides a real example of how the excited wave patterns can actually generate particles.

C.2 Many particle systems

We will now see the use of the occupation number representation to describe a system of many identical particles. Apart from changing the way the states are labeled, this method has the advantage of getting rid of the wave functions altogether.

In conventional quantum mechanics one labels each identical particle and its momentum is listed. In occupation number representation, instead of listing which particle is in which momentum state, one list the values of the momentum p_m and how many particles occupy each momentum state. In this occupation number representation the Hamiltonian is expressed as [5]:

$$\hat{H}\,|n_1,\,n_2,\,n_3\rangle = \left[\sum_m n_{p_m} E_{p_m}\right]|n_1,\,n_2,\,n_3\rangle. \tag{C.28}$$

Here the number of particles in each state is multiplied by the energy of that state and summed over all of the states. We can now move forward by drawing the following analogies with the harmonic oscillators problem [5]:

1. Quanta in oscillator problem correspond to the particles in momentum states in the many particle problem.
2. kth oscillator corresponds to mth momentum state p_m.
3. Total energy $E = \sum_{k=1}^{N} n_k \hbar w_k$ in the oscillator problem corresponds to $E = \sum_{k=1}^{N} n_{p_m} E_{p_m}$.

In the same way as was done for the many harmonic oscillators in the last section, a general state for many particle systems can be built by operating on the vacuum state $|0\rangle$:

$$|n_1,\,n_2,\,n_3,\,\ldots\rangle = |n_k\rangle = \prod_k \frac{1}{\sqrt{n_k!}}(\hat{a}^+)^{n_k}|0\rangle. \tag{C.29}$$

In the many particle system a creation (annihilation) operator $\hat{a}^+$ ($\hat{a}$) creates (annihilates) a particle in momentum state $|p_m\rangle$. However, this creation and annihilation of states must respect the exchange symmetry principles for identical particles. We recall that the exchange of two identical Bose particles or bosons results into the same state again, whereas the exchange of two identical fermions leads to a state, which is the original state with a minus sign. Suppose we put one particle in state p_1 and then another in state p_2, or do the same thing in the reverse order, we can write:

$$\hat{a}_{p_1}^+\hat{a}_{p_2}^+ = \lambda\hat{a}_{p_2}^+\hat{a}_{p_1}^+. \tag{C.30}$$

If $\lambda = +1$, then the wave-functions are symmetric under particles exchange, and the particles are boson. On the other hand, if $\lambda = -1$, then the wavefunctions are antisymmetric under particles exchange, and the particles are fermion. Thus, the two cases correspond to two different types of commutation relation for the creation and annihilation operators [5].

In the case of bosons with $\lambda = 1$, we get:

$$\hat{a}_{p_1}^+ \hat{a}_{p_2}^+ = \hat{a}_{p_2}^+ \hat{a}_{p_1}^+.$$ (C.31)

From this with generalization of the labels of the momentum states to i and j we can write then commutation relation:

$$\hat{a}_i^+ \hat{a}_j^+ - \hat{a}_j^+ \hat{a}_i^+ = [\hat{a}_i^+, \hat{a}_j^+] = 0.$$ (C.32)

This indicates that the creation operators for different particle states commute. In a similar way we can get the commutation relation fore the annihilation operator as $[\hat{a}_i^+, \hat{a}_j^+] = 0$. We can also write:

$$[\hat{a}_i, \hat{a}_j^+] = \delta_{ij}.$$ (C.33)

The commutation of different operators shows that:

$$\hat{a}_{p1}^+ \hat{a}_{p2}^+ |0\rangle = \hat{a}_{pd}^+ \hat{a}_{p1}^+ |0\rangle = |1p1, 1p2\rangle.$$ (C.34)

Thus the same state is obtained independent of the order the individual particles are being put in the momentum states. This is of course the characteristic of the bosons. In general one can write [5]:

$$\hat{a}_i^+ |n_1, n_2, \ldots, n_i, \ldots\rangle = \sqrt{n_i + 1} \, |n_1, n_2, \ldots, n_i + 1, \ldots\rangle$$ (C.35)

and

$$\hat{a}_i |n_1, n_2, \ldots, n_i, \ldots\rangle = \sqrt{n_i} |n_1, n_2, \ldots, n_i - 1, \ldots\rangle.$$ (C.36)

In the case of fermions, $\lambda = -1$. To distinguish it from boson operators, creation (annihilation) operator is usually denoted by $\hat{c}^+$ ($\hat{c}$). With $\lambda = -1$ we find that:

$$\hat{c}_i^+ \hat{c}_j^+ + \hat{c}_j^+ \hat{c}_i^+ = \{\hat{c}_i^+, \hat{c}_j^+\} = 0.$$ (C.37)

Thus, the Fermion operators anticommute, that is a minus sign appears when their order is changed. The curly bracket in the above equation indicates an anticommutator. If we put $i = j$ in equation (C.37) then:

$$\hat{c}_i^+ \hat{c}_i^+ + \hat{c}_i^+ \hat{c}_i^+ = 0 \quad \text{or} \quad \hat{c}_i^+ \hat{c}_i^+ = 0.$$ (C.38)

This indicates that two fermions cannot be put in the same state. This is the Pauli exclusion principle, which indicates each state can have a single fermion or it remains unoccupied.

In a similar way we can show an anticommutation relation between the annihilation operators:

$$\{\hat{c}_i, \hat{c}_j\} = 0$$ (C.39)

and finally the commutation relation:

$$[\hat{c}_i, \hat{c}_j^+] = \delta_{ij}. \tag{C.40}$$

It may be noted that since $\hat{c}_i^+\hat{c}_j^+ |0\rangle = -\hat{c}_j^+\hat{c}_i^+ |0\rangle$, it really matters in which order the particles are being put into the momentum states. The creation and annihilation operators for fermions are defined as follows:

$$\hat{c}_i^+|n_1, n_2, \ldots, n_i, \ldots\rangle = (-1)^{\Sigma_i}\sqrt{n_i - 1}\,|n_1, n_2, \ldots, n_i + 1, \ldots\rangle \tag{C.41}$$

and

$$\hat{c}_i|n_1, n_2, \ldots, n_i, \ldots\rangle = (-1)^{\Sigma_i}\sqrt{n_i}\,|n_1, n_2, \ldots, n_i - 1, \ldots\rangle \tag{C.42}$$

where

$$(-1)^{\Sigma_i} = (-1)^{(n_1+n_2+\cdots+n_{i-1})}. \tag{C.43}$$

This introduces a factor of -1 for every particle in the many particle system, standing to the left of the state labeled n_i in the state vector [5]. It may be noted that the the square roots are redundant in equations (C.41) and (C.42) since $n_i = 0$ or 1, and they can actually be omitted.

References

[1] Mattuck R D 1992 *A Guide to Feynman Diagrams in the Many-Body Problem* (New York: Dover)
[2] Fulde P 1995 *Electron Correlations in Molecules and Solids* (Berlin: Springer)
[3] Freerick J K 2006 *Transport in Multi Layered Nanostructures: The Dynamical Mean-field Theory Approach* (London: Imperial College Press)
[4] Mahan G D 2000 *Many-Particle Physics* (New York: Plenum)
[5] Lancaster T and Blundell S J 2014 *Quantum Field Theory for the Gifted Amateur* (Oxford: Oxford University Press)

Appendix D

Green's function and Hubbard model

In this appendix we will provide an introduction to Green's function and its application to the Hubbard model. We have used some of these results involving Green's function in chapters 3 and 5. This appendix is meant for the interested advanced readers who may like to know a bit more on the mathematical background leading to those results. For more details on these subjects the readers are referred to the excellent books by Martin, Reining and Ceperley [1], Fulde [2], Freerick [3], Mattuck [4] and Lancaster and Blundell [5], and the narratives below will closely follow some of those books.

D.1 Green's function

A detailed description of a many-electron system requires information on the time-dependent wave function of the whole system $\Psi(r_1 \cdot r_2, \ldots, r_N, t)$. This is a very difficult if not totally unsolvable problem. However, to get information on the important physical properties of a many-electron system it may not be necessary to know the detailed behaviour of each electron in the system, but rather just the average behaviour of one or two electrons. The quantities which can describe this average behaviour are the one-particle propagator and two-particle propagator respectively, and physical properties may be calculated directly from them [4]. These propagators are formally known as Green's functions.

The one-electron propagator or Green function is defined in the following way. If an electron (or quasiparticle) is put into an interacting lattice at point r_1 at time t_1 and let it move through the lattice colliding with the other electrons for a time (i.e. let it 'propagate' through the lattice), then the one-electron propagator or Green function is the quantum probability amplitude that the electron will be observed at the position r_2 at time t_2. Instead of putting the electron at a definite position of the lattice, it may be more convenient to put it in with definite momentum, say k_1, and observe it later with momentum k_2. The single-electron Green function directly gets the energies and lifetimes of electrons or quasiparticles. It can also give the electron momentum distribution, spin and density, and can also be used to calculate the ground state energy. In a similar manner, the two-electron Green function is the probability amplitude for observing one electron at the point r_2 at time t_2 and another at r_4 at time t_4 if one electron was put into the lattice at the position r_1 at time t_1 and another at r_3 at time t_3. This two-electron Green's function is also very useful, giving directly the energies and lifetimes of collective electron excitations, as well as the magnetic susceptibility, electrical conductivity, and a lot of other non-equilibrium properties.

To give some idea on Green's function let us start with a general looking differential equation formed from a linear differential operator $\hat{L}$ [5]:

$$\hat{L}x(t) = f(t). \qquad (D.1)$$

The Green function $G(t, u)$ corresponding to this linear operator $\hat{L}$ is defined by the equation [5]:

$$\hat{L}G(t, u) = \delta(t - u). \qquad (D.2)$$

It may be noted that the Green function $G(t, u)$ needs two arguments. The first is the variable t one is interested in (here it is the time t), and the second is u the variable that describes the position of the delta function $\delta(t - u)$ that is used to define the Green function via equation (D.2). The inhomogeneous part of the equation (D.2) is built up from a set of delta functions weighted by an amplitude function $f(u)$.

Let us now take the example of an oscillator of mass m and spring constant K evolving under the influence of a time-dependent force $f(t)$ [5]. The equation of motion is written as:

$$m\frac{d^2x(t)}{dt^2} + Kx(t) = f(t). \qquad (D.3)$$

The force function $f(t)$ can be represented by adding together a number of delta functions:

$$f(t) = \int_0^\infty du f(u)\delta(t - u). \qquad (D.4)$$

Here u is a dummy variable in terms of which the force function $f(t)$ is broken into delta functions, and this kind of superposition is allowed here since the differential equation is linear. Instead of solving the equation (D.3) straightaway, one can first solve the equation for just one of the delta functions, and this solution is called Green's function $G(t, u)$:

$$\left(m\frac{d^2}{dt^2} + K\right)G(t, u) = \delta(t - u). \qquad (D.5)$$

The complete solution to equation (D.3) then can be obtained by the integral of Green's function, weighted by the function $f(u)$:

$$x(t) = \int_0^\infty G(t, u)f(u)du. \tag{D.6}$$

In quantum mechanics we know that the equation that governs the change of a wave function $\psi(x, t)$ is the Schrödinger equation $\hat{H}\psi(x,t) = i\frac{i\partial\psi(x,t)}{\partial t}$, where $\hat{H}$ is an operator function of x. Now, following the same argument as above one can write:

$$\psi(x, t_x) = \int G^+(x, t_x, y, t_y)\psi(y, t_y)dy. \tag{D.7}$$

Considering the wave function representing an electron, Green's function here propagates the electron from the space-time point (y, t_y) to (x, t_x). Green's function G^+ is called time retarded Green's function and defined as $G^+ = G$ for $t_x > t_y$ and $G^+ = 0$ for $t_x < t_y$. This later condition prevents the electron going back in time. Similarly one can define the time-advanced Green function by $G^- = G$ for $t_x < t_y$ and $G^- = 0$ for $t_x > t_y$. From equation (D.7) $\psi(y, t_y)$ ($\psi(x, t_x)$) can be interpreted as the amplitude to find a particle at (y, t_y) (x, t_x). Then the propagator $G^+(x, t_x, y, t_y)$ is the probability amplitude that a particle in state $|y >$ at time t_y, ends up in a state $|x >$ at time t_x. This interpretation means that Green's function can be expressed as [5]:

$$G^+(x, t_x, y, t_y) = \theta(t_x - t_y)\langle x(t_x)|y(t_y)\rangle. \tag{D.8}$$

The symbol $\theta(t_x - t_y)$ is the unit step function, which equals to 1 for $t_x > t_y$.

Now if one starts from Green's function, what can one learn about the electron? This can be examined by considering yet another form of the Green function as a function of space but instead of time in the frequency or energy energy domain [5]:

$$G^+(x, y, E) = \sum_n \frac{i\psi_n(x)\psi_n^*(y)}{E - E_n}. \tag{D.9}$$

Considering equation (D.9) as a function of the complex variable E one can have two important observations:

1. Existence of singularities on the real axis when $E = E_n$, that is when the parameter E equals the energies of the eigenstates $\psi_n(x)$.
2. The residues at the poles are i times the wave functions.

Thus starting with Green's function of an electron it is possible to have access to the energies of the electron and its wave functions. To ensure causality it is more appropriate to replace E_n with $E_n - i\epsilon$, so that Green's function is written as:

$$G^+(x, y, E) = \lim_{\epsilon\to0^+} \sum_n \frac{i\psi_n(x)\psi_n^*(y)}{E - E_n + i\epsilon}. \tag{D.10}$$

The power of Green's functions is even more realized in the perturbation problems. The solvable part of the problem is treated in terms of a particle propagating from point to point, while the perturbation is considered as a scattering process that interrupts the propagation. Writing the perturbation problem by splitting up the Hamiltonian into two parts $H = H_0 + V$, where H_0 is the solvable part and V is the perturbing potential, Green's function in the presence of a perturbation is written as:

$$G = \frac{1}{E - H_0 - V}.$$ (D.11)

Here G is termed as the full propagator, whereas Green's function $G_0 = \frac{1}{E - H_0}$ for the solvable part of the Hamiltonian is termed as the free propagator. The full propagator G can be expanded as a function of free propagator G_0 as:

$$\begin{aligned} G &= \frac{1}{E - H_0 - V} \\ &= \frac{1}{E - H_0} + \frac{1}{E - H_0} V \frac{1}{E - H_0} + \frac{1}{E - H_0} V \frac{1}{E - H_0} V \frac{1}{E - H_0} + \cdots. \end{aligned}$$ (D.12)

In general one can calculate each of the terms on the right-hand side of equation (D.12) and hope that G may be well approximated by just the first few of these terms The full Green function thus can be expressed as an expansion in free propagators and potentials:

$$G = G_0 + G_0 V G_0 + G_0 V G_0 V G_0 + \cdots$$ (D.13)

The perturbation problem involving an infinite series of terms has now been converted into a clear picture of electron scattering. The amplitude G for an electron to go from y to x is a superposition of the amplitude G_0 for moving freely from y to x plus the amplitude $G_0 V G_0$ making the trip with a single scattering event at some point along the way, plus the amplitude $G_0 V G_0 V G_0$ making the trip with two scattering events and so on. The algebra of Green's functions thus helps to visualize the actual physical processes [5].

Green's functions are often most useful to know in the momentum and energy domain for the quantum mechanical perturbation problems. These are expressed in the following forms:

$$G^+(k, t_x, t_y) = \theta(t_x - t_y) e^{-iE_k(t_x - t_y)}$$ (D.14)

and

$$G^+(p, E) = \frac{i}{E - E_k + i\epsilon}.$$ (D.15)

Many a times energy E is expressed in terms of frequency w, and Green's function as $G(k, w)$. Another useful concept is 'spectral function', which is defined in terms of retarded single-particle Green's function as:

$$A(k, w) = -\frac{1}{\pi} \operatorname{Im} G^+(k, w). \tag{D.16}$$

The spectral function can be obtained from angle-resolved photo-emission spectroscopy (ARPES) for $w < 0$ and in inverse ARPES for $w > 0$ [1, 6].

D.2 Hubbard's approach in correlated electron systems using Green's function

Hubbard proposed in a series of papers [7–10] a formalism to study the interacting electron systems in terms of the one-electron Green function. Hubbard first considered isolated atoms (including intra-atomic) interactions and treated the coupling between the atoms perturbatively. This was formulated in terms of the one-electron Green function [1]:

$$G_\sigma(k, w) = \frac{1}{\left[w - E_0 - \sum_\sigma(w) \right] - \Delta E_k}. \tag{D.17}$$

Here,
1. E_0 is the independent-electron energy for the localized atomic-like state.
2. $\sum_\sigma(w)$ represents the effects of intra-atomic interactions, and
3. $\Delta E_k = E_k - E_0$ is the independent-particle dispersion associated with the inter-atomic hopping.

Hubbard considered the average of Green's functions for isolated atoms with fractional occupation of the lattice sites. This means the average of atoms that are occupied or not with opposite-spin electrons is represented by Green's function:

$$\bar{G}_\sigma^{atom}(w) = \frac{1 - n_{-\sigma}}{w - E_0} + \frac{n_{-\sigma}}{w - E_0 - U}. \tag{D.18}$$

This gives the average spectrum with poles at E_0 and $E_0 + U$ with weights at $1 - n_{-\sigma}$ and $n_{-\sigma}$. A self-energy corresponding to this Green's function can also be defined by:

$$G_\sigma^{atom}(w) = \frac{1}{w - E_0 - \sum_\sigma^{atom}(w)} \tag{D.19}$$

where,

$$\sum_\sigma^{atom}(w) = U n_{-\sigma} + U^2 \frac{n_{-\sigma}(1 - n_{-\sigma})}{w - E_0 - U(1 - n_{-\sigma})}. \tag{D.20}$$

A system of decoupled atoms will have massive degeneracy, and a coupling (no matter how small) between the atoms is required to determine the actual ground state. Hubbard then proposed the first approximation [7], often termed as 'Hubbard I', which assumed that the self-energy to be the same as in the case of isolated atoms even in the presence of dispersion, and ΔE_k remained unchanged even in the presence of interactions. This leads to Green's function for the crystal lattice:

$$G_{k,\sigma} = \frac{1}{w - E_0 - \sum_{\sigma}^{atom}(w) - \Delta E_k}. \tag{D.21}$$

Using the expression for $\sum_{\sigma}^{atom}(w)$ given in equation (D.20), which has a singularity unless $n_\sigma = 0$ or 1, it can be shown that for any interaction and for any partial filling of the band, the single band always splits into two parts with fractional weights. This is, however, not in accord with the experimental results and established theories for weak interactions. On the other hand, such splitting of one band into 'upper Hubbard' band and 'lower Hubbard' band is quite appropriate for a system with large Coulomb interaction U. As discussed in chapter 3, we recall here that the upper Hubbard band corresponds to adding an electron at a lattice site occupied by an electron with opposite-spin, and the lower Hubbard band represents sites with no other electron. In the next step forward [9], commonly known as the 'Hubbard III' approach, Hubbard formulated the problem in terms of the propagation of an electron in the lattice in the presence of other electrons. One is concerned here with the dynamics of an electron in the presence of only the electrons with opposite spin, since the interaction U is only between the electrons with opposite-spin. Now there are two separate approximations in the 'Hubbard III' approach. The first approximation is that the electrons with the opposite spin is treated as a random array of fixed scatterers like an alloy. Thus, an electron say with up-spin encounters an array of lattice sites either with no electrons that have an energy E_0, and other sites occupied by an electron with a down-spin that have an energy $E_0 + U$, and the same situation occurs for the electrons with down-spin. The second approximation is a simplification of this alloy problem, which is known as the coherent potential approximation (CPA). This was derived independently by Hubbard [9], and by Soven [11] and Velicky [12] for electrons in alloys. Within the framework of CPA, a translational invariant effective medium is defined with a coherent potential denoted by $\sum^{CPA}(w)$. The band energy is then replaced by a complex function $\left[E_k + \text{Re } \sum^{CPA}(w)\right] + i \text{ Re } \sum^{CPA}(w)$, which represents shifts in the energies and scattering due to disorder. The system is translational invariant, hence k is conserved and one can define the CPA Green function and the on-site CPA Green function as:

$$G_k^{CPA}(w) = \frac{1}{w - E_0 - \sum^{CPA}(w) - \Delta E_k} \tag{D.22}$$

and

$$G_{00}^{CPA} = \frac{1}{N_k} \sum_k G_k^{CPA}(w). \tag{D.23}$$

The determination of self-energy $\Sigma^{CPA}(w)$ requires that it should represent the average scattering in the static alloy. In the Hubbard model each lattice site is considered to be embedded in an effective medium with a Green's function:

$$G^{+,-}(w) = \frac{1}{w - E^{+,-} - \Sigma^{CPA}(w) - \Delta(w)}. \tag{D.24}$$

Here $(+)$ or $(-)$ represents sites with or without electron with an opposite-spin, with energies $E_0 + U$ and E_0, respectively. The function $\Sigma^{CPA}(w)$ is to be determined by the self-consistency. The CPA condition is that the effective medium is the same as the average over the sites [1] and

$$G_{00}^{CPA}(w) = n_{-\sigma}G^+(w) + (1 - n_{-\sigma})G^-(w). \tag{D.25}$$

Equation (D.25) is valid for electrons with either spin, where $n_{-\sigma}$ denotes the occupation of the site by an electron with opposite-spin. We may note here that it is the same as the average of Green's functions in the case of isolated atoms (see equation (D.18)), except that isolated atoms are replaced by renormalized atoms embedded in a medium chosen appropriately to represent the coupling to the rest of the crystal.

There is a close correspondence here with dynamical mean-field theory (DMFT). The essential difference lies in the meaning of the self-energy and how it is determined. In the case of the Hubbard model the Green function $\bar{G}(w)$ is a static average of the simple functions expressed in the equation (D.24). In comparison, in DMFT it is the solution of a difficult many-body problem of a site with interacting electrons embedded in a bath [1]. The CPA equations can be written in several ways and one form is:

$$\sum(w) = \bar{E} - \frac{\left(E_0 - \sum(w)\right)\left(E_0 + U - \sum(w)\right)}{\left(w - \sum(w) - b\Delta(w)\right)}. \tag{D.26}$$

This is solved self-consistently with equation (D.22) to find $\Sigma^{CPA}(w)$. The term b in the denominator of equation (D.26) is unity for the CPA, and $b = 0$ corresponds to the limit where G is the atomic resonance.

It is not quite correct to assume that the electrons with opposite spins are in fixed static positions at the lattice sites. All electrons should be in the same dynamical condition. The motion of the electrons of the opposite spins may actually lead to additional possibilities for hopping for a electron. To take into account of this effect Hubbard introduced a 'resonance broadening' correction. To this end Hubbard increased the hybridization function in the denominator of equation (D.26) by a factor $b > 1$. This is then solved self-consistently with equation (D.23), which does not contain the factor of b. Hubbard proposed a value of $b = 3$ for the case of half-filled band to take into account that the an electron can hop between two sites without an increase in energy U if an electron with opposite spin hops on either site, thus increasing the total hopping probability by a factor of 3. This was used to estimate the point at which a gap opens, which was interpreted as a metal–insulator transition.

The excitation spectrum in the Hubbard III approximation is correct in two limits: the atomic limit with split bands and the non-interacting electron limit with independent-electron bands. In between there is a gradual development of the upper and lower Hubbard bands and an opening of a gap. It is a qualitative improvement over the Hubbard I approximation for weak interactions which shows a gap for all values of $U > 1$. The feature of the gap opening is in semi-quantitative agreement with DMFT results [1]. The main drawback of the Hubbard III approximation is due to the replacement of the effects of electron–electron interactions by 'static disorder'. This misses the crucial dynamical aspects. These are covered at least approximately by the dynamical mean-field theory.

References

[1] Martin R M, Reining L and Ceperley D M 2016 *Interacting Electrons: Theory and Computational Approaches* (Cambridge: Cambridge University Press)
[2] Fulde P 1995 *Electron Correlations in Molecules and Solids* (Berlin: Springer)
[3] Freerick J K 2006 *Transport in Multi Layered Nanostructures: The Dynamical Mean-field Theory Approach* (London: Imperial College Press)
[4] Mattuck R D 1992 *A Guide to Feynman Diagrams in the Many-Body Problem* (New York: Dover)
[5] Lancaster T and Blundell S J 2014 *Quantum Field Theory for the Gifted Amateur* (Oxford: Oxford University Press)
[6] Damascelli A, Hussain Z and Shen Z-X 2003 *Rev. Mod. Phys.* **75** 473
[7] Hubbard J 1963 *Proc. R. Soc. Lond. Ser.* A **276** 238
[8] Hubbard J 1964 *Proc. R. Soc. Lond. Ser.* A **277** 237
[9] Hubbard J 1964 *Proc. R. Soc. Lond. Ser.* A **281** 401
[10] Hubbard J 1965 *Proc. R. Soc. Lond. Ser.* A **285** 542
[11] Soven P 1967 *Phys. Rev.* **156** 809
[12] Velicky B, Kirkpatrick S and Ehrenreich H 1968 *Phys. Rev.* **175** 747